Lecture Notes in Energy

Volume 23

For further volumes:
http://www.springer.com/series/8874

Evanthie Michalena · Jeremy Maxwell Hills
Editors

Renewable Energy Governance

Complexities and Challenges

Springer

Editors
Evanthie Michalena
ENeC Laboratory, UMR 8185
Paris-Sorbonne (Paris IV)
and CNRS INSHS
Paris
France

Jeremy Maxwell Hills
CTL Consult
Newcastle upon Tyne
UK

ISSN 2195-1284 ISSN 2195-1292 (electronic)
ISBN 978-1-4471-5594-2 ISBN 978-1-4471-5595-9 (eBook)
DOI 10.1007/978-1-4471-5595-9
Springer London Heidelberg New York Dordrecht

Library of Congress Control Number: 2013954027

Printed on acid-free paper

Springer is part of Springer Science+Business Media (www.springer.com)

Preface

Two decades after the take-off of the renewable energy (RE) systems and continuous increases in the electricity bills, customers start to seek answers. Answers related to the efficiency/effectiveness of RE systems, answers related to local benefits in relation to huge profits of some private RE developers, and answers related to tax being spent supporting the ballooning institutional architecture (Energy Regulators, Transport System Operators, Public Utilities, International RE Agencies).

From the point of view of the editors—two scientists who have followed the RE Market almost since its birth—the above issues have not been discussed sufficiently (if not at all) and answers have not been given. Indeed, the focus of academic and institutional studies has been on the race for RE, the technological advances, and the barriers and constrains which slow that race down. In the meantime climate pressures and "environmental problems related to energy" constitute the perfect "camouflage" for developers to become even richer, for customers to pay for unknown or "hidden" services, and for energy institutions to be created whose role remains dubious.

To approach some of those emerging questions, we have thought of suggesting this book, a book where examples of "rushing"/"not-well-planned" *Renewable Energy Governance* have been analyzed. As already mentioned, RE has been around for more than two decades now, which allows maturity and experience to lead to more secure conclusions.

This book is made from case-study interrogations of highly educated people that live (or work in) countries from around the world. Therefore, there is a collection of case studies from USA, Brazil, New Zealand, Australia, Kenya, China, Japan, and European countries (Belgium, France, Germany, Greece, Ireland, Spain, and UK). To keep things simple, this book is mainly emphasizing RE Governance for electricity purposes and not transport. Therefore, cases on biofuels are only referred at a limited extent in this book.

The choice of authors was not "accidental". The energy sector covers a range of aspects related to RE such as trade, production, employment, pricing, taxes, environmental regulation, etc. Selected academics were asked not only to cover the different disciplines surrounding RE, but also to be as illustrative as possible and to be critical, keeping an objective view over RE expansion. In short, this book offers an honest, fresh, and thorough look at RE Governance facts and/or mistakes.

The idea was that each chapter should be edited from a pair or academics, one being experienced (a Professor, an Associate Professor or equivalent) and the other a young Researcher (a Ph.D. candidate, a postdoc, a Lecturer, a research fellow or equivalent) or an Executive from the Market. This was done to add depth and freshness to the views under examination and move away from "ritual" explanations and deeply engrained path-dependent thinking. We have deliberately asked Market Executives to participate, as this would balance the academic perspective of this book. We judged this inclusion as necessary because we consider that the voice of the Market should definitely go hand in hand with the voice of academics on crucial issues that touch everyday life, such as energy.

This book is designed to provide lecture material as well as numerous case studies to be further considered in seminars and discussions related to energy governance. More particularly, the book is targeted at those who want to actively learn about renewable energy and the governance aspects that shape and define the sector. The book does not provide a recipe but a palimpsest of emerging RE ideas and examples from across the word which raise questions, create challenges, and require solutions. It is designed to provide background information and case studies to help with the development of lectures for specialist undergraduate modules or postgraduate Higher Education courses. It is also for postgraduate or research students who are from other sectoral backgrounds, such as engineering or finance, but who want to more broadly access the aspects which shape the RE sector. Finally, it is for professionals who have experience of energy utilities or the RE domain and are well founded in traditional expositions and descriptions of the sector, but who are not entirely convinced by contemporary rhetoric and want to probe deeper to see what lies beneath the smooth veneer of the RE sector.

Evanthie Michalena
Jeremy M. Hills

Contents

Part I
Introduction

Introduction: Renewable Energy Governance: Is it Blocking the Technically Feasible?

Evanthie Michalena and Jeremy Maxwell Hills

Abstract Answers on Renewable Energy (RE) governance have been sought through many initiatives, Directives and legislative instruments, but to what extent the needs of RE have been met through these developments or to what extent questions have been appropriately answered remains debatable. In a dynamic world which is affected by external drivers, such as financial crises and geopolitical manoeuvrings, private companies seem to be at the forefront of RE deployment but they also lag behind or just shut down: why is this happening in the massively expanding RE sector? National governments can back up the whole RE deployment system, or create an investment boom, but they employ different criteria and differences in approaches which maybe suggests there is no "best" model when it comes to governance of RE systems. In the middle of this game with differing rules are the energy consumers, the general public, to whom RE benefits are deemed to accrue, but not without cost. Questions are beginning to emerge from consumers about costs and benefits of their governments policy on RE. The most bespoke of questions is "what is RE finally for", where and to whom do the benefits really accrue? What do you see when you look behind the financial and institutional cloak of subsidised RE delivery?

Yes, world energy needs to increase and environmental concerns need to be accommodated. Yes, solutions are urgently required. Then let us exploit all the opportunities offered by RE.

Not so fast….

Rapidly switching investment to cleaner energy technologies from fossil fuels is an option but it cannot operate alone. The scale of the world's energy needs and

E. Michalena (✉)
ENEC Laboratory, Sorbonne University and CNRS, Paris, France
e-mail: evamich@gmail.com

J. M. Hills
CTL Consult Ltd , 7 Mitford Road, Morpeth NE61 1RG, UK
e-mail: J.Hills@CTL-Consult.com

E. Michalena and J. M. Hills (eds.), *Renewable Energy Governance*,
Lecture Notes in Energy 23, DOI: 10.1007/978-1-4471-5595-9_1,

the ability for renewables to fill the gap left from fossil fuels, if they are abandoned, is unbridgeable. Countries need energy now and in the next 5 years and solely RE cannot meet those needs.

Many cards are dealt on the table which will allow the "luxury" of managing the transition to cleaner energy technologies without energy shortages or economic crises. For example, the carbon capture and storage (CCS) method which could potentially make coal/lignite electricity generation more friendly in environmental terms. Still, such interventions are costly, which might undercut the economics of other solutions (such as renewables); climate change (and its everyday impacts) is also pushing toward more daring solutions. Aside from the opportunity cost (the investment going into fossil fuels would be better spent on renewables), there are some more arguments that favor RE politics: First, fossil fuels are depleting commodities, which means that the further we go down the route of using them without alternatives, the more dependent on them we become and the bigger the consequences of reliance on a finite resource. Second, impacts on the groundwater (for example from fracking) and other environmental sustainability issues should also be accounted for in decision making. Added to that, the International Energy Agency says we must keep two-thirds of all fossil fuels in the ground in order to keep temperatures below 2 °C.[1]

Therefore, recent decisions are that RE is now intensively served up on political plates and locked-in with mandatory targets. As a result, last year, wind energy was the number one source of new American electric generating capacity, accounting for 42 % of all new capacity. RE as a whole accounted for 55 % of all new American generating capacity. In Spain, wind power this winter just surpassed that from all other sources. Germany is on track to get 35 % of its electricity from renewables by the end of the decade. But Germany soon will be surpassed by China as the country with the largest installed base of solar. This year China will more than double its installed solar from 4 to 10 GW (http://www.economist.com/debate/days/view/936/).

Still, RE governance creates headaches and at local, national, and regional levels as well. The problem here is that the concept of "governance" does not only concern "technology." It also concerns actors and frameworks, economics and institutions. "Governance" is a complex issue to talk about and "RE governance" is an even more difficult one, as it is insidiously present and highly variable in the way in which it affects the plethora of spatial and temporal domains of RE.

At an EU level, the first Directive to liberalize energy markets was the 93/94/EC. Since then, many technological and institutional changes have occurred so as to prepare national economies for a new energy regime; and not without cost: Energy Regulators have been unveiled and expanded, unbundling processes have detached System Operators from Electricity Utilities, private investors have presented their plans for the construction of energy plants, including RE ones (Frantzeskaki et al. 2008), and the most daring ones or the ones who got quickly familiarized with the

---

[1] http://priceofoil.org/2012/11/12/iea-acknowledges-fossil-fuel-reserves-climate-crunch/

governmental mechanisms have entered the RE Market with high profits to follow. In Greece, for example, such an example is TERNA ENERGY, which is active in the Hellenic RE area since 2002, has demonstrated an impressive growth in its turnover going from 17.9 in 2007 to 32.8 M€ in 2011 and from 118 MW RE installed capacity in 2007 to 266 MW of RE in the first semester of 2013[2]; this is in times of recession not only for the GEK-TERNA as an Association, not only for Greece as a country, but also for the whole world and the RE sector in particular.[3]

For the rest of the Hellenic companies, even if the RE penetration has reluctantly started in about 2002, the RE uptake has been institutionalized with the first Directive 2001/77/EC and the Directive 28/2001/EC has turned RE goals into "mandatory" ones, their "uptake" remains on paper…

Therefore, almost 20 years from RE's appearance on the scene, interesting questions that should probably be discussed in national Parliaments and the answers to which are hidden in the chapters of this book, evolve around three main axes:

(a) *Transition Paths Toward RE*

- Which are the drivers toward RE implementation and how those differ between countries? How strong is the role of the central government in promoting RE? And on which aspects? [see "Institutional Factors that Determine Energy Transitions: A Comparative Case Study Approach"—Koster and Anderies]
- Why local level RE needs the national energy decision-making so as to bear fruits? [see "Renewable Energy: Urban Centres Lead the Dance in Australia? " (Hamilton and Kellett), "Germany: Challenges of a Full Transition to Renewable Energy" (Lauber and Buschmann), "Green Electricity Certificates in Flanders: The Gradual Extension of a Market-based Mechanism and Doubts over its Cost-efficiency " (Moorkens et al.)]
- Which energy policy (top-down or bottom-up) is best to promote RE whilst protecting at the same time local people's rights? Which institutional level or levels are necessary to have a functional energy policy? [see "Renewable Energy: Urban Centres Lead the Dance in Australia?"—Hamilton and Kellett]
- How self-sustaining communities can undertake RE risks and accumulate benefits? [see "Outliers or Frontrunners? Exploring the (Self-) Governance of Community-owned Sustainable Energy in Scotland and the Netherlands"—Frantzeskaki et al.] Why Northern European communities can do this, whereas

---

[2] Profits EBITDA have been up to 17.9 M€ in 2007 (67.1 % more than in 2006). Then in 2010 = 1.4 M€, 32.8 M€ in 2011 with further continuous increases in 2012 and 2013. The installed RE capacity has been increased from 118 MW in 2007 to 266 MW of RE in the first semester of 2013. *Sources* (http://www.capital.gr/tools/view_printer.asp?ID=402802), (http://www.euro2day.gr/news/enterprises/article/689531/terna-energeiakh-afxhse-53-ta-kerdh.html), (http://www.euro2day.gr/market_announcements/com_results/article-market-announcement/1100892/terna-energeiakh-apotelesmata-a-trimhnoy-2013.html).

[3] Bad-news-there-s-less-being-spent-on-green-energy-good-news-there-s-more-green-energy-20130417.

Southern Europe communities appear to be less adept? [see "Endogenous Tourism Development Through Renewable Energy Governance: A Questionable Challenge" (Parpairis and Lagos) and "Outliers or Frontrunners? Exploring the (Self-) Governance of Community-owned Sustainable Energy in Scotland and the Netherlands" (Frantzeskaki et al.)].

(b) *"Hidden" Problems Coming up from RE*

- Why money-motivation is not so obvious at a national level? Ok, RE has penetrated but what about the community's wellbeing? Are RE national targets good enough to satisfy everyday struggles? [see "Renewable Energy Governance in Kenya: Plugging into the Grid, 'Plugging into Progress'"—Mwangi et al.]
- Why a country would be indifferent in promoting RE? [see "Renewable Energy: Urban Centres Lead the Dance in Australia?" (Hamilton and Kellett) and "Renewable Energy in New Zealand: The Reluctance for Resilience" (Byrd and Matthewman)]
- What about punishment of governmental behaviors in the case of RE? Does this concept really exist? Does it work effectively? Who bears the blame for the Market distortions that remain unresolved? And what is the price that national economies and private investors pay? Who are the private investors who really had a share in the RE market pie? And who were left outside? Which were the reasons? How much does the citizen feel protected from corruptions and protective reforms toward "big interests," especially in countries hammered by financial crises? [see "The Development of Renewable Energy Governance in Greece. Examples of a Failed (?) Policy "—Metaxas and Tsinisizelis]
- While in other countries [see "Outliers or Frontrunners? Exploring the (Self-) Governance of Community-owned Sustainable Energy in Scotland and the Netherlands"—Frantzeskaki et al.], the governance of RE has already been taken into local hands, small countries still struggle to find ways to handle the issue of RE authorization. Is the basic concern the incompetent governments, or the fact that incompetent governments are too well embedded in the governance system? [see "Lost in the National Labyrinths of Bureaucracy: The Case of Renewable Energy Governance in Cyprus "—Fokaides et al.]
- Why farmers are "obliged" to become RE developers and are there any benefits out of this choice? What is the role of Mayors in asking for a bigger portion out of the pie? How RE Gov changes over time? What are the games between big developers and State/local authorities? [see "Champagne And Metal Flowers: Who Is Invited To The Wind Generation Party In France?"—Burger and Mancebo]
- As time runs by quickly and RE possible sites are rare, developers feel trapped and Governments feel confused. Is there any time to learn from past mistakes other approaches should be adopted? Is RE capable to cover alone large-scale needs? And under which circumstances? [see "Renewable and Conventional

Electricity Generation Systems: Technologies and Diversity of Energy Systems" (Kosmadakis et al.) and "The perplexed Technical Governance of Wind Turbines in Greek Islands " (Caralis et al.)]

- Which is (or can be) the fate of non-interconnected islands, when it comes to RE use? Can it really be affordable, or is it about an utopia dream? Why private investors do not rush to seal the deal on interconnections? Have countries economically benefited from their clean tech foresight? And if yes, how much (in relation to the direct and indirect costs created?). Which are the technical, social, and economic consequences out of REintermittency and failure of immediate delivery? How much high could be the cost of back-up facilities and hybrid solutions? [see "The perplexed Technical Governance of Wind Turbines in Greek Islands "—Caralis et al.]
- Are the criteria efficiency and effectiveness sufficient ones to assess a situation, or other criteria are useful as well, such as relevance, sustainability, stakeholders' involvement, etc.? Why Environmental Impacts Assessment is defective when used as a tool alone? And, if "defective" indeed, why the State has set it as a mandatory process for an RE project to be implemented? Which are the environmental problems to be created by RE? [see "Environmental Impacts of Renewable energy: Gone with the Wind?"—Kouloumpis et al.]
- What happens to countries with vulnerable people [see "Renewable Energy Governance in Kenya: Plugging into the Grid, 'Plugging into Progress'"—Mwangi et al.] and to countries into recession and what is needed to get them competitive in RE—if there is this need at all ? [see "Times of Recession: Three Different Renewable Energy Stories from the Mediterranean Region "—Boemi and Papadopoulos].

(c) *RE Governance—Further Food for Thought*

- Under which conditions the multi-aging system can create confusion? [see "Renewable Energy Governance challenges within a "puzzled" institutional map"—Kottari and Roumeliotis]
- Which are the geopolitical benefits of using RE instead of other technologies? For example, is the building up of a regionalization status, possible? Can RE create a new economic engine alongside significant fossil fuel exploitation? [see "Geopolitics, Climate Change and Energy Governance: a Grey Area in the Black Sea Region."—Hills and Michalena]
- Which is the relation of RE with other political interests and how RE does serve them? Is it possible that an RE energy success can constitute a notable political failure? Do RE State policies and tools reach effectively the public? And if not, what are the reasons? [see "Germany: Challenges of a Full Transition to Renewable Energy"—Lauber and Buschmann].

(d) *Considerations for a Good Future RE Governance*

- Which are the State policies that allow competition? Which are the main financial mechanisms chosen by States, why are they privileged and what were their consequences? When the RE system is effective but not efficient and how to create balance between the financial instruments and the Market? Alternative models and suggestions can be made to accelerate RE permits allowance, but at what cost? Can incentives boost RE sources which are limited in a specific area? In other words, how to stimulate investments in desirable technologies? And how to choose which technologies we should limit support? How to get a good feedback from the RE system and then, to manipulate it according to our needs and at a reasonable cost fairly distributed between customers and developers? [see "The Shadows Cast by Inadequate Energy Governance: Why More Sun Does Not Necessarily Mean More Photovoltaic Electricity" (McCormack and Norton) and "Green Electricity Certificates in Flanders: The Gradual Extension of a Market-based Mechanism and Doubts over its Cost-efficiency" (Moorkens et al.)]

- Why a country being rich in energy sources would be interested in including RE in its energy mix and which should be the best way to succeed in this? [see "Building on Norway's Energy Goldmine –Policies for Expertise, Export and Market Efficiencies"—Moe Skjolsvod et al.]
- How many agents are good enough for the implementation of an RE system? [see "Building on Norway's Energy Goldmine –Policies for Expertise, Export and Market Efficiencies"—Skanavis et al.]
- Which are the advantages of authoritarian political systems toward RE adoption? [see "The Political-Economic of the Green Industrial Revolution: Renewable Energy as the Key to National Sustainable Communities"—Clark and Li].

The main content of the book is structured as follows: After the introductory chapter that emancipates the most interesting—by experience—questions on what is blocking RE Governance, there is a first chapter called "Technical Introduction." This chapter introduces the reader to the variety of energy technologies and their main technical and economic aspects. This chapter will serve both as a technical entry to the world of energy technologies, but also as the first triggering of the concept "Governance" when this is related to energy. The rest of the chapters are classified around four Parts, the first part working around the efficiency of transition paths toward RE, the second part giving evidence on "hidden" problems that RE's exploitation creates, the third part instigating further food for thought on RE governance schemes, and a fourth part giving some hints on what might be considered to be a "Good RE Governance." The book finally closes with Answering the "Mega-What" Question: Who is Finally Renewable Energy For.

Reference

Franzeskaki N, Michalena E, Angeon V, Van Daalen E (2008) The on-going Greek energy transition to sustainability. In: The proceedings of the conference protection and restoration of the environment IX (PRE9), Kefalonia, Greece 30 June–3 July

Renewable and Conventional Electricity Generation Systems: Technologies and Diversity of Energy Systems

George Kosmadakis, Sotirios Karellas and Emmanuel Kakaras

Abstract In this chapter, the primary technical aspects of conventional and renewable energy systems are presented. The description focuses on commercial systems installed across the world, together with a brief introduction to some promising technologies currently under development, such as Carbon Capture and Storage (CCS). Conventional energy systems include power plants using fossil fuels (natural gas, coal, etc.), while renewable energy systems include solar, wind, geothermal, biomass, and small-hydropower applications. These technologies are briefly described accompanied by economic figures (installation cost, fuel cost, specific cost of electricity, etc.) and emissions data (where applicable). Some insight on the energy strategy in specific countries is provided and how this can be related to local conditions and electric power requirements.

1 Introduction

Lately, renewable energy systems are increasingly being used for electricity generation, either at small-scale decentralized systems with capacity in the kW scale or even medium-scale systems (often called utility-scale) with capacity of a few MW. However, the large-scale systems with capacity of some hundreds of

G. Kosmadakis (✉)
Internal Combustion Engines Laboratory, Thermal Engineering Department,
School of Mechanical Engineering, National Technical University of Athens,
Heroon Polytechniou 9, 15780 Zografou, Athens, Greece
e-mail: gkosmad@central.ntua.gr

S. Karellas · E. Kakaras
Laboratory of Steam Boilers and Thermal Plants, Thermal Engineering Department,
School of Mechanical Engineering, National Technical University of Athens,
Heroon Polytechniou 9, 15780 Zografou, Athens, Greece
e-mail: sotokar@mail.ntua.gr

E. Kakaras
e-mail: ekak@central.ntua.gr

E. Michalena and J. M. Hills (eds.), *Renewable Energy Governance*,
Lecture Notes in Energy 23, DOI: 10.1007/978-1-4471-5595-9_2,

MW are still using conventional technologies based on fossil fuels (natural gas, coal, oil, lignite, etc.). These very large plants operate at high loads (operation at a range of 50 % to 100 % of their capacity), having high capacity factors (operation of many hours annually), and covering the base load needs of the electricity grid. One of the most important disadvantages of conventional technologies is the environmental impact. The combustion of fossil fuels leads to the inevitable production of carbon dioxide (CO_2), while most of the times harmful emissions are produced, such as carbon monoxide (CO), nitrogen oxides (NO_X), sulfur oxides (SO_X), unburned hydrocarbons (HC), and solid particles.

Another critical disadvantage of conventional technologies is that they need continuous fuel supply to operate, which contributes to the operating costs. This cost depends on various local and global parameters, such as fuel availability and type, fuel purity, world economic conditions, local prices, etc.

On the other hand, renewable energy technologies do not require any fossil fuel during normal operation. Their operation is based on the exploitation of natural resources, such as the sun and wind, having relatively lower operating costs, although they still require some maintenance. The most important disadvantage of renewable energy technologies is the fluctuation of their power output, which depends on the intermittency of the particular natural resource (Kosmadakis et al. 2011), which can have a significant variation, even on an hour-to-hour basis. This aspect has brought many second thoughts and skepticism on their wider implementation when energy supply must be secured, irrespective of any possible fluctuations in renewable energy resources availability.

Additionally, the installation cost (in terms of cost per installed kW) of renewable technologies is usually much higher than that of fossil fuel fired plants, bringing some restrictions on their scale and their total capacities for electricity generation. Although this cost is rapidly decreasing during the recent years, securing a lower specific energy cost (cost per kWh), and slowly approaching grid parity (specific cost equal to the electricity price for an end-user), there are still many issues to be resolved, such as excess energy storage and their safe and efficient integration to the electricity grids (Steinke et al. 2013).

The following sections introduce the mainstream commercial conventional and renewable technologies and their main characteristics and features. Cost parameters are also presented, having in mind that the cost of electricity of each technology depends a lot on the prevailing local and national circumstances, such as climate, economic, market, and even political conditions. This aspect should be considered before any techno-economic comparison is made among the different technologies.

2 Renewable Energy Technologies Overview

The commercial renewable energy technologies for electricity generation are briefly described next. Their main features are provided, together with an insight into their common scale and real applications.

2.1 Solar Energy

One of the most widely developed renewable energy sources is solar energy. Solar energy applications are constantly increasing in the last few years, and they are considered perhaps the most promising that can significantly contribute to the total electricity generation. There are two main technologies involved in the exploitation of solar energy, which differ in the way that solar radiation is harvested and converted to electricity. These are the solar photovoltaic (PV) and the concentrated solar power (CSP) technology, which are presented in the following sections.

2.1.1 Solar Photovoltaic Technology

Solar photovoltaic technology (PV) is the most popular technology for capturing solar energy and converting it to electricity (the solar radiation is directly converted to electricity with the use of semiconductor materials). One reason for its popularity is the modular design/size of a PV unit which has no moving parts, permitting it to be installed even on building roofs with generation capacity starting from a few Watts.

The basic element of a PV is the solar cell. There are various technologies of such cells, having a large variety of efficiency and cost. The most common solar cells are the crystalline silicon ones (either mono- or poly-crystalline), while thin films are also increasing their market share, due to their low cost and sufficient performance. Multijunction solar cells are still at a very early stage of commercialization, due to their extremely high cost and are used only in special applications (e.g., in space) and in high-concentration PV plants.

A number of solar cells form a solar panel (or solar module) with common power output of some hundreds of Watts. A PV plant consists of many such modules arranged in arrays in order to produce the required power, ranging from a few kW up to a few MW. It should be mentioned that PVs produce direct current (DC) electricity and inverters are required to convert it to alternating current (AC), decreasing somewhat their efficiency. The maximum electric efficiency of such plants at real conditions is around 10–15 %, considering also their losses at the cables, cells, and their temperature effect (Kosmadakis et al. 2011; Skoplaki and Palyvos 2009).

PV cells are also used in concentrating photovoltaic (CPV) units, where lenses or curved mirrors are used in order to increase the direct solar radiation on the PV cell surface (Kosmadakis et al. 2011). In such units, a solar tracking device is used, which traces the Sun's movement during the day. Such trackers can be also used in flat PVs, which are not very common, because of the increased maintenance and installation costs.

In CPV units, the total PV cell area per kW produced is decreased, since the incident solar radiation on the PV cell area is significantly high, making it affordable to use high-efficiency cells, such as multijunction ones (Pérez-Higueras et al. 2011). The concentration ratio can even reach values of around 1,000 for the high-concentration configurations. A major issue in such cases is the over-heating of the PV cells, due to the high incident radiation. There are many methods to remove these large quantities of heat, either using passive ways (e.g., fins on the back of the module, applicable mainly on low-concentration systems), or active ways (e.g., using fans to cool the PV cells, for higher concentration ratios) (Kosmadakis et al. 2011; Pérez-Higueras et al. 2011). In some cases, this heat is effectively recovered and used for heating purposes, resulting to a CPV/Thermal system, which is usually installed at the kW scale and with low concentration ratios (Kosmadakis et al. 2011).

The most common way of specifying the cost of PV and any other energy production technology, is by calculating the specific cost (cost per kWh) or else *Levelised Cost Of Energy* (LCOE) (Hernández-Moro and Martínez-Duart 2013). This value for the PV technology depends on the local weather conditions, but reasonable values are in the range of 0.12–0.25 €/kWh, which are however rapidly declining the last years. The installation cost of PV plants is around 2,000–2,500 €/kW for the small-scale systems, decreasing to just 1,500 €/kW (or perhaps even below that) for the larger ones (Branker et al. 2011; Kosmadakis et al. 2011). In Fig. 1 the average cost of PV plants is depicted, which is decreasing during the last few years for both utility (of MW scale) and commercial scale (of kW scale), while the future prediction/projection includes some smaller price decrease (Branker et al. 2011; Hernández-Moro and Martínez-Duart 2013). The decrease of LCOE is actually the combined result of both price decrease and PV efficiency increase, so that its reduction is steeper and quickly approaching the grid parity in areas with high insolation, such as in the USA, China, and Spain, where the solar PV industry is also very active.

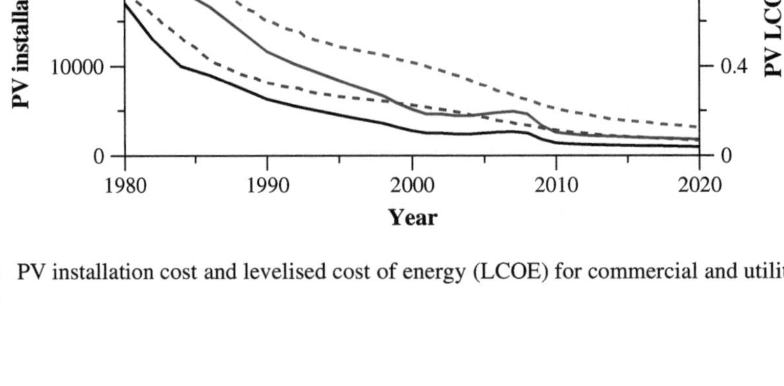

Fig. 1 PV installation cost and levelised cost of energy (LCOE) for commercial and utility scale plants

2.1.2 Concentrated Solar Power Technology

Solar thermal collectors are used to capture the solar radiation, convert it to heat and finally to electric energy with the use of a heat engine. Flat plate collectors are rarely used for electricity generation, since their maximum operational temperature is usually lower than 120 °C and the resulting efficiency is too low. The most common method for power production using solar thermal technology is the concentrated solar (thermal) power (CSP) technology.

This technology concerns lenses or mirrors, which concentrate the solar radiation on the receiver, in which a fluid is circulated, heated, and then driven to the heat engine for power production. The CSP plant configurations vary, depending on the geometry and the operation of the system, as well as the solar radiation harvesting method, which can be divided into parabolic trough collector, Fresnel collector, dish Stirling collector, and the solar power tower (Viebahn et al. 2011). In all these variations, the main concept is the same, whereas the heat engine and the concentration ratio can change, bringing a variation to the operating temperature as well. The most popular CSP technologies, together with their most common commercial plant scale and efficiency are depicted in Table 1.

The solar power tower shows the highest efficiency, due to the operation at high temperature and capacities. On the other hand, its installation cost is higher than the parabolic troughs, which are today the most common CSP technology.

CSP plants are mostly of large scale with common capacities of more than 10–20 MW, since the heat engine technology used is derived from conventional plants with some proper modifications. Usually water-steam is generated and expanded in a conventional steam turbine, while a thermal-storage unit can be also included in the plant configuration to stabilize the power production and operate during the night as well. This additional feature increases the installation cost, but provides flexibility to the power production and solves some electricity grid issues that could arise. In CSP plants with capacity lower than 1 MW, the heat engine is usually an organic Rankine cycle (ORC) engine, which is the ideal technology for small/medium-scale systems (Kosmadakis et al. 2011). Finally, the solar dish Stirling unit is of kW scale and has increased concentration ratio and temperatures, making its Stirling power cycle operate with high efficiency, having unfortunately high costs. The main advantage of the Stirling Dish technology is that no cooling system is needed, making this technology more applicable for areas, where no water is available (e.g. in deserts).

Table 1 CSP technologies and indicative plant scale and solar to power conversion efficiency

Solar thermal technology	Scale (MW)	Solar to power conversion efficiency (%)
Solar dish stirling engine	0.01–10	20–25
Linear fresnel collector	0.1–10	15–20
Parabolic troughs	1–150	20–25
Solar power tower	2–300	30–35

CSP plants are not very common and just lately their development has been expanded. They are mainly installed in areas, where there are significant incentives (such as in Spain and the USA) and high solar potential, preferably over 1,800 kWh/m^2/year of direct solar radiation. Their development started from the 80s with the Solar One and the SEGS projects in the USA (NREL 2012a) and continued with Planta Solar 10 in Spain (NREL 2012b). These two countries are the front-runners in the sector and have invested great resources on research facilities of this technology.

The LCOE of CSP plants is in the order of 0.15–0.20 €/kWh, showing a large range of installation costs of 2,000–6,000 €/kW (Hernández-Moro and Martínez-Duart 2013). These values depend on the technology used, its scale, and the possible use of a thermal storage unit, while the solar power tower and the solar dish Stirling have the highest costs.

2.2 Wind Energy

Another very common renewable energy source is wind. With the use of a wind turbine (most common is the horizontal axis turbine), the kinetic energy of wind is converted to mechanical power and then to electricity. Usually, several wind turbines are installed together and constitute a wind farm. Their capacity ranges from some kilowatts up to some megawatts (the large wind farms can even reach a capacity of 100 MW), while they can be installed either on-shore (not only lower installation cost, but also lower average wind speeds) or off-shore (higher costs, but much higher average wind speeds).

A common wind farm's capacity factor is around 20–30 % (in other words, how much time per year the system operates at maximum capacity), while their installation is advised only in areas with high wind potential (e.g., high-altitude areas). This technology is very mature, since many commercial units are installed every year, and with thousands of GW already installed around the globe. Their cost is steadily declining, although with a slower rate compared to PV, and is currently equal to around 1,000–1,500 €/kW for the large on-shore units of megawatts-scale. This cost is much higher for the smaller ones (Bolinger and Wiser 2012), due to the economies of scale and for the off-shore ones, since large infrastructure is required for installing them at deep waters.

The LCOE of wind farms takes into consideration the equipment cost, the operating and maintenance (O&M) cost and the electric energy produced, while it shows a large variety, depending on the size, the location, and the wind potential. Common values are around 0.10 €/kWh or even less for large wind farms, which is highly competitive and has pushed wind farms to a fully commercialization. However, there are still some relevant research activities, mainly dealing with the development of very large wind turbines and very small ones (of few kW, appropriate to be installed in the urban environment).

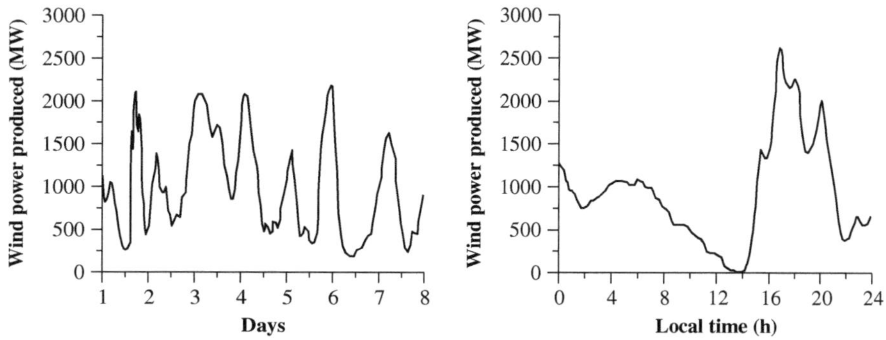

Fig. 2 Fluctuations of wind power produced during a representative week (*left*) and day (*right*)

Today, wind farms are installed in various areas of the world, reaching a total global capacity of 300,000 MW (Bolinger and Wiser 2012) (four times higher than the total PV capacity). The main reasons for such high penetration of wind energy plants are their low costs and mature technology. On the other hand, the cost-effectiveness of such plant is influenced from the selection of the installation location which should be critically evaluated. For a specific location, a large database of wind measurements over many years should be available, before any engineering work is implemented.

The most important disadvantage of wind power plants is the strong fluctuations of the power output and the great uncertainty on prediction of wind velocity. These two factors greatly contribute to the low capacity factor of such plants. The wind velocity and power produced can become unavailable during a very short time (even within an hour), which means that at this case there should be reserved power from other type of power plants or energy storage technologies, to cover the required energy. In Fig. 2 is shown a representative power fluctuation of wind plants during a week (left chart) and during a day (right chart). Such fluctuations are very strong and reduce the flexibility of an electricity grid, which should at all times be able to cover the energy demand.

The recent world leader of wind farms is China, who is rapidly progressing during the last few years, having large wind potential. Other important global players are the USA, Germany, Spain, India, Canada, France, and many other countries of the European Union (GWEC 2013; Hu et al. 2013), which show well-developed wind industry and applications. Denmark has currently the highest wind penetration worldwide, since around 20 % of its energy production comes from wind farms, many of those being off-shore (GWEC 2013), following political decisions to focus on wind power, in order to decrease their carbon emissions and coal use. Apart from that, the wind energy industry in Denmark is very strong, supporting the installation and operation of many wind farms.

2.3 Bioelectricity Generation

Biomass is a renewable energy source and refers to waste and residues from agriculture (including vegetal and animal substances), forestry and related industries, energy crops, as well as the biodegradable fraction of industrial and municipal waste that can be used as fuel for different scale power production. Its use for electricity production is "CO_2-free", or, in other words, "CO_2-neutral", since the amount of the CO_2 released during its utilization equals the amount, which has been assimilated from the plant during its growth (photosynthesis).

The potential of the so-far unexploited biomass for energy power, fuels, and chemicals from biomass is of increasing importance in addressing issues of global warming and sustainability. The total amount of primary bioenergy production in the 27 Members States of the European Union (EU-27) was 100.77 Mtoe in 2009 and 112.73 Mtoe in 2010 respectively with a continuous growing market, in order to meet the goals of 2020 (Jäger-Waldau et al. 2012). The bioelectricity predictions in EU-27 for 2020 are depicted in Fig. 3, where the expected bioelectricity production is shown in each EU country for solid, gaseous and liquid biomass.

Biomass can vary in composition and form according to fuel properties, cultivation, and harvesting period. The low energy content of biomass fuel imposes additional techno-economic barriers concerning availability, logistics, and replacement of food crops. The conversion of biomass can be realized with either thermochemical processes including combustion, gasification, pyrolisis, liquefaction, or biochemical processes, such as anaerobic digestion, fermentation, and enzymes.

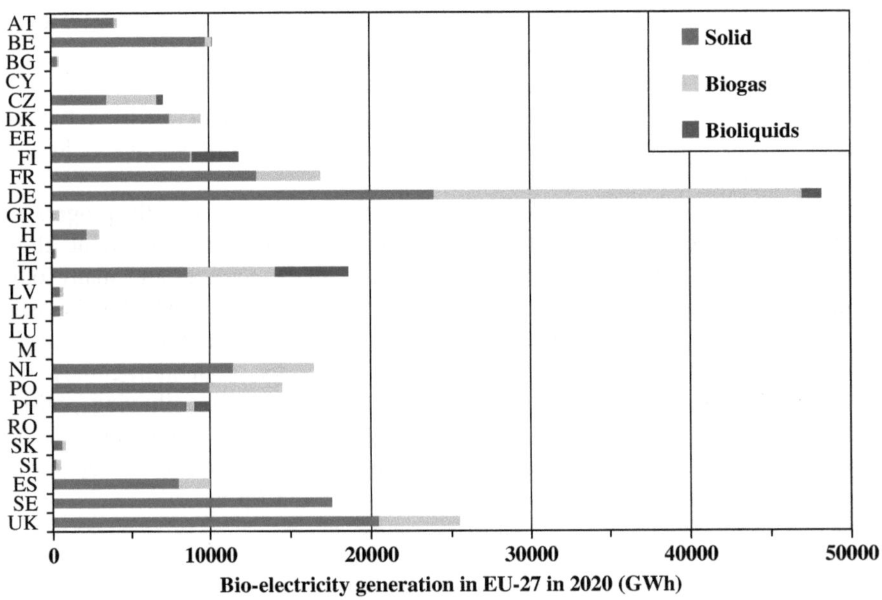

Fig. 3 Estimation of bioelectricity generation in EU-27 by 2020 (Jäger-Waldau et al. 2012)

Biomass combustion has gained increasing attention for heat generation in domestic boilers and with respect to availability issues can find application in larger scales for electricity generation. The combustion of solid biomass usually takes place in steam or organic Rankine cycle power plants with electric efficiency in the range of 15–38 %, while it can be also co-combusted in conventional power plants, replacing some fossil fuel and reducing its CO_2 footprint.

Biomass thermal gasification and anaerobic digestion have attracted the highest interest amongst the biomass conversion technologies as the solid biomass can be converted to a multidiversable energy carrier gas. The produced gas is called syngas (from thermal gasification) or biogas (from anaerobic digestion) and can be exploited in emerging technologies for heat and power production and realize the decentralized energy production concept. The total global installed capacity of electricity power plants possible to be fed with raw material of renewable origin was 25.8 GW in 2008 and 28.7 GW in 2010, showing an average annual increase of about 2 GW (Jäger-Waldau et al. 2012). The available small-scale technologies for converting the biomass derived gas to electric power include internal combustion engines (ICE), microturbines, and fuel cells. Small-scale efficient combined heat and power (CHP) plants based on biomass gasification can have electrical efficiencies from 20 to 30 % of the biomass fuel lower heating value (LHV), being capable to operate for many hours per year.

Solid biomass combustion is implemented in small to medium scale plants using the technology of organic Rankine cycle (ORC), Stirling, or steam engines which however still have high costs and low efficiencies (Rentizelas et al. 2009), despite their recent commercialization. The utilization of such plants is steadily increasing, since they can operate for many hours per year with high capacity factors (depending on the fuel availability) and are considered a reliable renewable energy source, since the power production fluctuations can be minimized and controlled. The conversion efficiency decreases with the reduction of the power output, which in reverse is restrained by fuel availability and price. For power output higher than 2 MW, biomass combustion plants are used with water steam cycle reaching electric efficiencies of up to around 38 %, depending on the fuel type and the steam parameters.

For scales over 20 MW, Biomass Integrated Gasification Combined Cycle (BIGCC) technology can be used, reaching efficiencies up to 40 % (Jäger-Waldau et al. 2012). In addition, biomass is the main renewable source for liquid fuel production (such as biodiesel) used predominately in the automotive sector.

The installation cost of biomass plants is rather high due to economies of scale issues. For large steam cycle biomass plants (above 10 MW) the specific investment cost can be higher than 2,000 €/kW. For technologies used for power output lower than 2 MW, probably including ORC technology, an installation cost higher than 5,000 €/kW is expected, whereas for gasification technologies the cost is around 6,500 €/kW (for turn-key projects). The resulting LCOE is about 0.20 €/kWh depending on the fuel cost and the technology used.

In the case of biogas plants, the installation costs, including the fermenter and the internal combustion engine, are in the range of 4,500 €/kW (Karellas et al. 2010). The resulting LCOE is about 0.20 €/kWh (Stürmer et al. 2011). Taking into consideration the power production control of such units, it is concluded that bio-energy power plants can be a highly competitive technology, especially if there is a thermal energy consumer (having a combined heat and power—CHP—plant at this case), for example a district-heating network.

Many bio-energy plants for electricity generation exist in areas where solid biomass and syngas/biogas are widely available at low cost and require little transportation. Such areas can be near forests (using wood residues) common in the USA and Scandinavian countries, agricultural areas (using agricultural waste) common in Latin America, and landfills, animal waste or energy crops for biogas production using an anaerobic digester and fuelling internal combustion engines for electricity generation, being very common in Germany (Gewald et al. 2012). It should be highlighted that although Canada has the highest solid biomass production in the world, the biofuel is usually converted to pellet and used for heating purposes and not electricity generation.

2.4 Geothermal Energy

The underground fluid thermal energy, called geothermal energy, can be recovered and used for electricity generation with geothermal plants. These include drilling, in order to use the hot fluids found some hundred meters below the surface of the earth. These plants are using either water at low-temperature range (over 100 °C but usually below 200 °C) and special steam turbine technology, or even a working fluid boiling at low temperature (such as pentane, R134a, R245fa, etc.), and organic Rankine cycle or Kalina technology (Campos Rodríguez et al. 2013). Due to their low-temperature operation, their efficiency is lower than 25 % depending on the technology used, but their capacity factor is high, reaching even 70–80 % (Dipippo 2004, 2012), since geothermal plants do not show significant power output fluctuations during the year, in contrast to solar and mainly wind plants. Geothermal plants are usually constructed with capacity of a few MW, although there are also some plants in the kW scale.

Their installation cost is site-specific, usually ranging from 2,000 up to 4,000–5,000 €/kW. Although these costs are high, their capability to operate for many hours per year enables them to have a low LCOE, usually lower than 0.15 €/kWh (Dipippo 2004, 2012), which is an important advantage, and can reach even values of 0.08 €/kWh for cases of high-enthalpy/temperature geothermal fields.

The low worldwide penetration of geothermal energy is mainly related to the scarcity of suitable places for construction of such plants. This fact can be witnessed by the low total installed capacity of around 12,000 MW (compared to the 300,000 MW of wind farms' capacity). USA is the world leader, while many plants also exist in Italy and Mexico (Dipippo 2012). In some areas such technology is very common for electricity and heat generation, such as Iceland, where around 25–35 % of the electricity consumption is produced by geothermal plants. Also, New Zealand and Philippines have a high geothermal energy penetration with many installed plants and increased percentage of the total electricity generation (Dipippo 2012).

In conclusion, geothermal plants can be installed in rather limited locations, and this is the main reason why their market penetration in the renewable energy pool is increasing very slowly. Also, their high installation costs can be sometimes restrictive. On the other hand, they are reliable plants, being able to operate as base-load plants as well, and are accompanied with very low operating costs and without any fuel cost.

2.5 Small Hydropower Plants

There is a strong debate as to whether large hydropower plants are considered a renewable energy source or not. During their operation, a large area is flooded using a dam, significantly changing the natural environment and relocating people. For this reason, the discussion in this chapter only deals with small hydropower plants, which do not involve the construction of a dam to change the natural flow of a river. Such plants usually have a capacity lower than 10 MW (Anagnostopoulos and Papantonis 2007), while they require a careful planning and design of the installations, since the available water quantities of the river could change over time and during a year.

There are various turbine types, which can convert the gravitational energy of water to mechanical power and then to electricity with the use of a generator. The most important ones are the Pelton, Francis and Kaplan hydroturbines. All of them are suitable for specific cases and are installed after careful investigation of the available water flow rates and head (in other words the water pressure).

The LCOE of such plants is usually very low, since the operational costs are extremely low, with common values around 0.10 €/kWh, being site specific. Moreover, their installation can be accomplished with moderate values of €/kW, normally using a custom-made hydro-turbine and a large piping infrastructure (with length up to few km). Although a dam is not used to form the water reservoir (in large hydro-plants the dam construction is a large proportion of the total cost), their installation cost is usually from 2,000 up to 4,000 €/kW (ESHA 2013).

An important advantage of such plants is that they have very short response time, which means that they can start or stop operating within very few minutes. Also, they can operate for many hours per year, depending on the water flow availability. Nevertheless, the annual produced electric energy can vary each year, depending on the intensity and duration of the rainfalls. This is an important restriction for such plants, since they can even stop operating during dry months, since hydroturbines cannot operate at extreme off-design conditions with low water flow rates. The indicative water availability and electric energy produced during a typical year, together with the indicative water resources annual fluctuation is shown in Fig. 4. It should be mentioned that hydropower production is almost a linear function of the available water quantities. The peak water flow rates observed during the summer months (months 6 and 7) are due to the snow melting.

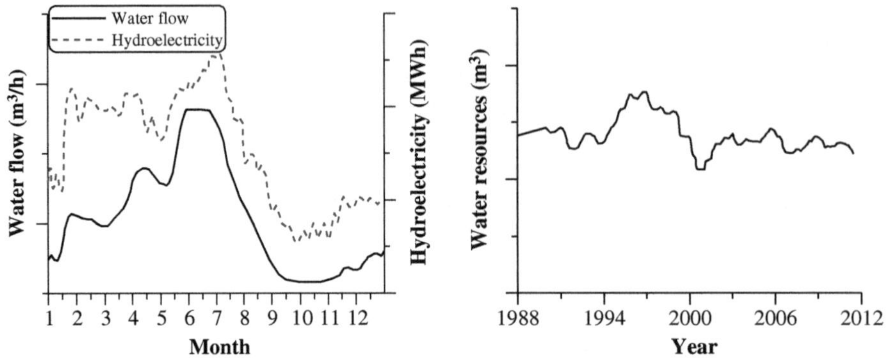

Fig. 4 Indicative available water flow and electricity during each month (*left*) and available water resources each year (*right*)

As far as the installed hydropower is concerned, China is the world leader in small hydropower plant capacity, having more than the half global capacity, with almost 100,000 MW. Japan is a major player as well, while the USA, Canada, and many countries of Latin America have also installed many small hydropower plants, due to their increased water resources. European countries, such as Italy, France, Switzerland, and Germany, have much lower installed capacities in the range of 2,000 MW each (ESHA 2013).

2.6 Other RES Ready to Get Commercialized

Ocean energy plants take advantage of the energy of waves, currents, tides or even the seawater temperature difference at different depths (most common as ocean thermal energy) and convert it to electricity with various methods. Such plants are usually off-shore and, up to now, they have been installed mainly in the UK and the USA, even at medium scale (of few MW).

Most of these projects are mostly demonstration ones, while such technology has just recently moved to commercialization. These plants are very promising, but there is still lot of research and development to be conducted, in order to decrease their costs and improve their reliability and efficiency.

3 Fossil Fuel Fired Power Plants

Electricity generation from fossil fuels in Europe and the rest of the world is expected to continue to play an important role in the international energy mix. Despite the increase of the renewable energy sources and their penetration in the

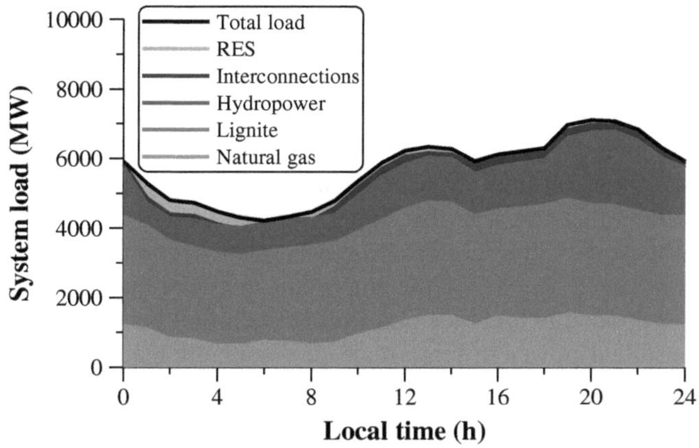

Fig. 5 System load and energy mix during a typical winter day in Greece (IPTO 2013)

global energy mix, fossil fuels cover more than 50 % of the total electricity production. The present energy mix during a typical winter day in Greece is depicted in Fig. 5, showing the technology brake-down of energy produced (IPTO 2013). It is observed that most of the electricity produced comes from fossil-fuel fired plants, whereas the contribution of renewables is very small.

Four of the most indicative technologies for power generation are presented in this section, namely the steam power plants, the open cycle gas turbine, the combined cycle gas turbine and the internal combustion engines.

3.1 Coal, Lignite-Fired Steam Power Plants

It is generally expected that coal will continue to play a key role in the future energy mix as it is the most abundant and cheapest fossil fuel source. The higher stability of the coal market compared to the oil and natural gas market, guarantees stable electricity costs, which could not be achieved without the utilization of coal as the main fuel for electricity production. Such solid fossil fuels are combusted in steam power plants, where the power cycle is based on the steam-Rankine thermodynamic cycle, using a steam turbine. These plants are widely installed in many countries, since it is the most common conventional technology.

The increase of power plant efficiency and capacity is one of the main goals of the coal industry and is directly related to the steam parameters in the boiler, seeking to increase the steam temperature and pressure. Progress in boiler design throughout the twentieth century resulted in tremendous improvements in both coal and lignite combustion and advanced steam cycles. The optimum boiler design is usually a balance between breakthroughs in combustion and metallurgy. However, the reliability of these new developments has to be also proven, since the large-scale power plants have normally a life time of at least three decades. Furthermore, advances in emission control processes and equipment have managed the reduction of the environmental impact of the combustion process especially in the large-scale equipment. It should be noted that since the early applications of pulverized coal-fired boilers, emissions' rates have been reduced more than 300 times on a kilowatt-hour basis. Ongoing improvements are anticipated that will further improve this figure to more than 700 times.

The development of the large-scale power plant technology in the twentieth century is considered as one of the important engineering achievements. Coal, lignite pulverization, and combustion of pulverized fuel contributed to the drastic increase of installed power in large-scale power plants leading to a reduction of specific capital costs. At the same time, the progress in material science and steel manufacturing methods facilitated the development of steam boilers operating in supercritical steam parameters (above the critical conditions: 220 bar, 374 °C). This in turn led to a considerable increase of the electric efficiency and to a decrease of the electricity generation cost. Pulverized fuel firing has also contributed to the reduction of labor costs in steam power plants, increasing the flexibility of operation and allowing use of an extremely wide range of fuels, such as lignite, brown coal, hard coal, etc. The alternative, fluidized bed combustion technology, allows even higher fuel flexibility (e.g., petcoke, slurry, lignite, anthracite, etc.) with excellent environmental performance in terms of NOx and SOx emissions.

The plants' efficiency increase in the last few years was based on the progress of new high temperature resistant materials. The development of the 9 % Cr steels P91 and P92 in the late 1980s and 1990s was the result of an international effort and allowed the increase of supercritical steam parameters to the range of 300 bar and 600 °C, showing a satisfactory power plant performance in terms of reliability, flexibility, efficiency, and economy (Bugge et al. 2006). The requirements on further increase of plant efficiency and minimization of CO_2 emissions lead to the current research effort on the development of component materials for a 700 °C ultra supercritical power plant and on the steam cycle optimization. Brown coal pre-drying technology may play an important role toward the realization of this goal. It is estimated that the optimization of the drying process in future brown coal power plants may lead to an efficiency increase of 4–6 % points. Besides, the development of an efficient brown coal drying process is a necessary step towards the implementation of oxy-fuel firing in future generations of brown coal power plants (Agraniotis et al. 2012). One of the most advanced brown coal pre-drying technologies, currently under development, is the fluidized bed drying concept with waste heat utilization (WTA-Drying) developed by RWE (2008).

The expected plant efficiency increases through the application of lignite pre-drying technologies and supercritical parameters with 700 °C live steam temperature is shown in Fig. 6, where it is depicted that the future plants can reach an electrical efficiency of 50 % and therefore, a reduction of the CO_2 emissions is expected.

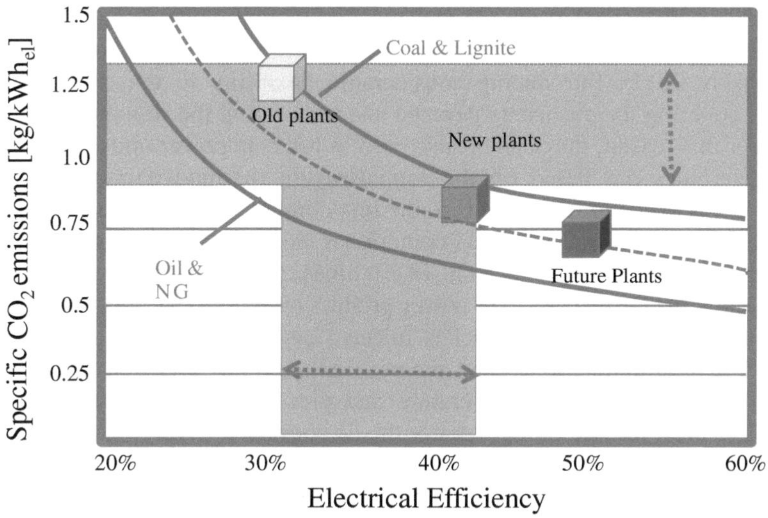

Fig. 6 Specific CO_2 emissions as a function of power plant's electrical efficiency

Regarding the CO_2 emissions there are research activities and demonstration plants for testing the three main Carbon Capture and Storage (CCS) technologies, namely: (a) Pre-combustion capture, (b) Oxy-fuel combustion and (c) Post-combustion capture. The NO_X emissions are in the range of 3,000 kg/GWh, and SO_X emissions around 6,700 kg/GWh, while particle emissions around 9,000 kg/GWh (Spath and Mann 2000). These emissions are controlled with the optimization of the combustion system (air–fuel staging), DeNOx systems, desulfurization systems, and particle filters.

The electricity generation cost of such plants is calculated as the sum of the following costs: (a) annualized investment cost, (b) fuel cost, (c) fixed operating and maintenance cost, (d) variable operating and maintenance cost, and (e) costs for CO_2 allowances. The latter is also regarded as variable cost. An overall installation cost of 2,000 €/kW is a representative value for a state of the art Greek lignite power plant (Agraniotis et al. 2012). It should be noted that, due to different reasons related with logistics and infrastructure, the installation cost of a similar lignite plant in Central Europe may be considerably lower ranging from 1,500–1,800 €/kW. The resulting LCOE of such plants has a large variation, depending a lot on the fossil fuel costs and quality, but is usually in the range of 0.04–0.07 €/kWh.

3.2 *Gas Turbine Plants*

Gas turbines operate with gaseous or liquid fossil fuels, usually natural gas, reaching efficiency values up to around 30–35 % (De Sa and Al Zubaidy 2011). Their capacity can range from a few (e.g., 0.05 in the case of microturbines) up to hundreds (e.g., 250) of MW and they are a modular technology. An important advantage of gas turbines is their possibility to commence operation within a few seconds (De Sa and Al Zubaidy 2011), introducing considerable flexibility to the electricity grids, quickly covering the electricity demand and performing the so-called "peak shaving". For this reason, they can be also used as back-up power units.

Gas turbines are based on the Joule/Brayton thermodynamic cycle. Atmospheric air is driven through vents to the first component of the gas turbine, which is the axial compressor. The axial compressor significantly raises the air pressure through a number of stages to about 13–16 times the atmospheric value (De Sa and Al Zubaidy 2011) for base-load power production units. The pressurized air then enters the combustor, where fuel is injected and the combustion process takes place. The high-temperature and pressure exhaust gas then enters the turbine, where it is expanded. Its temperature and pressure drop produces mechanical power and with the use of a generator this is converted to electric power, feeding the grid.

Modern gas turbines have sustained modifications aiming at their efficiency increase. Such modifications include conducting the compression and the expansion of the working medium in two separate stages and inter-cooling or reheating it respectively in between. Another common method of increasing the gas turbine efficiency is the use of a heat exchanger to partially recover heat from the exhaust gas leaving the turbine and using it to preheat the low-temperature gas before the combustion. This process is called regeneration and can significantly reduce fuel consumption, thus increasing the overall thermal efficiency. It is mainly used in small-scale plants using single stage compressors that can be found in the so-called microturbines.

The most important factor affecting the efficiency of a gas turbine is the temperature of the exhaust gas leaving the combustion chamber and entering the turbine. The value of this temperature is usually limited by the turbine blade cooling system and the materials used, since it is necessary to avoid overheating and excess thermal stresses. For state-of-the-art technology turbines, it can reach a maximum value of about 1,300 °C.

The installation cost of gas turbines is very competitive, and is lower than 1,000 €/kW, while common values are from 400 up to 800 €/kW (Horlock 2003). Nevertheless, such plants require fossil fuels of high quality and purity for their operation, resulting to significant operational cost. The LCOE of such plants, which takes into account all kinds of costs (installation, operational, O&M, etc.), is around 0.07–0.12 €/kWh (Horlock 2003), having a large variation, depending on the fuel price, engine scale, and ambient conditions, since their performance is sensitive mostly to air temperature.

Gas turbines are being installed everywhere in the world, being most common in countries, where there is available natural gas (using preferably advanced methods of drilling and using perhaps hydraulic fracturing, known as fracking, to decrease costs and environmental concerns). They can be integrated in an electricity grid as base or auxiliary units, due to their capability to adjust their capacity to varying loads. They are very common in the Middle East and in countries where there is scarcity of water supply, since they do not require cooling unit. Due to the combustion of high-quality fuels, gas turbines have relative low specific emissions, while the production of nitrogen oxides and sulfur oxides is minimum. The exhaust gases are mainly composed of vapor water and carbon dioxide, while some small traces of carbon monoxide are also produced, due to incomplete combustion. Typical concentrations of gas turbines emissions include CO_2 emissions of about 0.7–0.8 kg/KWh, 100 kg/GWh for NO_X, 2 kg/GWh for SO_X, 44 kg/GWh for CH_4 and 27 kg/GWh for CO (Spath and Mann 2000). These values change according to the fuel properties and total plant's efficiency and performance.

3.3 Combined Cycle Plants

The exhaust gas from the gas turbines usually exits the turbine with a temperature higher than 500 °C, and can be used as heat input in a bottoming steam power cycle. In this case, a Combined Cycle (CC) is formed, where the gas turbine cycle is acting as a topping cycle and its exhaust heat feeds the heat recovery steam generator, fully or partially substituting the required fossil fuel of the bottoming steam Rankine cycle.

The efficiency of a combined cycle is actually a combination of the efficiency of the two cycles, as described next. According to the second law of thermodynamics, the efficiency of these power plants is limited from the so-called Carnot efficiency. This efficiency depends on the maximum and the minimum temperature of the power cycle. Usually, the minimum temperature is the environment, which serves as a heat sink. The maximum temperature is linked with technological operation and limitations, dealing mainly with the development of materials. The efficiency of the steam power plants, gas turbines, and combined cycle plants is presented in Fig. 7, as a function of the range of maximum temperature, depicting the Carnot limit as well.

This combined cycle efficiency depends on many parameters of the cycles, such as the combustion chamber temperature, the exhaust gas temperature from the gas turbine, the gas turbine efficiency, the live steam parameters, the condensation temperature, and the heat recovery steam generator efficiency. The ambient temperature also plays a significant role, since higher ambient temperatures lead to lower gas turbine power output and efficiency and to lower Combined Cycle Power output. Therefore and especially in hot climates, the combined cycles have air chillers which cool the inlet air temperature down to 10 °C (Kakaras et al. 2006).

The efficiency of a combined cycle can reach very high values, up to 59 %, offering high flexibility on the power contribution of the top/bottom cycles. For natural gas fueled CC plants, the installation cost ranges from 500 to about 900 €/kW (Kaplan 2008) and the cost of electricity production is around 0.06–0.09 €/kWh (Black 2010), depending on the natural gas price. The combined cycle power plants have also the possibility for additional combustion in the Heat Recovery Steam Generators (HRSG), offering the option of using solid fuels as well.

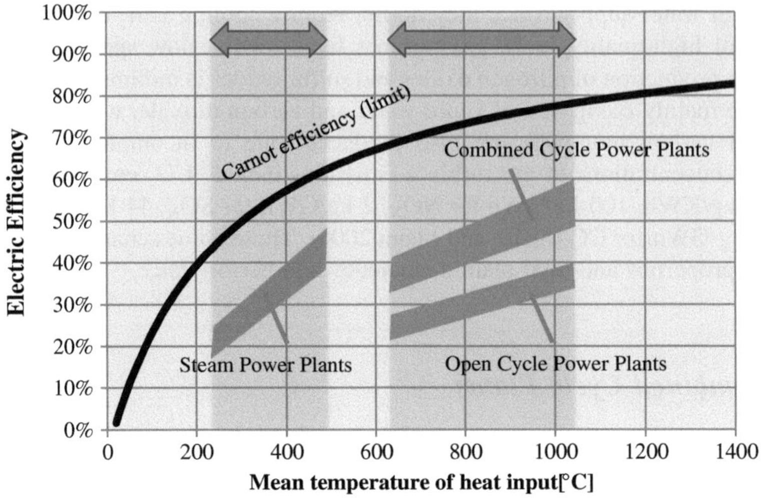

Fig. 7 Efficiency of the three fossil-fired technologies as a function of mean temperature heat input

Most of the new natural gas fired power plants built are based on the combined cycle principle, since it has some important advantages, which were mentioned previously. Apart from the high conversion efficiency, such plants show low values of specific emissions, according to previous Fig. 6, in the range of 0.4–0.5 kg CO_2/kWh (Spath and Mann 2000).

Finally, it should be mentioned that hybrid solutions for the use of solar thermal energy in combined cycles can be applied in many cases. Such plants are called "Integrated Solar Combined Cycles" and solar energy is used for the production of steam, saving some amount of fossil fuel during daytime. Usually, the solar thermal power plant performs in a solar dispatching mode, where the gas turbine always operates at full load, depending only on ambient conditions, whereas the steam turbine is somewhat boosted to accommodate the thermal hybridization from the solar field (Montes et al. 2011). Integrated Solar Combined Cycle System Technology (ISCCS) is widely regarded as a "technology bridge" between fossil fuels and solar thermal energy production with some operating projects already, while many others are planned.

3.4 Reciprocating Internal Combustion Engines

Internal combustion engines (ICE) are very reliable machines that are also used for electricity generation. They are usually fuelled with diesel oil, heavy oil, natural gas, or even biogas/syngas. Their capacity is usually up to some megawatts, but with the installation of multiple units, the capacity of such plant can reach several megawatts of electric power.

They are mainly installed in isolated or autonomous areas (noninterconnected grids), including small islands. They are very popular in the Greek islands, in Middle East, and in rural areas of Canada. They are the most common prime mover in all kinds of ships, and in building installations as cogeneration plants (CHP), or even as back-up plants.

Their efficiency can be very high, reaching even values of 45–54 % for the two-stroke low-speed large engines using diesel oil, which are also turbo-charged (Henningsen 1998). The engines running on gaseous fuels have lower capacity and also lower efficiency, around 30–44 %, since they operate with lower compression ratio (EPA 2008). The installation cost of such engines can be extremely low, reaching even a cost of 300–500 €/kW for the large plants, while their LCOE is around 0.10–0.15 €/kWh (Torrero 2003). Nevertheless, for the smaller plants with capacity lower than 1 MW, these cost figures rapidly increase, due to their lower efficiency and higher cost per unit output.

A significant disadvantage of internal combustion engines for electricity generation is the high emissions produced (EPA 2008). One reason for this is the fuel used, which sometimes can be of low quality (especially for the case of heavy oil). Except from the inevitable production of carbon dioxide (around 0.5–0.7 kg/kWh), the combustion of fuel in ICEs leads to the emission of carbon monoxide (around 1 kg/MWh) sulfur oxides (around 10 kg/MWh), nitrogen oxides (around 12 kg/MWh), and most importantly to the production of unburned hydrocarbons and soot (Henningsen 1998). For most of these emissions, standards and limits have been produced (Tier standards), which are mainly applicable for maritime large-bore engines, but they are also used for on-shore large diesel engines for power production (IMO 2013). The power output and the total efficiency of the plant can be increased, when coupling the ICE engine with an ORC waste heat recovery system. In this case an increase of the power output by 10 % is expected (Gewald et al. 2012). Such systems are usually called ICE Combined cycles.

4 Discussion and Conclusions

The mainstream technologies for electricity generation have been presented, beginning with the renewable energy plants. Then, the conventional plants have been analyzed with specific focus on their cost, the fossil fuels utilized, and their emissions.

Although there is a common belief that RES can be applied at large scale and replace conventional power plants using fossil fuels, it has been shown that only a small fraction of the electric energy consumption is covered by renewables. Although their use is steadily increasing during the last years, their takeup is not expected to be seen the next few years. There are many issues yet to be overcome, such as the lack of cost-effective energy storage for large-scale utilities (aiding in the security of supply) and their high installation costs, even if the latter are rapidly decreasing. To this context, great resources are given on research facilities, with a final aim to decrease the specific costs and increase the renewable energy sources penetration, not only for electricity generation, but also for heating/cooling purposes (co- /tri-generation systems). New materials, techniques, and system designs are investigated, which can further increase the efficiency. After all, the utilization of wind, solar and biomass energy is more than welcomed, which could control the produced emissions from fossil fuel fired plants in future energy scenarios.

On the other hand, the use of fossil fuels allows the generation of cheap electric energy, with good power balance management over the electricity grid. However, since carbon taxes have been finally introduced for conventional plants, and most importantly, given the uncertainty of the availability of the unexploited fossil fuel resources, the game of technologies and prices has moved toward a game of fuels governance. This game can even include efforts on increasing the conversion efficiency and the further development of carbon capture and storage technology (or even both), which can improve the environmental performance of fossil fuel plants and decrease the fuel consumed per kWh produced.

In either case, given the pure technological status of both types of technologies (renewable/conventional), the takeup of renewables does not look instant during the next few years. Still, renewables can surely play a supportive role in the energy mix, which in most countries still depends a lot on conventional power plants. And perhaps hybrid solutions (using mostly solar energy) can aid in their further introduction at larger scale, which can rapidly increase their energy share, toward the greenhouse gases emissions reduction and to a more sustainable energy future.

References

Agraniotis M, Koumanakos A, Doukelis A, Karellas S, Kakaras E (2012) Investigation of technical and economic aspects of pre-dried lignite utilisation in a modern lignite power plant towards zero CO_2 emissions. Energy 45(1):134–141

Anagnostopoulos JS, Papantonis DE (2007) Optimal sizing of a run-of-river small hydropower plant. Energy Convers Manage 48(10):2663–2670

Black J (2010) Cost and performance baseline for fossil energy plants. Volume 1: bituminous coal and natural gas to electricity. US Department of Energy, NETL REPORT-2010/1397

Bolinger M, Wiser R (2012) Understanding wind turbine price trends in the U.S. over the past decade. Energy Policy 42:628–641

Branker K, Pathak MJM, Pearce JM (2011) A review of solar photovoltaic levelized cost of electricity. Renew Sustain Energy Rev 15(9):4470–4482

Bugge J, Kjær S, Blum R (2006) High-efficiency coal-fired power plants development and perspectives. Energy 31(10–11):1437–1445

Campos Rodríguez CE, Escobar Palacio JC, Venturini OJ, Silva Lora EE, Melián Cobas V, Marques dos Santos D, Lofrano Dotto FR, Gialluca V (2013) Exergetic and economic comparison of ORC and Kalina cycle for low temperature enhanced geothermal system in Brazil. Appl Therm Eng 52(1):109–119

De Sa A, Al Zubaidy S (2011) Gas turbine performance at varying ambient temperature. Appl Therm Eng 31:2735–2739

Dipippo R (2012) Geothermal power plants: principles, applications, case studies and environmental impact, 3rd edn. Butterworth-Heinemann, USA
Dipippo R (2004) Second Law assessment of binary plants generating power from low-temperature geothermal fluids. Geothermics 33(5):565–586
Environmental Protection Agency (2008) Energy and environmental analysis. technology characterization: reciprocating engines. Combined heat and power partnership program. http://www.epa.gov/chp/documents/catalog_chptech_reciprocating_engines.pdf
European Small Hydropower Association (2013) SHP in figures. http://www.esha.be/about/about-small-hydropower/small-hydropower-in-figures.html
Gewald D, Siokos K, Karellas S, Spliethoff H (2012) Waste heat recovery from a landfill gas-fired power plant. Renew Sustain Energy Rev 16(4):1779–1789
Global Wind Energy Council (GWEC) (2013) Global status overview. http://www.gwec.net/global-figures/wind-energy-global-status/
Henningsen S (1998) Chapter 14: Air pollution from large two-stroke diesel engines and technologies to control it. In: Handbook of air pollution from internal combustion engines. ISBN: 978-0-12-639855-7
Hernández-Moro J, Martínez-Duart JM (2013) Analytical model for solar PV and CSP electricity costs: present LCOE values and their future evolution. Renew Sustain Energy Rev 20:119–132
Horlock JH (2003) Advanced gas turbine cycles. Pergamon Press, Pergamon
Hu Z, Wang J, Byrne J, Kurdgelashvili L (2013) Review of wind power tariff policies in China. Energy Policy 53:41–50
International Maritime Organization (IMO) (2013) Air pollution and greenhouse gas emissions. http://www.imo.org/OurWork/Environment/PollutionPrevention/AirPollution/Pages/Default.aspx
Independent Power Transmission Operator (IPTO) of Greece (2013) Interconnections net flows. http://www.admie.gr/en/operations-data/system-operation/real-time-data/reports/
Jäger-Waldau A, Monforti-Ferrario F, Banja M, Bloem H, Arantegui RL, Szabó M (2012) Renewable energy snapshots 2012. JRC Scientific and Technical Reports, European Commission, EUR 25756 EN, JRC 77772. http://publications.jrc.ec.europa.eu/repository/handle/111111111/27434
Kakaras E, Doukelis A, Prelipceanu A, Karellas S (2006) Inlet air cooling methods for gas turbine based power plants. J Eng Gas Turbines Power 128:312–317
Kaplan S (2008) Power Plants: characteristics and costs. CRS Report for Congress, RL34746
Karellas S, Boukis I, Kontopoulos G (2010) Development of an investment decision tool for biogas production from agricultural waste. Renew Sustain Energy Rev 14(4):1273–1282
Kosmadakis G, Manolakos D, Papadakis G (2011) Simulation and economic analysis of a CPV/thermal system coupled with an organic Rankine cycle for increased power generation. Sol Energy 85(2):308–324
Montes MJ, Rovira A, Muñoz M, Martínez-Val JM (2011) Performance analysis of an integrated solar combined cycle using Direct Steam Generation in parabolic trough collectors. Appl Energy 88(9):3228–3238
National Renewable Energy Laboratory (NREL) (2012a) Concentrating solar power projects in the United States. http://www.nrel.gov/csp/solarpaces/by_country_detail.cfm/country=US%20%28%22_self%22%29
National Renewable Energy Laboratory (NREL) (2012b) Concentrating solar power projects in Spain. http://www.nrel.gov/csp/solarpaces/by_country_detail.cfm/country=ES
Pérez-Higueras P, Muñoz E, Almonacid G, Vidal PG (2011) High Concentrator PhotoVoltaics efficiencies: present status and forecast. Renew Sustain Energy Rev 15(4):1810–1815
Rentizelas A, Karellas S, Kakaras E, Tatsiopoulos I (2009) Comparative techno-economic analysis of ORC and gasification for bioenergy applications. Energy Convers Manage 50(3):674–681

RWE (2008) The WTA technology, an advanced method of processing and drying lignite. Information brochure. http://www.rwe.com/web/cms/en/77254/rwe-power-ag/power-plant-new-build/wta-plant/

Skoplaki E, Palyvos JA (2009) On the temperature dependence of photovoltaic module electrical performance: a review of efficiency/power correlations. Sol Energy 83(5):614–624

Spath PL, Mann MK (2000) Life cycle assessment of a natural gas combined-cycle power generation system. National Renewable Energy Laboratory. Task No. BP00.1030. http://www.nrel.gov/docs/fy00osti/27715.pdf

Steinke F, Wolfrum P, Hoffmann C (2013) Grid versus storage in a 100% renewable Europe. Renewable Energy 50:826–832

Stürmer B, Schmid E, Eder MW (2011) Impacts of biogas plant performance factors on total substrate costs. Biomass and Bioenergy 35(4):1552–1560

Torrero E (2003) Costs of utility distributed generators, 1–10 MW: Twenty-Four Case Studies. Electric Power Research Institute (EPRI). Technical update. http://www.epri.com/abstracts/Pages/ProductAbstract.aspx?ProductId=000000000001007760

Viebahn P, Lechon Y, Trieb F (2011) The potential role of concentrated solar power (CSP) in Africa and Europe: a dynamic assessment of technology development, cost development and life cycle inventories until 2050. Energy Policy 39(8):4420–4430

Part II
Transitions Towards Renewable Energy Systems

Institutional Factors That Determine Energy Transitions: A Comparative Case Study Approach

Auriane Magdalena Koster and John Martin Anderies

Abstract In this chapter, we consider energy transitions as important elements of attaining development goals. In the development context, energy transitions pose both challenges and opportunities. Obstacles to transitions include (1) an existing, centralized, complex energy-grid system, whose function is opaque to most users, (2) coordination and collective-action problems that are path dependent, and (3) difficulty in scaling-up RE technologies. Because energy transitions rely on both technological and social innovations, we are interested in how institutional factors can be leveraged to surmount these obstacles. This research attempts to address the question of what constellation of institutional, biophysical, and social factors are essential for an energy transition. Our objective is to derive a set of "design principles" that we term institutional drivers for energy transitions, analogous to Ostrom's institutional design principles. This chapter will analyze energy transitions using the Institutional Analysis and Development Framework (IADF) to conduct a comparative case-study analysis. The comparative case-study analysis allows us to uncover recurring patterns across cases that help to identify institutional factors associated with energy transitions.

A. M. Koster (✉)
Mary Lou Fulton Teachers College, PO Box 37100, Mail Code: 3151, Phoenix, AZ 85068, USA
e-mail: auriane.koster@asu.edu

J. M. Anderies
Arizona State University, School of Sustainability, 800 S Cady Mall, Tempe, AZ 85281, USA
e-mail: m.anderies@asu.edu

E. Michalena and J. M. Hills (eds.), *Renewable Energy Governance*, Lecture Notes in Energy 23, DOI: 10.1007/978-1-4471-5595-9_3,

1 Introduction

Institutions are critical for the functioning of complex social systems. They facilitate exchange, reduce transaction costs, and structure repeated interactions between people. They also play a critical role in promoting investment and innovation. This is especially true in the case of common-pool resources and shared infrastructure. Energy systems are intimately related to common-pool resources—e.g., the climate system—and are comprised mainly of shared infrastructure—i.e., large power utilities. Promoting change in this enormous, complex infrastructure system will require careful consideration of institutional change. Comparative case-study analysis is an essential tool for attempting to extract basic design principles from complex coupled human-environmental systems (Ostrom 1990).

This chapter takes a comparative case-study approach across energy systems to begin to uncover some basic features related to how they are governed. The goal is to exploit this structure to help facilitate transitions to renewable energy (RE) systems when appropriate. The chapter characterizes what a unsuccessful transition may look like by looking at some "successful" (at least relatively speaking) transitions. It is difficult to observe unsuccessful transitions because by virtue of them being badly planned, they do not occur. Therefore, an unsuccessful transition can be identified as lacking the characteristics that we argue are necessary for success.

Although RE technologies can be very difficult to implement, investment in new RE is higher than it ever has been in the past (Fig. 1). In 2011, $257 billion was invested in RE.[1] The top RE users in the world have increased their RE investment between 2010 and 2011 (Fig. 2). However, modest the recent increases may seem, these increases occurred during the worst financial crisis since the Great Depression.

For the first time in history, in 2011, solar investments surpassed those of wind (Fig. 3). This jump in investment was largely due to rooftop solar photovoltaic (PV) installations in Germany and Italy, small-scale solar projects in China and the UK, and an increase in large-scale solar-thermal electricity generation projects in Spain and the US. The drop in wind investments was partly due to lower turbine prices, European policy uncertainties, and a slowdown in the original fast-paced growth of wind installations in China (McCrone 2012). While government support and industry investments continue to rise, the price of RE continues to drop. The per megawatt (MW) price of solar PV has dropped by 76 % and the price of wind power by 18 % since 2008 (McCrone 2012). These drops reflect not only fierce competition in the supply chain but fierce competition with fossil-fuel-based energy sources. In 2011, RE investments (excluding large hydro) surpassed net fossil-fuel investments (McCrone 2012)

---

[1] Investors include corporate R&D, government R&D, VC/PE, public markets, small distributed capacity, and asset finance. Small distributed capacity and asset finance have always been the RE investment leaders, far surpassing the other four categories.

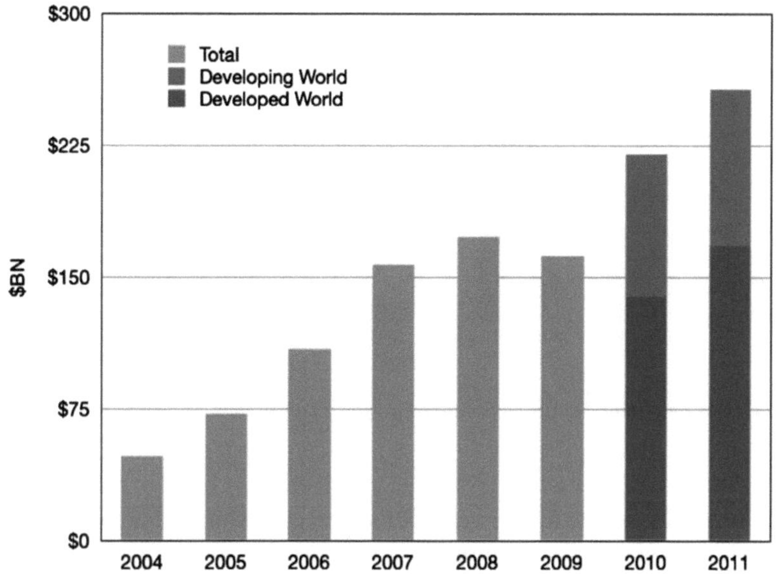

Fig. 1 Global renewable energy investment. Data source: McCrone (2010, 2012)

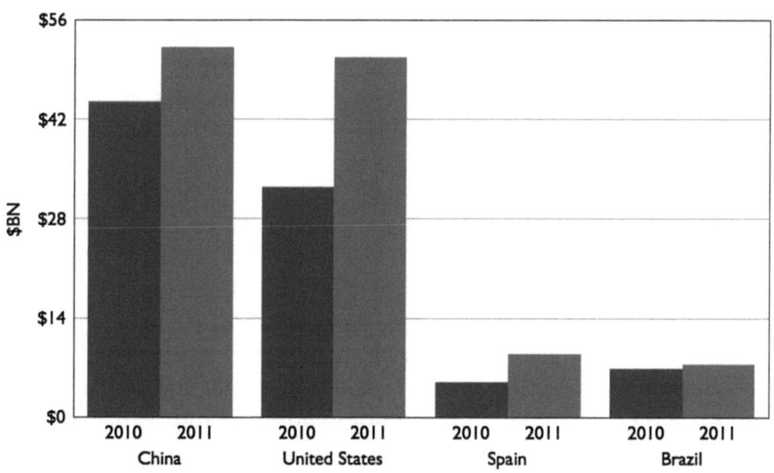

Fig. 2 Renewable energy investment by country. Data source: McCrone (2011, 2012)

These developments show the extent to which RE has developed, but what progress has been made with energy transitions around the world? Over what time scales are energy transitions feasible? "In a manifestly complex world dominated by hegemonic ideologies of neoliberal capitalism, global finance, and commodity

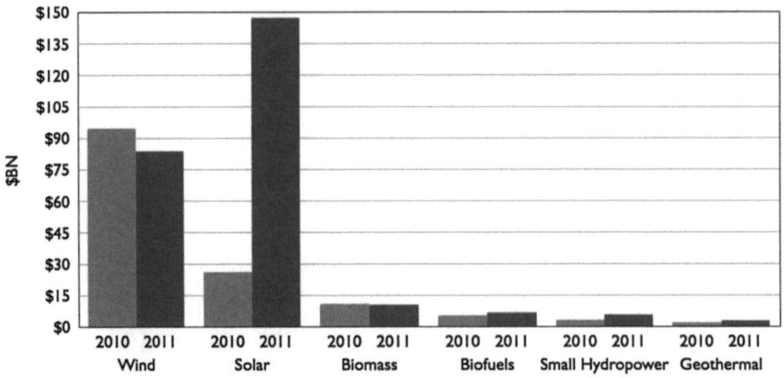

Fig. 3 Global renewable energy investment by type. Data source: McCrone (2011, 2012)

flows is it really possible to intervene and deliberately shift technologies, practices, and social arrange (not to mention their systemic interaction and interdependencies) onto an altogether different, altogether more sustainable track?" (Shove and Walker 2007, p. 763).

The goal of a transition is to dislodge currently dominant socio-technical regimes and replace them with new configurations (Shove and Walker 2007). Shove and Walker argue that in order to successfully transition, "most recommend the deployment of multiple methods and tools for intervention, also arguing for processes of governance (rather than government), for the involvement of diverse actors and knowledge, and for explicit recognition of the uncertainties and limitations of science-based expertise" (p. 764). The goal of this chapter is to explore what these methods and tools are for specific countries.

With current consumption rates and without considering environmental impacts, it has been estimated that there are only 70 years of oil, 72 years of natural gas, and 230 years of coal left (General Electric 2013; Murphy 2012). Eventually, there must be a shift away from our dependence on these finite resources. Wang et al. (2010) argue that there are three primary ways to reduce carbon emissions: slow economic growth, reduce energy intensity, and develop RE. This chapter will describe strategies and implementation of the third way in the most progressive countries. The countries chosen represent four of the top RE users: Brazil (3.8 % share of total), Spain (6.5 % share of total), China (9.1 % share of total), and the United States (23.2 % share of total) (BP 2012) (Fig. 4).

When selecting case studies, we endeavored to include high RE users while also ensuring geographical and political diversity. The European Union (EU) has taken great strides toward RE implementation and each country within the EU could be used as a case study. However, only one top RE user was chosen: Spain.

Otherwise, each of the other case-study countries represent a major geographical area, with the exclusion of Oceania which does not have a nation in the top ten RE user list. An analysis of these cases will help to uncover some

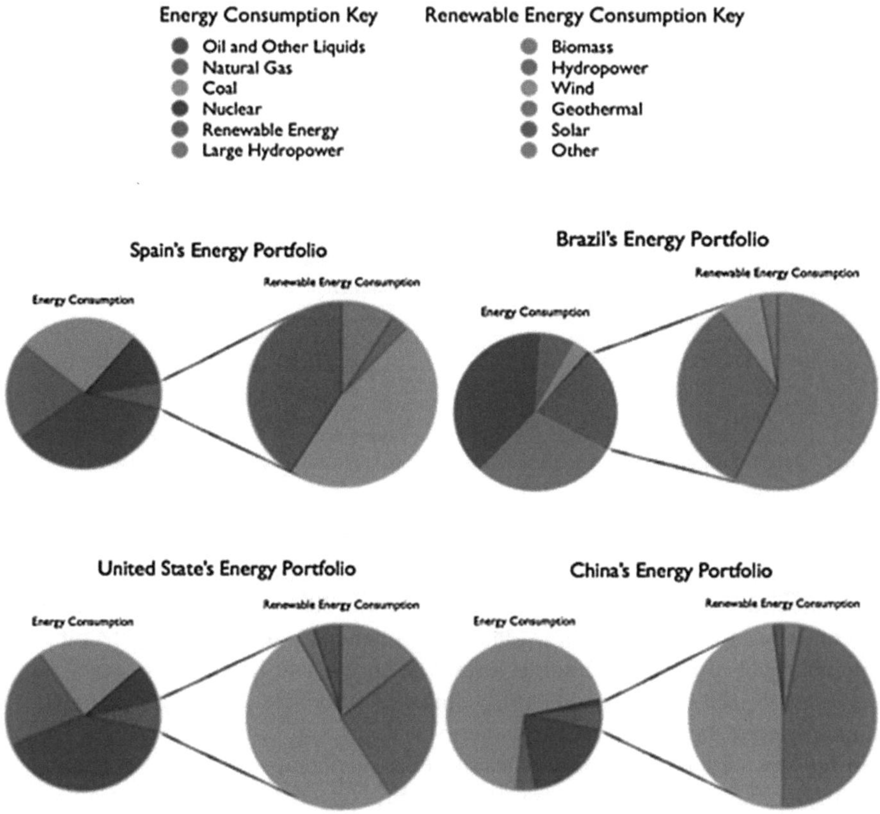

Fig. 4 The renewable energy portfolio constitutes an outcome of the interactions of the exogenous variables and decision-making environments in the IADF

institutional similarities among them that may lead to successful energy transitions. Based on these similarities, we derive a list of institutional drivers that may lead to successful energy transitions.

2 Institutional Analysis and Development Framework and Energy Transitions

The Institutional Analysis and Development Framework (IADF) is a tool developed by Elinor Ostrom (2005) and her colleagues to facilitate institutional analysis. The intent of the IADF is to help structure thinking about elements that influence decision making in collective action situations and provide insight on

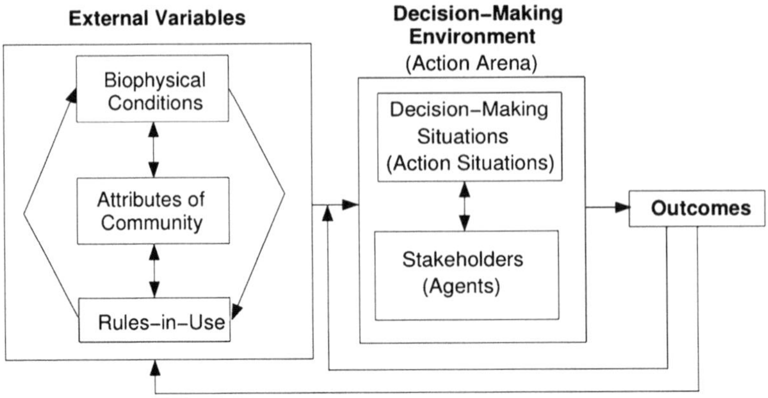

Fig. 5 Institutional and Analysis Development Framework (AIADF). *Source* adapted from Ostrom (2005). Original labels for the IAD appear in parenthesis where appropriate

how people behave in social dilemmas.[2] The IADF has been most-often used to study the management of common-pool resources because they "are among the core social dilemmas facing all people" (Ostrom 2005, p. 219). The Framework has been used to develop large databases related to studying a wide range of common-pool resources including forests, fisheries, and irrigation systems.

Although it has been used to study some commodities, the IADF has never, to our knowledge, been used to study energy systems. Ostrom's IADF can be applied to organize key issues and themes of energy management and dynamics by identifying key decision-making factors (Fig. 5). This chapter uses Ostrom's (2005) Institutional Analysis Development Framework (IADF) to examine energy transitions and better understand the energy-implementation decision-making process of the case-study countries.

Energy management occurs at many different scales, from the individual, to the utility, to the country scale. As such, countries face the choice of whether to invest in large-scale RE projects and/or to implement renewable-energy policies that aid in the implementation of such projects. Complicating this decision is the fact that within governments, energy decision making can either be highly centralized, and be the sole responsibility of one ministry or department within the government, or it can be decentralized and spread throughout ministries and departments. At the other end of the spectrum, individuals also have the opportunity to invest in renewable-energy projects, but at a much smaller (and perhaps more appropriate) scale. Does whether energy policy is administered in a centralized or decentralized fashion impact the transition to RE? Should countries rely on individuals to invest

[2] A collective action is an action that aims to improve the standings of a certain group of people. A collective action dilemma arises when multiple people would benefit from a certain action, but it is impossible for any one person to act alone. Thus, the group must work together to solve the problem for the benefit of everyone.

and lead the transition? Do different institutional arrangements (defined precisely below) have any impact? Here, we apply the IADF to address some of these questions and analyze strengths and weaknesses in the institutional arrangements involved in different approaches to supply an essential commodity, energy, from common-pool resources such as solar radiation and wind. The application of the IADF to the case studies yields several key features that are important for the success of collective action and governance for energy transitions:

a. Credible commitment, in this case the government;
b. Polycentric governance, defined as a system where citizens are able to organize not just one but multiple governing authorities at differing scales, and where self-organized resource-governance systems may be special districts, private associations, or parts of a local government (Ostrom 2005);
c. Stakeholder engagement;
d. Information rules promoting transparency;
e. Scope rules leading to innovation;
f. Formal rules that force a country to follow the Kyoto Protocol;
g. Constitutional rules impacting infrastructure and grid connectivity;
h. Collective choice rules leading to informal and formal monitoring.

Ostrom used the IADF to generate eight design principles that were associated with successful collective action. Using a similar method, our research analyzes energy transitions to develop eight institutional drivers that stem from the list above. We define an institutional driver as a certain combination of biophysical conditions, community attributes, and rules that generate a decision-making environment in which an energy transition occurs. For the purpose of this chapter, the decision-making environment of interest is the energy production/consumption system of a country. Given her political science background, Ostrom tended to focus more on understanding rules in use and strategic interactions. Here we focus more on the interaction between technology, rules, and people. In each of the cases, the analysis attempts to draw out the exogenous variables and decision-making environments that have been most influential in the country's RE development. Before turning our attention to the details of each case study, we summarize some of the key exogenous variables that relate to energy transitions.

Biophysical Conditions. One of the key biophysical conditions that influence energy use is the location of energy sources and users. A geographical analysis shows that natural RE resources are spread throughout the country for all of the case studies (Table 1). This increases the country's ability to take advantage of these RE resources without necessarily the need for a centralized grid system and, at the same time, exploit a diverse energy portfolio. If a centralized grid system is seen as a necessary component of an energy system, then perhaps not all RE sources could be taken advantage of. This suggests that in order to maximally exploit all available energy resources, a combination of centralized and decentralized governance regimes are necessary. This is the essence of a polycentric governance approach.

Table 1 Renewable energy potential by geographical location within the case-study countries (Biophysical Conditions)

Country	Geographical distribution by energy type				
	Biomass	Wind	Solar	Hydropower	Geothermal
China	Countrywide	N, NE	W	SE, NE	SW
US	Countrywide except C	C, NE	SW	NW	W
Brazil	N, NW	NE coast, E	Small potential in EC	Countrywide	S
Spain	Countrywide	S, C	S	Countrywide	E

Attributes of the Community. The analysis of the attributes of the communities in each of the case-study countries suggests that *all of the case study countries have many opportunities to increase RE use*. However, there are some obstacles. Two common obstacles in all of the case studies is price tolerance and infrastructure lock-in. Currently, the cost of RE is much higher than nonrenewable energy. *Without institutional change, RE will most likely never be price competitive with nonrenewable energy sources.* Many people place value on the environment and, by extension, using RE. Thus at the individual scale, there may be willingness to pay some premium for RE. Individuals may be willing to invest, but given the nature of energy as an essential good, few individuals can tolerate the cost of investing in RE.

Obviously, there are economies of scale in energy provision. Thus, joint provision of RE, just as with conventional energy sources, may reduce the burden on individuals and make RE more attractive. Unfortunately, all of the case-study countries have a relatively old grid infrastructure system that can primarily only support traditional energy sources, locking them into nonrenewable energy use. Thus, many countries are stuck between a rock and a hard place: Although RE technologies can be effective when implemented by individuals, it is too costly. Reducing this cost requires some centralization, but path-dependent infrastructure development has generated sunk costs that prevent this from happening. Because RE is caught between decentralized (too costly at the individual scale) and a centralized model (too costly to refit large-scale infrastructure), RE technologies are difficult to implement and investors are less likely to support them. Energy systems are extremely complex so the attributes of the community involved (everyone!) are far more numerous and diverse than, for example, a small fishing community or a small irrigation system of the type studied by Ostrom using the IADF. Thus, rather than attempt to characterize the attributes of the community, we summarize some key characteristics related to opportunities and barriers to RE implementation in the case-study countries (Table 2).

Rules-in-Use. So it is the only hope that a situation in which the environment becomes so degraded that conventional energy prices increase and make RE more attractive, or do we just need better regulation? It turns out that all of the case-study countries have a variety of rules in use to overcome this RE implementation dilemma, especially scope and payoff rules (Table 3). Most of the countries rely on a

Table 2 Opportunities and barriers to renewable energy adoption by country

Opportunities	Barriers
China	
- Largest investor in RE	- Minimal stakeholder involvement
- National and provincial RE incentives/ standards	- Largest carbon emitter in world (e.g. critical need to maintain economic growth for stability)
- Hierarchical approach to RE	- Lacking grid connectivity of renewable energy
- Cooperation mechanisms with organizations and other governments	
-Innovative governmental decision-making	
US	
- Goal of national security and energy independence	- Weak national policies
- Strong state incentives for renewable energy implementation	- Policy direction changes with political values of President
- Renewable energy positively impacts tourism	- Only reactive during times of crisis
	- Innovation is mainly demand-side driven
	- Noncompliance determined on state-by-state basis
Brazil	
- Democratic government	- One of the fastest growing GDPs in world
- Good relations with neighboring countries	- Inequitable income distribution
- Most industrialized country in Latin America	- Most solar PV systems are off-grid
- Strong governmental push for renewables	- Minimal stakeholder investment
- Off-grid solar implementation in rural communities	
- Ethanol price is market-driven	
Spain	
- Investigation, development, and innovation within energy companies and the government	- Minimal adoption of solar energy even with the natural resource and technology innovation
- Low population density, so little public opposition	- Major debt problem
- National, provincial and city mandates	- Complex permitting process for renewable energy
- Pioneer of photovoltaic investigation	- Consumers rarely involved in decision- making process
- Job creation from renewable energy implementation	
- Looking for ways to reduce debt	
- Renewable energy generation information relayed immediately to citizens via website	

Renewable Portfolio Standard (a scope rule in the language of the IADF) along with a multitude of financial incentives (payoff rules in the language of the IADF) to increase RE use. Our analysis suggests that *it is a diversity of rules, and not the reliance on just one type rule, that allows a country to undergo an energy transition.*

An analysis of all three classes of exogenous variables is required to understand a country's energy transition. In addition, each exogenous variable impacts and is impacted by the other exogenous variables. The rules in use and attributes of the community allow for a country to take advantage of the biophysical conditions to increase RE use. Rules in use can also remove barriers related to the attributes of the community, such as an inability to pay the high costs associated with RE, by, for example, decreasing the price of RE. The biophysical conditions which determine rules in use will have the greatest impact on RE use. For example, if a country has local RE resources (biophysical condition), then a net metering law (rules in use) would effectively increase RE penetration into the grid, increasing the likelihood that individuals will invest in RE (attributes of the community associated with risk tolerance). The different types of rules employed by different countries are quite diverse. Again, we provide an overview of the most common rule types in Table 3.

Key Definitions

Renewable Portfolio Standard (RPS): "An RPS is a policy that ensures that a minimum amount of RE (such as wind, solar, biomass, or geothermal energy) is included in the portfolio of electric-generating resources serving a state. RPS regulations generally impose obligations that increase over time" (Yin and Powers 2010, p. 1140).

Table 3 Renewable energy policies and mandates by country (Rules-in-use). *Source* KPMG International (2011); The Pew Charitable Trusts (2012)

Rules-in-use	China	US	Brazil	Spain
Important acts, plans, and laws	Renewable energy law	EPA[a] 2005; EISA[b] 2007; ARRA[c] 2009	Ten year expansion plan	Renewable energy plan
RPS	20 % by 2020	State-by-state	75 % by 2030	20.8 % by 2020
Government subsidies/ rebates	Yes	National and by state	Yes	Yes
Carbon cap	Working towards implementation	No	Yes	Yes
Tax incentives	Yes	Yes	Yes	Yes
Feed-in tariffs	Yes	State by state	2002–2007	Yes
Loans	Yes	National and by state	Yes	Yes
Procurement	Potential in future	Yes	Yes	Yes
Renewable energy credits	Yes	State by state	No	Yes
Net metering	No	State by state	No	Working towards implementation

[a] EPA = Energy Policy Act
[b] EISA = Energy Independence and Security Act
[c] ARRA = American Recovery and Reinvestment Act

Renewable Energy Credit (REC): "Most RPS policies are enforced through a credit-trading mechanism. When electricity is generated from a renewable source in states that have a renewable energy credit program, there are two resulting products—the electrons that are fed into the grid, and the environmental attributes associated with producing reduced-carbon or carbon-free electricity. In most states, these environmental attributes are accounted for in the form of renewable energy credits, or RECs. Each REC represents one MWh of electricity generated from an eligible renewable energy resource" (Yin and Powers 2010, p. 1143).

Feed-in-tariff (FIT): "The feed-in-tariffs guarantee a price that the grid company will pay to the renewable generator, in order to ensure that developers of more expensive renewable generation sources can recover their costs and earn an appropriate profit" (Schuman and Lin 2012, p. 8).

Text Box 3.1: Definitions for different institutional arrangements for promoting RE

3 Case Studies

Now that we have a sense of the IADF and some of the features of energy systems of potential interest, we will now describe each of the case studies to provide a foundation for the comparative discussion in Sect. 4.

3.1 Spain

Most of the RE used in the world is used in the EU. The EU's executive body, the European Commission (EC), played a major role in drafting the Kyoto Protocol and developed the first multi-country emissions-trading system (Potocnik 2007). Consumer buy-in has had significant impact on successful implementation. A 2006 poll indicated that 55–80 % of Europeans favored RE (Potocnik 2007). Increasing RE use can spur economic growth, create jobs, and mitigate the impacts of climate change. If the EU uses about 13 % less energy than it did in 2007, it could save €100 billion and reduce carbon dioxide emissions by nearly 780 metric tons each year (Potocnik 2007). An increase of 12 % in RE use would create an estimated 500,000 jobs (Romero et al. 2012). In 2007, the wind-energy sector alone provided 154,000 jobs (Romero et al. 2012).

Both the EU and its member countries are using a variety of mechanisms to successfully implement RE. For example, individual countries, most notably Germany and Spain, have used feed-in-tariffs (see Text Box 3.1 for a definition) (Schuman and Lin 2012).

Spain is a leader in several RE sectors in terms of MW installed capacity, both in Europe and the world. It is first in Europe in solar thermal, second in wind power and solar PV, and third in hydropower. Worldwide, Spain ranks second in solar thermal and PV, and fourth in wind power (Romero et al. 2012). Natural and demographic factors have contributed to the country's successful RE implementation. Spain has abundant wind and solar energy, which is the second most mountainous region in the European area (behind Switzerland), and is one of the least densely populated countries. Therefore, there has been little public opposition to the installation of wind and solar farms in the country.

Both the national government and private energy companies have invested in innovation, research, and development of RE technologies (Romero et al. 2012). Spain's Renewable Energy Plan, which implemented a FIT policy, is the most important RE implementation law to date (Gamboa et al. 2005). Unfortunately, in 2012, the Spanish government responded to the then ongoing economic crisis by suspending the FIT policy. As a result, Spain is considering net metering[3] as a replacement for the FITs to encourage consumer-based RE projects. Net metering projects promote the implementation of small-scale, grid-connected RE projects, where most of the cost is placed on the consumer rather than the government.

Three groups of stakeholders make decisions about RE in Spain: the government, which provides subsidies; the banks, which provide investment funds; and the development companies, which install the technology (Cristobal 2011). The role of private individuals has been insignificant compared to other countries, and only a handful of visionary entrepreneurs have increased the development and adoption of wind power (del Rio and Unruh 2007). Lack of public participation can lead to the failure of RE projects. For example, the Catalan wind parks project nearly failed because the lack of public involvement in the decision-making process led to public opposition to the project (Gamboa et al. 2005).

Wind dominates Spain's RE market because wind-technology and energy-generation costs are low. Economies of scale, research and development efforts, mass production, and technological improvements all allow for a relatively high return on investment (del Rio and Unruh 2007). Spain is beginning to build off-shore wind farms in Cantabria, Catalonia, and the Canary Islands (Romero et al. 2012). Until recently, 85 % of wind projects have been concentrated in Galicia (the leader), Castilla-La Mancha, Castilla, Leon, and Aragon; all of these are on-shore wind projects.

Although Spain is a leader in wind-energy development, it has some of the best insolation in Europe. Despite this great potential (as compared to Germany, for example), solar energy has not been widely used due to its high cost. "The result is a vicious circle where the technology is not adopted because it is expensive, and it is expensive because it is not adopted" (del Rio and Unruh 2007, p. 1507). Nonetheless, Spain has recently begun to focus on increasing implementation of

[3] Net metering is an electricity policy where consumers invest in small-scale, grid-connected RE projects and any excess electricity that they generate is sold back to their utility.

solar PV. Many large cities have made it mandatory to install solar PV on new buildings, and some provincial plans prioritize the use of solar PV. Such plans make economic sense: PV is more cost-effective if installed during construction rather than on existing buildings (del Rio and Unruh 2007, p. 1509).

Interestingly, while implementation of solar PV has been weak, technology development has been strong. Spain is the leading manufacturer of solar PV in Europe. del Rio and Unruh note that "The Spanish PV cell and PV installation industry is highly competitive and recognized for its quality, flexibility, innovativeness and commercial dynamism" (2007, p. 1509). It is the leading European manufacturer and exports 85 % of its production. This represents 40 % of European and 7 % of world production. del Rio and Unruh also point out that 4,000 jobs are currently linked to the PV sector in Spain (2,500 are direct) and the socioeconomic benefits of PV make it especially attractive for local communities (2007).

In fact, Spain is a world leader in RE innovation. Its Renewable Energy Technology Center is working on the hybridization of solar with biomass energy, something that has never been accomplished (Romero et al. 2012). Also, Spain has developed an innovative information system that relays immediate wind power information to the public every 12s (Romero et al. 2012; Schuman and Lin 2012).[4]

If Spain continued with its original RE plans, then RE implementation could help Spain with its current debt problem. If Spain reached its goal of 20 % renewables by 2020, an estimated half million jobs would be created (Romero et al. 2012). These include both direct and indirect jobs, stemming from more than 1,000 companies in the industry. Spanish wind developers and manufacturers are competitive on the international market, and there are more than 75 industrial wind-technology development centers in the country, 18 of which are wind turbine assembly lines (Romero et al. 2012).

Spain's governmental institutional arrangements generate the one notable barrier to RE implementation. The authorization procedures for construction, connection to the grid, and initiating production of RE are complicated. A project must go through several application processes and obtain several permits at the national, regional, and municipal levels (del Rio and Unruh 2007). For example, "...the implementation of wind farms is affected by 60 different regulations involving 40 different procedures between different administrative levels and causing lead times of 4–8 years" (del Rio and Unruh 2007, p. 1505). With such a complex permitting procedure, "different groups of decision-makers become involved in the process, each group bringing along different criteria and points of view, which must be resolved within a framework of understanding and mutual compromise" (Cristobal 2011, p. 498). It may be that this cumbersome permitting process is one reason for the fact that, although Spain is a RE innovation leader, it does not have high adoption rates locally.

[4] This information is relayed to the public via the following website: https://demanda.ree.es/eolica.html.

3.2 Brazil

Brazil is the most industrialized country in Latin America with a GDP of around $2 trillion. At the time of writing, it is the world's second fastest growing economy, and is estimated to have an annual GDP growth of 5.1 % over the next 10 years (Pereira et al. 2012). Much of Brazil's GDP is produced by energy-intensive industries such as aluminum and steel production, and by the growth in residential and commercial energy services.

The Brazilian government has been a strong advocate for RE use, especially biofuels, as is evident in its policies. Brazil has the world's most competitive program of development and production of biofuels and has the second largest share of the world ethanol production market (Pereira et al. 2012). Brazil accomplished this feat by initiating the Brazilian Ethanol Program (i.e., Proalcool) to offset petroleum-based fuel use with biofuels (Pereira et al. 2012). Brazil began investing in ethanol after the oil shocks of the 1970s to reduce its vulnerability to the global petroleum-based energy market (Geller et al. 2004). At the time, oil imports consumed half of Brazil's hard-currency export income. The program required that gasoline include up to 25 % ethanol, and encouraged car manufacturers to make engines that could run on 100 % hydrated ethanol. The program was accepted by all the key stakeholders from consumers to farmers to car manufacturers despite the fact that it was initiated by a military regime (Goldemberg 2007).

While most of Brazil's electricity comes from hydropower, the country is known worldwide for its ethanol production. Ethanol provides economic, environmental, and social benefits. Of course, cost is always a key consideration in the choice of energy sources. In 1980, the cost of ethanol was about three times the cost of gasoline, so Brazil's government provided $30 billion over 20 years to subsidize ethanol production. The subsidy was more than recompensed by $50 billion saved on gasoline imports during this time. Although initially the government had to subsidize ethanol, by 2004, the subsidies were no longer necessary, and ethanol became fully competitive with gasoline on the international market (Goldemberg 2007; Pereira et al. 2012). According to the World Bank, Brazil can make ethanol for 50 % less than it costs internationally to make gasoline.

Ethanol use has also had important environmental impacts. Conversion to ethanol removed lead additives and MTBE (methyl tertiary butyl ether) from gasoline and reduced sulfur, particulate matter, and carbon monoxide emissions as well as helped to efficiently mitigate greenhouse gas emissions by having a net positive energy balance (RE output versus fossil fuel inputs) (Goldemberg 2007). However, there has been global controversy over using land to produce biofuels instead of food. Brazil's experience demonstrates that controversy can be avoided if biofuel is sourced and used efficiently, especially in a country with the right climate for biofuel production. Sugarcane for ethanol production requires about 3 million hectares. Another 2.6 million hectares of land is used to produce sugar. These uses account for about 10 % of total cultivated land and 1 % of total arable land. If Brazil expanded ethanol production by a factor of 10–30 million hectares, it could produce enough ethanol to replace 10 % of gasoline consumed globally (Goldemberg 2007).

Obviously ethanol makes sense in the biophysical and geopolitical context in which Brazil operates. This likely has gone a long way in promoting the production and use of ethanol without having to make a strong case about the environmental benefits. In other situations, where incentives are not a strong, the Brazilian government has taken positive steps. In 2009, the Brazilian government initiated the 10-Year Plan for Energy Expansion (2010–2019) which included investing $28.2 billion in small hydropower, wind energy, and biomass, and $39 billion in biofuels (Pereira et al. 2012). Separately, the Program for Energy Development of States and Municipalities installed 5,700 off-grid solar PV systems, mainly in northern and north-eastern Brazil (Geller et al. 2004; Pereira et al. 2012).

Brazil's RPS of 10 % RE by 2020, set by the Programme of Incentives for Alternative Electricity Sources, was recently updated to 75 % by 2030. Brazil has made it easier for small-scale RE producers to enter the market. For example, the Regulatory Electric Power Agency allows for energy generated by small hydropower plants to enter the grid for free. The plants are also exempt from having to pay municipalities and state governments for the use of water resources. In 2005, Brazil also launched a program to encourage small-scale biodiesel production (Pereira et al. 2012).

Brazil has a progressive RE auction system which is similar to the one used in France (up until 2000), Ireland, and the UK. This system consists of establishing a total number of alternative sources of renewable electricity to be installed and uses several auction sessions to select the projects with the lowest costs (Pereira et al. 2012). Allowable sources include biomass, biodiesel, wind energy, and small and large hydropower plants. It encourages as many bidders as possible, allows for transparent use of energy sources, and sets a fair price for electricity (Pereira et al. 2012).

Most of Brazil's renewable electricity comes from hydropower. Brazil is the world's second highest user of hydropower, behind China.[5] It has more than 400 large- and medium-scale hydropower plants, which generate about 93 % of Brazil's electricity. About 80 % of new hydropower plants will be installed along the Amazon River (Pereira et al. 2012). Brazil's use of hydropower reflects the continent's historical use and is tightly linked to its biophysical context, i.e., the landscape and location of Brazil with its numerous, large rivers.

In order to promote other types of RE not so strongly driven by biophysical factors as ethanol and hydropower, Brazil has focused on influencing the attributes of the stakeholder community and rules in use. For example, Lambrides notes that in Brazil, "Both the Biomass Users Network-Central America and The Renewable Energy and Energy Efficiency Partnership are trying to create a culture of entrepreneurs in the energy sector" (2006, p. 79). Brazil is trying to increase its use of both solar thermal (for heating) and solar PV (for lighting, pumping, and

---

[5] Brazil uses 374 TWh, or 12.1 %, and China uses 485 TWh, or 15.8 %.

communication) (Pereira et al. 2012). Currently, most solar PV systems are off-grid, but Brazil is trying to increase its use of on-grid solar energy through a leasing structure (Lambrides 2006).

Brazil started using wind power in 1992 on Fernando de Noronha, an island off the northeast coast (Geller et al. 2004). With a total potential of 143 GW of wind energy, Brazil reached 931 MW in 2010. This represents almost half of Latin America's total wind-energy capacity (Pereira et al. 2012). Multinational companies have established themselves in Brazil to manufacture and/or assemble wind-power equipment and provide services. This has increased employment in the wind-energy sector and has encouraged universities and technical colleges to provide courses on wind energy (Pereira et al. 2012).

Brazil faces fewer obstacles to RE development than other Latin American countries, which must deal with policy and financial barriers "...including arrangements that favor low upfront costs and continued fuel costs (fossil fuel) over high upfront costs and low fuel costs (RE)" (Lambrides 2006, p. 78). Most Latin American countries focus on short-term energy prices and invest in projects with short construction time and low initial investment costs (Etzkowitz and Brisolla 1999; Organization of American States 2004). These countries must focus their attention on providing their citizens with power and may not have the luxury bearing the high initial investment cost of RE projects. Thus, they do not implement large-scale RE projects.

Much like the other case-study countries, Brazil is a democracy and has good relationships with its neighbors. However, among the case-study countries, Brazil is a bit of an outlier in that it considers a mid-developed country based on its per capita income. Brazil has a per capita income of around $10,500, but income distribution is highly inequitable. In northern areas, about 50 % of families earn less than $150 per month (Geller et al. 2004). Brazil's continued economic growth aimed at improving living standards has distinct implications for the development of its energy sector. Between 2005 and 2009, Pereira notes that "16 million Brazilians rose above the poverty line [and] around 12 million Brazilians obtained access to electricity" (Pereira et al. 2012, p. 3787). To sustain such growth, Brazil will need to invest $564 billion in the energy sector over the next 10 years (Pereira et al. 2012).

3.3 China

China is currently the largest carbon dioxide emitter in the world, slightly ahead of the United States. In 2007, 93 % of China's energy came from fossil-fuel resources and approximately 75 % of that energy was from coal. China has the largest population in the world at 1.4 billion in 2011 (19 % of world total) and a GDP development goal of 7.2 % per year between 2000 and 2020 (Wang et al. 2010). Taken together, these will dramatically increase energy needs. To begin to curb its fossil-fuel use, China invested $45.5 billion in RE in 2011 and set an RPS of 20 % by 2020.

The Chinese government has implemented several important RE policies; the most important of which is the Renewable Energy Law (REL) passed in 2005. Its purpose is "… to promote the development and utilization of RE, increase energy supplies, improve the energy structure, guarantee energy security, protect the environment and realize economically and socially sustainable development" (Schuman and Lin 2012, p. 2). The REL includes four key mechanisms: (1) a national RPS and a development plan; (2) net metering; (3) a FIT system where the price for RE stays above the wholesale electricity price for desulfurized coal-fired power; and (4) a surcharge on electricity consumption to pay for FITs, grid-connection projects, RE grids, research on renewables, pilot projects, rural use of renewables, and renewable-resource assessments. Each province and all four provincial-level municipalities also have RPSs (Schuman and Lin 2012).

China's Renewable Power Quota Regulations require that generators and grid companies meet their RPS targets by the target date, not on an annual basis. However, they must report progress toward these targets to the local branch of the National Energy Administration on a monthly basis. Although the reporting rules encourage information transparency, there is no penalty for noncompliance (Schuman and Lin 2012).

China has chosen a policy framework similar to that of Germany: it emphasizes priority connection for RE, a purchase policy for renewable generators, and FITs. However, "China appears to have the most unified, top-down approach to implementing RE policies and programs, whereas the EU and US systems have a greater degree of autonomy and diversity among their member states in terms of setting renewable policies" (Schuman and Lin 2012, p. 11). Having said that, China is taking the first steps toward developing a carbon market—a very decentralized, incentive-based approach to promoting RE use. Following the EU, China has setup a cap-and-trade system for greenhouse gas emissions in the cities of Beijing, Tianjin, Shanghai, Chongqing, and Shenzhen, and the provinces of Hubei and Guangdong (Marshall 2012). By setting this cap, China hopes to meet the carbon dioxide reduction goals stated in the Twelfth 5-year plan to a 17 % reduction by 2015, compared to 2010 levels (Marshall 2012; Wang et al. 2010).

China also initiates and funds significant RE research and development. In 2007, China implemented a cooperative RE program to demonstrate its commitment to RE use and greenhouse gas reduction through the creation and exchange of technological innovations. China wants to establish a dialog and a mechanism for cooperating with foreign governments, research institutions, and enterprises (Zhao et al. 2011). By May 2010, China had established 63 such bilateral or multilateral cooperation mechanisms.

China and the EU began cooperating in 1994 with the China-EU Energy Cooperation Forum. This relationship has led to environmental protection, technological exchange, industrial cooperation, and research and development. Cooperation between China and the US began in the 1990s and has led to demonstration projects, technological cooperation, and the establishment of a dialog mechanism. China has also developed a relationship with Japan that has significantly improved wind-power technology in both countries. The relationship has also led to environmental protection, energy conservation, and emissions reduction. Relationships between China and Germany and China and the Netherlands have led to the implementation of small-scale projects in rural areas. Germany has helped to bring small hydropower plants to Tibet, and the Netherlands helped to install 78,000 solar PV systems in remote homes in China (Zhao et al. 2011).

China has actually surpassed the US as the country with the largest amount of installed wind-capacity in the world. In 2010, China met its 2020 wind-power goal of 30 GW (Schuman and Lin 2012). Although in 2005 the Chinese government mandated that 70 % of wind-power equipment must be produced domestically, this requirement was dropped in December 2009 to provide new opportunities for foreign manufacturers (Zhao et al. 2011). China has set wind-turbine implementation standards and all turbines must undergo stringent testing before connecting to the grid. All turbines that are already in place must be retrofitted to meet the standard (Schuman and Lin 2012).

As with several of the case studies, grid connectivity is one of the major barriers to China's RE implementation. Grid enterprises have almost no incentive to build or expand grids to accommodate for RE. Four factors disincentivize grid expansion: (1) most RE plants are located in rural areas far from the existing grid; (2) grid enterprises will bear part of the higher cost of installing RE infrastructure; (3) since RE still accounts for only a small share of total electricity generation, grid enterprises will have little opportunity to recoup their investment in infrastructure; and (4) since RE is intermittent and sensitive to seasonal and climatic changes, it may cause grid instability and increase the complexity of grid management (Wang et al. 2010).

3.4 The United States

The US is home to about 4.5 % of the world's people who consume about 25 % of global fossil-fuel resources (Institute for Energy Research 2012) and consistently produces at least 25 % of global GDP (US Department of Agriculture 2012). The country is well-known as a major user of nonrenewable-energy sources, generating a large share of the world's emissions. However, the US continues to make the implementation of sustainable energy systems a priority, and in 2011 it led RE investment with $48 billion. Between 2008 and 2010, RE accounted for more than 50 % of new electricity capacity in the US, with wind accounting for 90 % of electricity from renewables (Wang et al. 2010). One feature that distinguishes the US from other case studies is the relatively strong separation between policy at the federal and state levels. Renewable energy policies and projects are not the exception.

3.4.1 Sustainable Energy Implementation at the Federal Level

Since the oil crisis of the 1970s, energy independence has been an important US goal. President Nixon initiated Project Independence in 1973 and President Ford continued this project and called for an increase in domestic RE use. President Carter then took the initiative a step further by heavily funding the development of new renewable technologies, and setting a timetable for their entry into the market (Grossman 2009).

The Energy Independence and Security Act of 2007 established federal energy-management goals and requirements. It updated the 1978 National Energy Conservation Policy Act. The Act sets standards in a number of areas including:

- Energy-reduction goals for federal buildings;
- Facility management/benchmarking;
- Performance and standards for new building and major renovations;
- High-performance buildings;
- Energy savings performance contracts;
- Metering;
- Energy-efficient product procurement;
- Office of Management and Budget reporting and;
- Reducing petroleum/increasing alternative fuel use (US Department of Energy 2010).

Energy independence is necessary to achieve the national security goals, so it has been promoted from both sides of the political spectrum as an important reason for focusing on domestic energy production (Grossman 2009; Salameh 2003). This common purpose has helped promote RE to some extent.

Although there are few federal mandates, some federal financial mechanisms promote RE generation, most notably the production tax credit (PTC) and the investment tax credit (ITC). The PTC, first established in the Energy Policy Act of 1992, provided project owners in 2010 with tax credits for wind, geothermal, landfill gas, qualified hydroelectric, municipal solid waste, marine and hydrokinetic power, and both open- and closed-loop biomass. The ITC was created by the Energy Policy Act of 2005 with the goal of reducing federal income taxes for RE project owners based on the capital investment value of the project. In 2010, the ITC provided a 30 % tax credit for solar, small wind, and fuel cells, and a 10 % tax credit for geothermal, combined heat and power, and micro-turbines. With no long-term policy and tax credit goals set in place, both the PTC and the ITC have undergone changes on an annual or biannual basis, resulting in *inconsistent policy support for renewables* (Schuman and Lin 2012). This can ultimately lead to the failure and/or lack of implementation of RE projects.

In 2009, the Federal Government enacted the American Recovery and Reinvestment Act, which created a new 1603 Treasury Grant program.[6] During the economic crisis of 2008, banks stopped taking advantage of the PTC due to profit

[6] The purpose of this program is to reimburse eligible applicants for a portion of the cost of installing specified energy technologies. This is in lieu of tax credits.

instability which led to a reduction in available tax equity capital.[7] The Act provides qualified project developers with cash grants up to 30 % of capital costs of a project in lieu of the PTC and ITC (Schuman and Lin 2012).

The Department of Energy (DOE) launched by President Jimmy Carter in 1977 was another response to the energy crisis of the 1970s. The DOE centralized the responsibilities of the Federal Energy Administration, the Energy Research and Development Administration, the Federal Power Commission, and other energy-related government programs into a single cabinet-level department. The current mission of the DOE is "to ensure America's security and prosperity by addressing its energy, environmental and nuclear challenges through transformative science and technology solutions" (US Department of Energy 2012, para 1). The DOE also created the Database of State Incentives for Renewables and Efficiency which provides easily accessible and transparent information about how to finance a RE and/or energy-efficiency project.

An issue to RE implementation in the US at the federal level is that RE has been prioritized primarily in association with a response to crises. Unfortunately, when projects are implemented in times of crisis or too quickly they are more likely to fail (Grossman 2009). Fortunately, at the state level, governments are more consistently innovative in the energy sector than is the federal government.

3.4.2 Sustainable Energy Implementation at the State Level

A fact that sets the US apart from the other case-study countries is that federal policies are fewer and less innovative than state policies. The fact that most federal energy policies must be periodically renewed reflects a lack of long-term commitment. The US has yet to set a national RPS, mandatory national RE targets, or FITs. Yet most states have an RPS and some states and cities are experimenting with FITs (Schuman and Lin 2012). For example, in California, "investor owned utilities are required to procure an additional 1 % of retail sales per year from renewable sources…and all utilities are required to achieve 33 % of their electricity from renewables by 2020" (Schuman and Lin 2012, p. 13). State-level RPS targets in the US depend on a number of variables including natural resources, political influence, and the organization of the RE industry within the state. States regulate their RPS differently, with some states imposing a financial penalty for noncompliance (Yin and Powers 2010) while others do not.

In addition to an RPS, states have used a number of other policy instruments to encourage RE development. One such instrument is the mandatory green power option: a utility company provides consumers with the option to buy green power in which the utility must then either generate through direct RE implementation or by purchasing RECs to equal the amount of green power purchased by the

[7] Tax equity capital is a sum of money that investors, as partners, give to invest in a certain renewable energy project. The investors are able to take advantage of tax benefits.

consumers. Some states have created a public benefit fund either by charging consumers a small amount or by having utilities pay to support local energy efficiency and RE projects. Many states have also enacted net metering laws and interconnection standards, both of which facilitate the implementation of small-scale consumer-generated RE projects (Yin and Powers 2010).

The strength of state-level policies is determined by a number of variables related to the attributes of the community and biophysical context. Currently, prices are higher when a greater proportion of power generation is composed of RE sources. As a result, higher income states are more likely to develop RE projects. Similarly, there will be a stronger demand for RE projects in a state where environmental values are stronger. A state with fewer natural fossil-fuel resource, for example, may also be more likely to develop RE in order to diversify its energy portfolio and reduce its dependence on outside sources of energy (Yin and Powers 2010).

4 Case Study Comparisons: Institutional Drivers

Through an application of the IADF to the four case studies, we have attempted to identify key institutional drivers associated with energy transitions. We acknowledge that the institutional lens is just one way to view energy transitions. Our intent in using the IADF was to try to understand the interaction between policies (i.e., rules-in-use), biophysical factors (i.e., natural resource endowments, technology), and attributes of the community (e.g., pro-environmental values, wealth) and energy transitions. We discuss eight such drivers below.

Government Commitment. Obviously, success with renewables requires government commitment. In all four countries, governments have prioritized RE implementation by setting standards and providing financial incentives. Present conditions simply do not support RE transitions based completely on energy value (price) (Fouquet 2010; Menanteau et al. 2003; Rotmans et al. 2001). Thus, government commitment is required to translate nonmarket considerations such as environmental values and energy security through institutional change to create the necessary market conditions to favor RE. On the other hand, government commitment alone is not sufficient. This is certainly true in our case studies and there are other examples where top-down (e.g., government- and donor-led) efforts have failed (Martinot et al. 2002; Nieuwenhout et al. 2001; you need to add some others). The question is where such commitment is most productive.

The case studies provide some hints. Except for the US, all of the case-study countries have set a national RPS. In the US, 29 states (58 %) have set an RPS. "RPS goals emphasize the importance of establishing long-term, consistent targets and price signals for RE to encourage sustained investment in RE projects" (Schuman and Lin 2012, p. 14). Further, the establishment of an implementation guide that outlines key steps and/or benchmarks is another kind of commitment that characterizes successful energy transitions (e.g., China, Spain, and Brazil).

Such a guide acts as a transition plan for change at all levels, from national to local. Local-level plans are important because they can incorporate local knowledge about what kinds of changes are most likely to succeed in a given community. Previous research suggests that transition plans need to be flexible and adaptable (Young 1982).

The case studies also suggest that no single financial mechanism will lead to high RE use—too much reliance on a single mechanism in a system as complex as the energy sector is likely to fail (Haya 2007; Menanteau et al. 2003). Governments must create an array of financial incentives, rebates, and subsidies as the case-study countries each have. The most common are tax incentives and government subsidies, rebates, and loans. In Brazil, the market, and not the government, drives the price for renewables. Therefore, sustainable financial schemes should aim ultimately to replace government support with market-based support.

The case studies support the idea that innovation and proactivity at local and provincial/state levels can promote a transition in a country that may face barriers to innovation at the national level. This is especially true in the US which allows state governments to establish their own RE standards and incentives. Following the advice of Schuman and Lin: "The central government should not prevent a province from taking such proactive measures should it choose to pursue a more aggressive RE policy" (2012, p. 17).

In at least three of the four case-study countries, formal rules accommodated consumer needs and preferences. For example, in Spain, there has been little opposition to RE projects and policies due its low density. In the US there have been many nongovernmental organizations pushing and helping to develop RE policies. In Brazil, the citizens have pushed for the use of biofuels, and in turn, the government has developed a number of policies that subsidize its use. In these countries, strong citizen commitment to alleviating climate change is reflected in extremely ambitious governmental policies.

All of the case-study countries have multiple organizations acting to promote RE. This supports Ostrom's (2005) argument that intentional change should be a multi-agency effort. Ministries and departments need to communicate with each other to coordinate implementation and avoid overlap. *Multi-agency efforts increase research and development as well as the successful deployment of RE technology*. Finally, the case studies demonstrate that national-level efforts necessary to implement RE projects that depend on economies of scale such as wind and/or solar farms, hydropower plants, and/or biogas plants. *It is these large-scale projects that may have the greatest potential to increase a country's RE use.*

Polycentricity. A self-governed, polycentric system is one in which there are many centers of decision making and different levels of organization where participants make many, but not necessarily all, rules that affect the use of a resource system (Ostrom 2008). As such, polycentricity can impact the scale at which different energy sources are used, as well as the rate at which technologies are implemented. All of the case studies allow for some form of polycentricity in RE implementation.

For example, the EU is emphasizing bottom-up energy management. In the 1990s, the European Commission initiated the SAVE and SAVE II Programmes, focusing on regional and urban energy management. Both Programmes encourage local and regional action, resource use, and sustainable development. In 2004, SAVE co-funded the creation of autonomous Energy Management Agencies at the local and regional levels, under the framework of the Intelligent Energy Europe-Programme. By 2013, about 80 of these energy agencies had been created (European Commission: Intelligent Energy Europe 2013; ManagEnergy 2013). One example is the Spanish Association of Renewable Energy Producers created under the SAVE program. Many industry groups belong to this organization which has enabled the government, banks, and industry to work together in RE implementation in spite of weak citizen engagement.

A wide range of stakeholders have been very supportive of RE policies and projects in Brazil, especially the Proalcool program. While lacking in citizen-led organizations, Brazil does have three notable market facilitation organizations that are supporting the growth of RE markets: Biomass Users Network Brazil, Brazilian Renewable Energy Companies Association, and Winrock Brazil, which is focusing on empowerment and civic engagement.

Although China's RE policies are very top-down, there are examples of polycentricity. For example, a greenhouse gas emissions cap-and-trade system has been deployed in the cities of Beijing, Tianjin, Shanghai, Chongqing and Shenzhen, and the provinces of Hubei and Guangdong. Also, in 2000, China established the China Renewable Energy Industries Association (CREIA), with a membership of over 200 from industry, organizations, academies, and individual experts. CREIA promotes the deployment of RE technologies by developing and disseminating studies and surveys, workshops, and expert groups (Europe-China Clean Energy Centre 2013). In addition, the China Sustainable Energy Program, established in 1999, consists of scientists, analysts, policy makers, and business leaders who help deploy RE policies and projects through reports, workshops, and grants (The China Sustainable Energy Program 2013).

The US allows for polycentricity in a number of different arenas. At the federal level, RE lobbyists support the missions of their organizations. These organizations can lobby to politicians for the use of RE technologies and the implementation of RE policies. The top four RE lobbyists in 2010 were the American Wind Energy Association, the National Rural Water Association, Growth Energy, and Renewable Fuels Association, spending nearly $5.4 million (Lacey 2010). The US also allows for polycentricity at the state, and even city, level since governments at this level can lead their own initiatives in setting RE policies and projects. There are also a number of organizations at various levels promoting RE development, including: the National Renewable Energy Laboratory, the only federal laboratory dedicated to the research, development, commercialization, and deployment of RE; the United States Renewable Energy Association, a volunteer advocacy group; and the American Council on Renewable Energy, a 501(c)(3) non-profit organization focusing on RE education.

Polycentricity is strongest in countries that allow for a variety of self governance regimes at different scales. The US is arguably the most polycentric in terms of RE among the case study countries. It allows for citizen engagement through a variety of nongovernmental organizations. The US also has RE lobbyists at the political level and umbrella organizations for RE industries. While Spain, Brazil, and China have industry groups, none of these countries have powerful citizen-led organizations. Polycentricity can also have negative impacts, most notably making coordination across scales and levels of organization more difficult. Given the complexity and multi-scale nature of energy systems, institutional arrangements for coordination across operational units will be necessary.

Stakeholder Participation and Community Building. All of the case-study countries have engaged a variety of stakeholders, although not all have involved all of them, and all have continually revised their RE laws and/or plans. Small-scale and rural RE projects in all of the countries emphasize the use of local knowledge. When stakeholders are engaged, the community benefits. A shared sense of community increases the likelihood that individuals will participate in collective efforts to improve the community and build social capital (Anderson and Milligan 2006). In a democratic setting, local communities with high levels of social capital can influence national policy. This power is exemplified in the US, where progressive state policies have preceded and served as models for national policies (e.g., California's emissions standards, Massachusetts' health-care reform). It is likely that state-level RE policies will have a similar impact.

Transparency of Information. Good consumption choices require sufficient product information. Given the nature of energy systems, providing meaningful product information is challenging. Transparency, facilitated by stakeholder engagement and careful documentation, is thus very important. All of the case-study countries have mechanisms to improve RE projects' transparency. Examples include the wind-information website in Spain that provides real-time information on RE generation and use and feedback on household-level generation, and consumption in the US. There are also policies to make large-scale RE implementation transparent. Every government in the case studies provides an annual report on energy production and consumption, including RE. Some research has concluded that making information available to the public may build public support and funding for RE projects and encourage additional investment in RE by providing investors with confidence and predictability regarding the expected development of RE in a particular region (Schuman and Lin 2012). Further, encouraging greater public discussion and analysis of the renewable development plans may help to improve RE policy as well (Schuman and Lin 2012). China has demonstrated that trading information and ideas helps promote energy transitions by creating collaborative networks and makes more people aware of both current and potential RE implementation. Trading and sharing information promotes more effective project monitoring, which in turn increases the likelihood of project success.

Pilot Programs and Technology Innovation. Again, due to the nature of energy infrastructure with multiple stakeholders, legacy infrastructure, multiple-infrastructure owners, and the diversity of energy production technology, we should not expect a smooth evolution of energy infrastructure. High risk pilot programs are thus an essential ingredient for change to occur in the energy sector, and multiple failures should be expected. All of the case-study countries have made use of pilot projects to initiate RE implementation. They also invest in technology innovation programs. Using pilot programs to test innovative technology helps a country evaluate the feasibility of larger-scale projects.

Compliance with the Kyoto Protocol. Because the transition to RE is, in part, a response to the global commons dilemma associated with atmospheric carbon concentration, there must be some institution at the global level to provide incentives to cope with that commons problem. Frankly, without the threat of global climate change, the incentives for transitioning toward RE, at present, would be very weak. We use the Kyoto Protocol as a label for such an institutional arrangement. All of the case-study countries signed the Kyoto Protocol and, with the exception of the US, also ratified it. The Kyoto Protocol requires countries to reduce their greenhouse gas emissions 5 % below 1990 levels between 2008 and 2012. Unfortunately, Kyoto acts as a norm, there is no enforcement mechanism that would transform it into a rule, so although it creates an incentive for the case-study countries to implement RE policies and projects, these incentives are not as strong as they could be.

Grid Connectivity. One of the key features in the IADF is the interaction between biophysical context and rules-in-use. We emphasize 'grid connectivity' because policies directed and facilitating energy transitions must be conceptualized as integrated plans that combine policy with practical physical considerations. In the case of energy, "policy" cannot be approached as mere making rules and expecting compliance. Policies must consider the existing system and the possible trajectories of change given the existing system. The importance of the inertia of the existing system is a common issue to RE implementation in the case-study countries. In all of the countries, it is still difficult to feed RE into the grid. All of the case-study countries will either have to retrofit existing grids or create new grid systems. Policies must focus on practical mechanisms to transform existing infrastructure not only, for example, on RE innovation. New technology must 'fit' existing infrastructure.

Monitoring. To achieve RE targets, countries need both formal and informal institutions that monitor compliance. Formal institutions include fines and other strong disincentives for violating laws and failing to meet RPSs that deter future noncompliance (Schuman and Lin 2012). The Kyoto Protocol mentioned above is a good example of a norm with no enforcement mechanism that, as a result, has been less effective than hoped. An institution, like Kyoto, can be formally written and agreed upon, but lacks formal operational mechanisms. Informal institutions, such as monitoring and sanctioning mechanisms, include self-policing and gossip (Anderson and Milligan 2006). "Social preferences such as shame, guilt, and reciprocity may allow coordination of the actions of large numbers of people in their mutual interest...[and] most successful communities do not rely entirely on good will, but supplement it with mutual monitoring and punishment for transgression of norm" (Bowles 2006, pp. 131, 148). Such informal mechanisms work well in small communities, but cannot be relied upon at large scales. Formal monitoring and enforcement mechanisms will be required to generate energy transitions.

5 Conclusions

In this chapter we have attempted to introduce an institutional perspective into the energy transitions discourse. The institutional perspective focuses on how groups of people come together to solve collective problems. As such, it has emphasized rules and decision making processes. The energy sector, on the other hand, involves extraordinarily complex interactions among rules (policy), decision making, technology, and physical resources. We have thus extended the institutional perspective to focus more on these interactions than the rules themselves.

In so doing, we have identified eight such interactions that we term 'institutional drivers' related to successful transitions in four case studies. The most important of these drivers seems to be government commitment. Among the case-study countries, the goals for RE share range from 20 % by 2020 in China and Spain to 75 % by 2030 in Brazil. To meet these goals, every country has created laws, policies, regulations, and financial mechanisms as well as engaged in technology innovation. The case studies suggest, however, that government commitment is insufficient.

First, most programs are 'norms', or enabling institutions, that encourage energy transitions. Reliance on such norms may be effective but will require the engagement of a large stakeholder base. Without such broad engagement of stakeholders, government commitment alone will likely only be effective in a top-down, centralized government context such as China. In other contexts, RE transitions will rely heavily on cultural norms and values and how they intersect with government programs. Thus, without the capacity to force RE transitions driven by the global commons dilemma of climate change, nations must leverage policies based on norms that are carefully integrated with cultural values and norms using polycentric governance designs.

More challenging is the fact that for energy systems, RE policy must start with considerations of existing infrastructure. Innovative RE technologies that cannot connect with existing infrastructure will have less impact, at least in the short run. This difficult challenge may reduce the effectiveness of policies based on incentives and stakeholder values simply because of the scale of the problem and the associated financial demands to transition existing infrastructure to accommodate new RE technologies. Thus, it may be that countries which adopt a top-down approach with disincentives for noncompliance may end up being the RE leaders in the short run (e.g., China). It would be interesting to explore this idea further by studying countries *with strong top-down leadership, and government-lead RE policy making* (e.g., Cuba, Qatar, Saudi Arabia, North Korea, and the United Arab Emirates).

In summary, taking an institutional view of energy transitions suggests that they face, in essence, the same fundamental problems of all social dilemmas: the benefits of over-harvesting from the commons (the capacity of the atmosphere to assimilate carbon) are too high. There are three ways to solve such commons dilemmas: top-down regulation, privatization of the commons, or a governance regime that relies on a rich set of institutions (rules and norms) integrated with the biophysical context and the relevant stakeholder communities. In the case of energy systems, the first two options likely will not work in the long run. The third way—a polycentric governance regime for environmental resources—involves elements that have been well studied for biological and water resources (e.g., Ostrom 1990, 2005) but less so for energy systems. We have taken an initial step in characterizing features of successful energy transitions summarized in Sect. 4. Unsuccessful attempts at transitions will be likely be missing one or more of the elements described herein.

References

Anderson AA, Milligan S (2006) Social capital and community building. In: Fulbright-Anderson K, Auspos A (eds) Community change: theories, practice, and evidence. The Aspen Institute, Washington, DC, pp 21–58

Bowles S (2006) Microeconomics: behavior, institutions, and evolution. Princeton University Press, Princeton, pp 127–166 (Chap. 4)

Cristobal JS (2011) Multi-criteria decision-making in the selection of a renewable energy project in Spain: the Vikor method. Renew Energy 36:498–502

del Rio P, Unruh G (2007) Overcoming the lock-out of renewable energy technologies in Spain: the cases of wind and solar electricity. Renew Sustain Energy Rev 11:1498–1513

Etzkowitz H, Brisolla SN (1999) Failure and success: the fate of industrial policy in Latin America and South East Asia. Res Policy 28(4):337–350

Europe-China Clean Energy Centre (2013) CREIA. http://www.ec2.org.cn/en/about-us/associates/creia-chinese-renewable-energy-industries-association. Accessed 9 March 2013

European Commission: Intelligent Energy Europe (2013) Local and regional energy agencies. http://ec.europa.eu/energy/intelligent/in-action/local-and-regional-energy-agencies/index_en.htm. Accessed 19 Mar 2013

Fouquet R (2010) The slow search for solutions: lessons from historical energy transitions by sector and service. Energy Policy 38(11):6586–6596

Gamboa G, Munda G, Russi D (2005) Tackling local conflicts caused by renewable energy sources: lessons learned from real-world case studies. In: European Regional Science Association Conference, pp 1–23

Geller H, Schaeffer R, Szklo A, Tolmasquim M (2004) Policies for advancing energy efficiency and renewable energy use in Brazil. Energy Policy 32:1437–1450

General Electric (2013) Natural gas. http://visualization.geblogs.com/visualization/gas/. Accessed 13 March 2013

Goldemberg J (2007) Ethanol for a sustainable energy future. Science 315:808–810

Grossman PZ (2009) US energy policy and the presumption of market failure. Cato J 29(2):295–317

Haya B (2007) Failed mechanisms: how the CDM is subsidizing hydro developers and harming the Kyoto Protocol. International Rivers, Berkeley

Institute for Energy Research (2012) Fossil fuels. http://www.instituteforenergyresearch.org/energy-overview/fossil-fuels/. Accessed 12 July 2012

KPMG International (2011) Taxes and incentives for renewable energy. Energy Nat Res

Lacey S (2010) Top 25 US energy lobbyists of 2010. Renewable Energy World. http://www.renewableenergyworld.com/rea/news/article/2010/12/top-25-u-s-energy-lobbyists-of-2010. Accessed 12 July 2012

Lambrides M (2006) Growing wealth promotes clean energy in Latin America. Energy Sustain Develop 10(3):78

ManagEnergy (2013) Energy agencies at a glance. http://www.managenergy.net/energyagencies.html. Accessed 19 March 2013

Marshall M (2012) China set to launch first caps on CO_2 emissions. New Sci. http://www.newscientist.com/article/dn21361-china-set-to-launch-first-caps-on-co2-emissions.html#.Uk8nMGT5kfc. Accessed 12 July 2012

Martinot E, Chaurey A, Lew D, Moreira JR, Wamukonya N (2002) Renewable energy markets in developing countries. Annu Rev Energy Environ 27(1):309–348

McCrone A (ed) (2011) Global trends in renewable energy investment 2011. Frankfurt School UNEP Collaborating Centre for Climate and Sustainable Energy Finance

McCrone A (ed) (2012) Global trends in renewable energy investment 2012. Frankfurt School UNEP Collaborating Centre for Climate and Sustainable Energy Finance

McCrone A, Usher E, Sonntag-O'Brien V (eds) (2010) Global trends in sustainable energy investment 2010. United National Environmental Programme

Menanteau P, Finon D, Lamy ML (2003) Prices versus quantities: choosing policies for promoting the development of renewable energy. Energy Policy 31(8):799–812

Murphy T (2012) Fossil fuels: I'm not dead yet. http://physics.ucsd.edu/do-the-math/2012/02/fossil-fuels-im-not-dead-yet/. Accessed 13 March 2013

Nieuwenhout FDJ, van Dijk A et al (2001) Experience with solar home systems in developing countries: a review. Prog Photovoltaics Res Appl 9:455–474

Organization of American States (2004) Policy reform for sustainable energy in Latin America and the Caribbean. Unit for Sustainable Development and Environment, vol 5

Ostrom E (1990) Governing the commons. Cambridge University Press, Cambridge

Ostrom E (2005) Understanding institutional diversity, 1st edn. Princeton University Press, Princeton

Ostrom E (2008) Polycentric systems as one approach for solving collective-action problems. Indiana University: School of Public and Environmental Affairs: Department of Political Science, pp 1–22

Pereira MG, Camacho CF, Freitas MAV, da Silva NF (2012) The renewable energy market in Brazil: current status and potential. Renew Sustain Energy Rev 16:3786–3802

Potocnik J (2007) Renewable energy sources and the realities of setting an energy agenda. Science 315:810–811

Romero SR, Santos AC, Gil MAC (2012) EU plans for renewable energy: an application to the Spanish case. Renewable Energy 43:322–330

Rotmans J, Kemp R, van Asselt M (2001) More evolution than revolution: Transition management in public policy. J Future Stud Strateg Thinking Policy 3(1):15–31

Salameh MG (2003) Quest for middle east oil: the US versus the Asia-Pacific region. Energy Policy 31:1085–1091

Schuman S, Lin A (2012) China's renewable energy law and its impact on renewable power in China: progress, challenges and recommendations for improving implementation. Energy Policy 51: 89–109

Shove E, Walker G (2007) CAUTION! Transitions ahead: politics, practice, and sustainable transition management. Environ Plann 39:763–770

The China Sustainable Energy Program (2013) About us. http://www.efchina.org/FHome.do. Accessed 19 March 2013

The Pew Charitable Trusts (2012) Who's winning the clean energy race? 2011 edn: Country fact sheets. http://www.pewenvironment.org/news-room/fact-sheets/whos-winning-the-clean-energy-race-2011-edition-country-fact-sheets-85899381104. Accessed 13 Oct 2012

US Department of Agriculture (2012) Economic research service: GDP shares by country and region projections. http://www.ers.usda.gov/data-products/international-macroeconomic-data-set.aspx#.Ubt3gfb5kfc. Accessed 19 Nov 2012

US Department of Energy (2012) History of the US Department of Energy. http://www.lm.doe.gov/land/sites/oh/fernaldorig/aboutfernald/dhist.htm. Accessed 21 Sept 2012

US Department of Energy: Energy Efficiency and Renewable Energy (2010) Energy independence and security act. http://www1.eere.energy.gov/femp/regulations/eisa.html. Accessed 21 Sept 2012

Wang F, Yin H, Li S (2010) China's renewable energy policy: commitments and challenges. Energy Policy 38:1872–1878

Yin H, Powers N (2010) Do state renewable portfolio standards promote in-state renewable generation. Energy Policy 38:1140–1149

Young O (1982) Regime dynamics: the rise and fall of international regimes. Int Organ 36(2):277–297

Zhao ZY, Zuo J, Feng TT, Zillante G (2011) International cooperation on renewable energy development in China: a critical analysis. Renewable Energy 36:1105–1110

Renewable Energy: Urban Centres Lead the Dance in Australia?

Cathryn Hamilton and Jon Kellett

Abstract Australia provides great potential as a case study for renewable energy governance. It is a large continent with a comparatively small and highly urbanised population. It possesses enormous mineral wealth and is a major exporter of fossil fuels, but it also has huge potential for the exploitation of renewable energy. Politically, it is a country divided between those who support large-scale exploitation of fossil fuels and those who advocate that the nation should grasp the opportunity of its rich renewable resources to become a world leader in this field. The potential for renewable energy development has been recognised in some areas, with large-scale wind energy development in particular. But the great distances between energy sources and users suggest that the urban centres themselves should be examined as sources of renewable energy. Governance is complex, with three levels, Federal, State and Local, each exercising power and capable of influencing energy concerns. The key question which is addressed in this chapter is, *in a regime with multiple layers of government, at what level is renewable energy development best promoted*? We address the politics of energy in the context of Australia's economy and governance arrangements. Drawing data from previous research carried out by the authors, we examine the potential of urban areas to generate and supply their own power from renewable energy. Using Hammer's (2009) capacity to act theory we examine the capacity of local government to develop urban renewable energy. We seek to identify hesitations towards RE adoption in all levels. A critical question concerns whether bottom-up or top-down action is preferable.

1 Background

Australia is unusual when its geographical statistics are compared against Europe and many parts of Asia. It is a very large country in terms of its land mass (7.6 million km^2) with a small population for its land area (22 million) giving an

C. Hamilton · J. Kellett (✉)
School of Natural and Built Environments, University of South Australia, Adelaide, Australia
e-mail: jon.kellett@unisa.edu.au

E. Michalena and J. M. Hills (eds.), *Renewable Energy Governance*,
Lecture Notes in Energy 23, DOI: 10.1007/978-1-4471-5595-9_4,

average population density of 2.9 persons per km^2. The vast majority of Australia is barely inhabited, since the bulk of the population is settled on the coast around the edge of a large desert continent. The Commonwealth of Australia is made up of six States and two Territories, most of which are larger than many European countries. The population is overwhelmingly urban (88 %) and 60 % of the population live in cities of a million or more. The five large cities, Brisbane, Sydney, Melbourne, Adelaide and Perth are strung out along a coastline thousands of kilometres in length and sufficiently distant from each other to make air travel the natural linking mechanism. A national electricity grid stretches from Cairns in northern Queensland to Port Lincoln in South Australia (SA), a distance of some 5,200 km and a separate grid operates in Western Australia (WA).

In Australia, the responsibility for energy governance is shared between the Federal and State governments. The Federal government based in Canberra, sets national policy directions across a range of issues including energy, but State government exercises considerable powers including over environment, transport, land use planning, water allocations and new capacity for energy systems. Over the past decade the key utilities including electricity, which previously were State owned, have been totally or partially privatised. The Australian energy market was deregulated progressively across the States, starting with Victoria in the mid-1990s. Whilst the financing of new capacity is now a market-driven decision of private electricity generating companies, the State governments are responsible for issuing operating permits and development consents and can clearly influence the location and type of primary energy consumption through these mechanisms. In Australia, Local government exercises powers devolved from the State level. Given the focus on urban living and the large distances across which electricity networks stretch, two key questions emerge.

Key questions: *In a regime with multiple layers of government, at what level is renewable energy development best promoted?* In a second question that reflects our own research, we also enquire: *What governance issues need to be addressed for Australian cities to provide a model for the successful application of renewable energy generation technologies within urban areas?*

2 Introduction

2.1 Australia: Primary Energy Producer and Consumer

Australia is well provided with fossil fuel resources, particularly coal, though oil and gas were discovered in the 1950s and continue to be produced today. However, the widespread occurrence and the ease with which Australia's coal measures can be exploited have created a situation where around 75 % of Australia's electricity is derived from coal (Australian Government 2012, p. xi). Queensland, New South Wales, Victoria and to a lesser extent Western Australia, all have active coal mines. In Victoria a large proportion of the fuel used to run base load power stations is lignite or brown coal with a high ash and water content, which in terms of air quality and greenhouse gas (GHG) emissions, can only be viewed as problematic.

Australians rank close to the top of the world league table when primary resource use and ecological footprint measures are considered. Australia's ecological footprint exceeds seven global hectares per person (the carrying capacity of the earth is calculated to be 1.7 global hectares per person). Per capita, Australians rank 15th in the world in terms of energy use (ABARE 2010) and GHG emissions exceeding 18 t per person contrast with an average of 10.3 t in the fifteen European countries making up the EU-15 (EU 2009). Australian cities are some of the most dispersed and lowest density in the world and as a result their populations are extremely car dependant. Australian housing is also large with the median floor area of a new house in 2009 totalling 240 m^2, the largest in the world. All of these factors suggest that environmentally, Australia is living beyond its means but politically, the Federal government has met resistance from the coal industry to suggest any change which might impact its dominant position as the world's largest coal exporting country. The Labour government has made some moves towards improving Australia's environmental performance, notably by its decision to sign the Kyoto protocol in 2007 and has introduced a carbon tax, but its room for manoeuvre is strictly limited by an electorate which appears almost evenly split on these issues.

2.2 *Breaking the Bonds of Fossil Dependency*

While energy efficiency is widely promoted in Australia for reducing carbon emissions in the residential sector (Harrington et al. 2006; Pears 2007), there has been a growing recognition of the opportunities that switching to renewable sources of electricity may offer to reduce carbon emissions (Beatley 2000; Lowe 2005; Pears 2007). As well as ample fossil fuels, Australia has significant renewable energy resources, although at present, nationally, only around 7 % of electricity is sourced from renewables (ABARE 2010).

In the post-war period significant hydro-electric schemes were developed in Tasmania and New South Wales and these continue to provide an important, though proportionally small contribution to national energy needs. Little potential exists for further development of hydro-electric schemes due to public opposition to building large dams in Australia. However, the potential of wind and solar energy is significant and wave and tidal energy along an extensive coastline also offer promise. Australia has the highest level of solar radiation per m^2 of any continent in the world, with enormous potential for solar energy development particularly in the central and northern latitudes, where very large areas receive more than 14 $MJ/m^2/day$. Wind regimes which offer wind speeds greater than 7 m/s also extend over much of the southern regions of the continent. In addition, large tracts of the inner continent are underlain by geothermal basins and Australia possesses 38 % of the world's known uranium reserves. Based on these resources, Australia has set mandated Renewable Energy Targets (RET).

In early 2010, the Federal government increased its RET from 5 to 20 % by 2020 so that "the equivalent of all household electricity comes from renewable sources by 2020" (Wong 2009). In a recent National Energy White Paper (Australian Government 2012), the Federal government set out the main energy issues which must be addressed over the next decades. Whilst it is encouraging to see significant discussion of carbon reduction, the role of renewable energy technologies and the need for a transition to clean energy, the importance of the fossil fuel export market remains a key policy driver. Energy cost and competitiveness within the domestic market also suggest that low cost power sources, notably coal, will remain important in future energy supply, particularly for base load.

In recent years, retail electricity prices in Australia have increased significantly despite domestic electricity demand falling. These rises are mainly the result of a peak demand problem. As summer heat-waves increase in intensity and duration, the gap between peak electricity demand and base load demand has increased. Generators have invested in new capacity in order to cater for this shortfall. However, this extra capacity remains idle for much of the year, whilst its investment cost is recouped through increasing electricity prices to consumers.

The contrasting attitudes of different Australian State governments in respect of the primary energy mix are clearly demonstrated when considering Victoria and SA. The former has extensive reserves of brown coal and derives much of its base load power from three coal burning power stations located in the Latrobe valley close to these fuel sources. Despite the introduction of the carbon tax, the Victorian government appears wedded to this source of fuel. Prior to the more recent Federal renewable energy targets, Australian States have been variously setting their own targets for renewable energy. SA legislated a target of 20 % of electricity consumed in the State to be generated from renewable sources by 2014. It was also the first Australian State government to introduce climate change legislation in 2007, which set mandatory targets for carbon emissions reduction, thus providing the context for both energy efficiency measures across all sectors and encouragement for renewable energy developments across a range of technologies.

Whilst Victoria had previously set a renewable energy target of 10 % by 2016 and has overseen the development of a number of wind farms in the past decade, recent changes to the Victorian planning rules now require much larger separation distances between turbines and residential properties than hitherto. Wind energy developers have reacted to these changes by looking elsewhere, particularly to neighbouring SA, for opportunities. SA already has the largest proportion of wind energy developments in Australia with around half of the nation's installed capacity and 24 % of its own power usage sourced from its 15 operating wind farms. Scope exists for significant development of more wind energy and the SA State government has recently changed its planning rules to reduce the scope for third-party appeals against wind farms and acknowledges that the rural character of SA includes the presence of wind farms. Amongst the Australian States, SA therefore is the prime example of the State government choosing to back renewable energy and shift its primary fuel mix away from fossil fuels in the interest of carbon emission reduction. The development of wind energy in SA is almost all in rural areas, despite there being potential for urban wind developments in periurban metropolitan Adelaide and along the coast (Girardet 2003).

The potential of urban areas to supply a proportion of their own energy needs has been the subject of numerous commentaries (Diesendorf 2007; Droege 2006; Hatfield-Dodds and Dennis 2008). Federal government programmes such as "Solar Cities" and rebates for solar PV installations have seen a significant uptake of roof mounted solar installations over the past 5 years, initiating a decentralisation of energy supply. This trend has been accelerated by the introduction in SA and elsewhere, of a Feed-in-Tariff (FiT), which offers householders significant saving on their electricity bills by buying back surplus electricity from solar PV panels at favourable rates. The uptake by householders has been enthusiastic and has been assisted by falling costs of the technology as rebates have progressively been removed and FiTs reduced. For the present, large-scale application of a policy for urban areas to provide a proportion of their own energy needs from renewable energy remains more speculation than regulation. Examples of planning rules (e.g. the Merton Rule) have been developed in the UK that require all new developments over a certain size (normally 1,000 m^2) to provide at least 10 % of their own energy from renewable sources (New Rules 2010). In the Australian context of large isolated cities with a significant range of potential energy options, such a concept presents an intriguing possibility. Further investigation of this potential is supported by the rapidly increasing price of electricity and the peak load problem.

2.3 What Research Acknowledges

Giddens (2009) states that humans may know that action needs to be undertaken to address climate change but that putting action into practice is problematic. Whilst Federal and State governments in Australia have been bickering over appropriate carbon reduction mechanisms and targets, Local government has been working on carbon reduction since the 1990s through the Cities for Climate Protection (CCP) program of the International Committee for Local Environmental Initiatives (ICLEI). The focus of Local government has been on its own operations and facilities, and it has widely adopted the corporate aspects of the CCP program as Bulkeley and Betsill (2003) noted in their study of climate change governance. They also noted that action to reduce carbon emissions at the community level was more challenging.

Many researchers consider that local policy responses that address the causes of climate change are important (Bai 2007; Betsill and Bulkeley 2004; Burch 2010; Coenen and Menkveld 2002). However, Bai (2007) concluded that managers in Local government struggled with the global framing of the climate change issue.

Betsill and Bulkeley (2004) observed that processes of policy learning within climate change programmes are neither rational nor straightforward, and instead take place in discursive struggles as different actors seek legitimacy for their interpretations of what local climate protection policies should mean. Coenen and Menkveld (2002) identified two paths for Local government action. The first relies on expanding the competence and scope for action by Local government, in particular, to solve problems that may be obstacles to taking action to reduce carbon emissions. The second path was for Local government to better use its existing capacity to act. They observed that options for carbon reduction were at times not used even though municipalities had the competence to use them. Coenen and Menkveld (2002) identified roles for Local government including regulation, spatial planning, sustainable housing and sustainable production and consumption within its community.

Research addressing Australian Local government's broader planning provisions for climate change mitigation has previously noted "limited evidence of tangible planning requirements to achieve these goals", with "only three local plans referring to climate change mitigation" (Gurran and Phibbs 2008). In a review of 18 Local government authorities across South East Queensland (SEQ), Burton (2007) found that 16 had planning schemes and of these, only seven mentioned climate change mitigation. A similarity in strategies undertaken to mitigate climate change at the local level was observed across the SEQ region with the main focus on energy efficiency through building design or orientation. Burton observed that there were no attempts to adopt renewable energy targets and become less reliant on the main coal-powered electricity grid, mainly due to economic reasons. In our previous study of metropolitan Local government in Victoria and SA undertaken in 2008, only one Victorian Local government authority mentioned the use of planning policy for promoting renewable energy, whilst none of the SA Local government authorities that responded mentioned action in this area.

In our previous research, we also found that Local government thought leadership, appropriate planning policy, mobilising the community and using renewable energy were ideal actions to address climate change. Our study noted, however, a number of obstacles to implementing such ideal action. These obstacles included a lack of political leadership from higher levels of government (State and Federal) which, in the Victorian Local governments surveyed, was as frequently mentioned as a lack of finances. In Australia, Local government is yet to be recognised in the federal constitution, and is reliant on State jurisdictions for defining its responsibilities and for approving funding, particularly the level of rates and fees that Local government can collect from its citizens to deliver services. The pressure on Local government in Australia to deliver more services on behalf of the State government and cost shifting from higher levels of government has been noted by others (Dollery et al. 2008).

According to Giddens' (1993) theory of structuration, systems and structures that have been established within communities are locked in by the agency or actions of citizens and institutions. To enable changes to the systems including those that will allow for decentralised energy supply, conscious and deliberate action is needed. Hammer (2009) considers that capacity to act on energy services is reflected in the levers that Local government can apply, and is manifested through its policies and programmes. He identified five levers:

- rule-making;
- regulatory oversight;
- financial incentives;
- direct expenditure and procurement and;
- information gathering and dissemination; convening; facilitation and advocacy.

We postulate that, if Local government is to be an appropriate level to act on climate change and, if increasing the local uptake of renewable energy is an ideal mitigation mechanism, then the presence or absence of policies and programmes at Local government level should reflect its capacity to act to increase the uptake of renewable energy across its spatial area of responsibility. Using the five levers described by Hammer (2009), we analyse data from an in-depth survey of three Australian urban Local government cases for which we have also undertaken renewable energy resource assessments.

3 The Urban Cases

Some information and statistics about each case are provided in Table 1 to gain an appreciation of some of the physical, social and economic issues faced by each Local government authority. In Table 2, we provide a summary of the average energy demand for sectors within each case and the solar energy resource (SER) that we have estimated from renewable energy resource assessments. (from Hamilton et al. 2008a; Sunter et al. 2010). From these studies, it appears that there is potential to generate electricity or heat water using energy from the sun, sufficient to match the energy demand of the residential sector in each of the SA cases. For Manningham, the SER has potential to contribute approximately 85 % of the energy demand for the residential sector.

Table 1 Key statistics of three Australian local government cases (*Sources* ABS 2007, 2012 and council strategies)

Name	Area km^2	Resident population		Dwellings		Expected growth % by date
		2006	2011	2006	2011	
Playford, South Australia	346	69,418	79,115	27,243	32,375	47–64 by 2018
Onkaparinga, South Australia	518	155,919	159,576	61,169	67,867	26 by 2028
Manningham, Victoria	114	115,702	111,300	38,305	42,570	17.5 by 2031

Table 2 Summary of energy demand and solar energy resource (SER) for each case[a]

Sector	Average energy used in playford in 2006 PJ per year	Estimated SER in playford PJ per year	Average energy used in Manningham in 2006 PJ per year	Estimated SER in Manningham PJ per year	Average energy used in Onkaparinga in 2006 PJ per year	Estimated SER in Onkaparinga PJ per year
Residential	1.24	1.37	3.39	2.98	3.49	4.06
Commercial	0.57	0.05	1.09	0.13	1.75	0.22
Industrial	4.59	0.14	2.7	0.01	11.23	0.12
Transport	3.38	na	6.34	na	8.81	na
Total	9.78	1.56	13.52	3.12	25.28	4.40

[a] PJ PetaJoules per year

4 Analysis of Local Capacity to Act

The following discussion summarises the policy and programmes that reflect Local government's capacity to act in respect of the five levers proposed by Hammer (2009) together with the obstacles and opportunities we identified in our research.

4.1 Local Rule-Making

Local government has specific powers to facilitate or obstruct renewable energy development by the exercise of planning and building control. Under the Development Act 1993 (South Australia), new development is managed through a Development Plan (DP), a statutory document which is reviewed every 5 years. The DP is produced by the relevant Local council after a process including consultation with its community and other stakeholders and is approved by the State Minister for Planning and Development. New DPs for both the City of Playford and the City of Onkaparinga were consolidated in 2012 (Government of South Australia, 2012a, b). A review of these documents indicates that standardised wording exists for both cases. In recent years, an attempt has been made to standardise policy statements using a planning policy library that Local government is required to adopt. For example on-site renewable energy generation is included in a section on Energy Efficiency. In addition, the SA State Residential Development Code allows the fixing of PV panels flush to roof surfaces subject only to building consent and exempts them from planning consent, unless the dwelling is listed on a heritage register.

In the State of Victoria, Local government is required to have a Planning Scheme (PS) and a Municipal Strategic Statement (MSS) which builds on the State Planning Policy Framework (SPPF) for Victoria (Government of Victoria 2008). These Planning Schemes operate for 5 years. There is also a requirement on each

Local government authority to encourage land use and development that is consistent with the efficient use of energy, to minimise GHG emissions (SPPF Clause 15.12) and to promote energy efficient building and subdivision design (SPPF Clause 15.12–1) (Government of Victoria 2008). Manningham's Planning Scheme addresses, amongst other things, energy efficiency, urban design and alternative energy sources. In respect of alternative energy sources, most relevant to renewable energy, the objective in the MSS (Clause 21.07) is to reduce air pollution and limit GHG impact by progressively reducing dependence on non-renewable energy. Strategies to achieve this objective include: promoting the benefits and the business case for applying appropriate alternative energy options for a range of different circumstances and promoting and encouraging the incorporation of alternative energy installations including solar photovoltaics (PV), solar water heaters (SWH) and grid-interactive power generators into the design of buildings (MSS Clause 21.07).

The State government policy in both States standardises processes for solar PV installation for all dwellings, but stops short of mandating developers to install renewable energy in new buildings. Its policies remove planning controls over solar PV panel installation in most instances. Also whilst policy has advanced in respect of taking solar access into account, the current policy statements are rather general in nature. Overshadowing of PV panels, even partially, can have significant impacts on the performance of most PV systems. Kellett (2011) argues for a detailed review of policy in respect of solar access rights and suggests that the application of planning controls may not always be the most effective mechanism. However, such arguments appear to be well in advance of current policy.

4.2 Local Regulatory Oversight

State government legislation enforcing the National Construction Code requires Local government to ensure that new housing complies with the relevant building standard in each jurisdiction. The minimum standard for all new housing is specified in the National Construction Code (Volumes One and Two of the Australian Building Code). A proliferation of different building standards for housing across a city, however, is not desirable and may create problems with training and compliance for the building industry.

The current standard for energy efficiency is six stars on a scale of one to ten, where ten represents passive heating and cooling. At six stars, renewable energy is not mandated, although a solar water heating system is one option required for meeting the standard for an energy efficient water heater under the building code. An energy efficient heat pump system is also able to achieve the six star standard required. As capturing solar energy in the form of electricity to offset the electricity of operating such a system is not mandated, the building code is limited in its ability to enhance the uptake of renewable energy.

The percentage of new houses built each year is small compared to the existing housing stock. Trends of design and methods of construction of houses influence the style of housing, including the style of roof on which panels are placed. Using the number of North Facing Roof (NFR) facets as an indicator of roof complexity our renewable energy resource assessments for the three cases found that, since 1960, the complexity of roof profiles of houses has increased (Hamilton et al. 2008b). Our early research noted that while the area of NFR was sufficient to accommodate a SWH system and 1,000 Watt of PV panels for approximately 80 % of houses in each Local government area, in reality, the roof areas may comprise two or more NFR facets per dwelling and are not always of rectangular shape to easily accommodate PV panels. In the Onkaparinga, renewable energy resource assessment (Sunter et al. 2010) estimated that a 3,000 W PV system would offset the energy demand of an average household. Depending on the capacity of each PV panel, such a system requires approximately 24 square metres of useable roof area. A recent review of aerial photography of each case study area reveals that, in many instances, the size of PV installations has been increasing and solar PV panels are indeed being installed on multiple roof facets of houses as was predicted by our earlier research (Hamilton et al. 2008b).

Ideally, new house design should maximise roof area for efficient collection of available solar energy. A review of aerial photography of recently built housing in the joint SA State and Local government Playford Alive urban regeneration project reveals complex roof profiles. In marketing and house design documents which provide information about encumbrances on dwellings in the Playford Alive development, it was noted that while SWH systems were mandated for each dwelling, PV panels for generating electricity locally were not. The building regulations and consequent house designs are failing to cater for the future uptake of potential renewable energy available at a local level. House design energy specifications could be identified at spatial scales similar to climate zones and be administered relatively easily by Local government which already undertakes building approvals.

4.3 Financial Incentives

While all three of the cases studied were, at various times, involved in bulk purchasing schemes for residents, none had provided financial incentives directly to residents for installing renewable energy technology. Comments from Playford staff reflected an increasing community demand for extra services and concern that a lack of funding was a key issue to implementing ideal action to address climate change. Both Playford and Onkaparinga are located on the metropolitan fringe and cover a relatively large spatial area. The capital and maintenance costs for infrastructure to connect dispersed and expanding communities were noted as issues for both SA Local government cases. Our previous survey of SA and Victorian Local government conducted in 2008 found that three of the 18 SA councils had provided financial incentives in the form of interest-free loans to residents to install SWH systems, while one SA council was providing interest-free loans for PV systems. Local government clearly had some flexibility in its ability to provide financial incentives, although competing demands may make it difficult to provide such assistance at scale.

4.4 Direct Expenditure and Procurement

Policies and programmes which involved direct expenditure and procurement focused on increasing use or capture of renewable energy through recycling of wastes and use of biomass wastes and landfill gas rather than on solar or wind as renewable sources of energy for households. The expansion of services in both Playford and Onkaparinga to enable organic wastes, including food and green wastes to be collected and recycled, was evident. Local government had capacity to implement such programmes through its existing responsibility for waste management at the local level. Costs for these services were recovered from households through council rates or taxes. A long history of reporting GHG emissions for waste exists in Australian Local government through the ICLEI Cities for Climate Protection program.

In Onkaparinga, climate change has been an important issue due to the extensive coastal development in the area and the pressure for further growth along the coastline. The local impacts of climate change such as sea level rise have resulted in a programme of education aimed at Onkaparinga's elected members and senior staff. Onkaparinga has also developed a Climate Change Strategy as part of its Community Plan (City of Onkaparinga 2008). An important focus of the Climate Change Strategy is the Community Owned Renewable Energy Project (City of Onkaparinga 2009) which has established a community fund for implementing renewable energy projects. Onkaparinga is monitoring the uptake of solar PV systems by households and has the top three postal codes in SA for the number of PV installations (Rec Agents Association 2012). In addition, the Council has ambitious plans for a demonstration renewable energy park (Futurtec), incorporating wind, solar and biomass technologies to power a new 20 ha low carbon industrial development. The intention is that the energy park will serve as a tangible example to the community and to developers of what can be achieved using these technologies. Different approaches to household energy efficiency have developed in each Australian State. SA Local government cases appeared to have adopted self-help programmes with 'do-it-yourself' energy audit kits for testing energy efficiency of household appliances and fittings while Victorian Local government focused on providing assistance to retrofit equipment or technology for improving energy efficiency. Of our three cases, Manningham had actively partnered with other councils to form the Northern Alliance for Greenhouse Action (NAGA) which was developing and implementing a grassroots programme to involve communities in promoting sustainability including renewable energy.

4.5 Information Gathering and Dissemination; Convening; Facilitation and Advocacy

Of the three cases studied in depth, Manningham has been the most proactive in providing information to its residents, running seminars on various aspects of sustainability and facilitating the bulk purchase of insulation, PV and SWH systems for residents. Many actions of Manningham through NAGA, however, were focused on lobbying State government for information and autonomy to act on energy services. Playford partnered with SA State government and a small number of other Local government authorities in the 'Adelaide Solar Cities' programme, established by the Federal government and promoted information to its community about options to purchase renewable energy technology at a rebated price. Local government staff at Playford expressed frustration that the programme provided little feedback to local government about the actual uptake of PV installations from the programme in its area. More recently, and following the renewable energy resource assessments conducted for each council, both Playford and Onkaparinga joined an initiative titled Solar Councils, which facilitates bulk purchase of PV systems for residents. A review of a recent community discussion paper about the environmental issues that are important in planning Playford through 2043 (City of Playford 2012) noted the need to improve energy efficiency for new development. Whilst the discussion paper alluded to carbon emission reduction as a key area for action and included a brief reference to using solar power, there were no statistics included to raise awareness of the potential renewable energy available on rooftops nor renewable energy targets the community could aim for. In addition, the link to designing housing to maximise the opportunity for capturing solar energy was not being discussed with the community. Local government staff at Onkaparinga commented that obtaining information about the actual number of PV installations in their area from the Federal government, who maintained the database, was difficult. Providing easy access to information for Local government and cooperation across levels of government is an ongoing concern.

5 Discussion: A Role for Local Government

From the analysis of capacity to act using these three cases, it may be concluded that Local government's role to support its community to mitigate climate change through switching to renewable energy is not well defined. While there is clearly potential to provide energy supply from locally installed PV panels on rooftops, widespread concern was expressed about Local government's ability to effectively influence decisions at a household level as it is just one actor in the energy services system. Where Local government was taking action, it had overcome obstacles by using networks across Local government and lobbying higher levels of government for decisions on policy or increased autonomy to enable local programmes to be funded or established. With the standardisation of planning policy evident in both South Australia and Victoria, and mechanisms such as the Residential Development Code in South Australia, it appears that government policy stops short of mandating developers to install renewable energy in new buildings. However, it has taken steps to simplify the installation of renewable energy technology on buildings by removing planning controls over solar panel installation in most instances. Also, whilst it can be observed that policy has advanced in respect of taking solar access into account, the current policy statements are rather general in nature. Despite the finding that Local government has a limited capacity to mandate the uptake of PV technology, it is variously promoting renewable energy options to its residents and facilitating bulk purchasing opportunities. These Local government initiatives result in increasing awareness across the community and, along with reduced costs of installation of rooftop systems, a significant increase in the uptake of SWH and PV systems has been observed in both Onkaparinga and Playford in South Australia.

The importance of the impact of housing design on fitting rooftop systems has been noted and a potential future role for Local government is identifying the limitations that current and planned housing design have on maximising local renewable energy generation. Such information should be communicated to residents and provided to other stakeholders such as State Government. Local government has an important role and has clearly the best opportunity to involve its community in discussions about the potential to supply energy from the built environment. Local government needs to lobby higher levels of government to establish regulations that mandate renewable energy in new dwellings, which it could then enforce at the local level. Local community support is important to encourage elected members at State and Federal Government level to commit to seeking changes to building regulations that maximise the efficiency of solar panels fitted to roof facets. Unless there is such leadership being demonstrated at the Local government level, the current systems at higher levels of government are unlikely to be challenged.

There is still much work to do at Local government level to raise the awareness of local communities about the potential for capturing renewable energy locally. Kellett (2003) previously noted that the community should be involved in making decisions about incorporating renewable energy at the local level. A more strategic approach could be adopted by Local government to identify and agree outcomes for carbon reduction including the role that renewable energy can play in achieving targets. In the UK, the Merton Rule has been influential in promoting the potential for renewable energy to provide some of the energy demand of new development. However, prior to such planning mechanisms being proposed, Local government could be communicating the outcomes of renewable energy resource assessments to its citizens. While strategic planning by Local government recognises the need to adapt to climate change, these plans should also aim to maximise the uptake of renewable energy to reduce carbon emissions. The authors have previously proposed that renewable energy resource plans, which support and complement development plans, could be adopted by Local government (Kellett and Hamilton 2009). Such plans would establish the potential that exists across an area, identify the pathways to achieve this potential, estimate costs and infrastructure requirements and provide a basis for evaluating progress towards achieving agreed targets.

6 Conclusions

Often described as "The Lucky Country", in respect of its energy options, Australia would appear to be just that. Its abundance of natural resources offers significant choice and its future energy security appears to be assured. Politically, Australia can choose between continuing a high carbon footprint at low initial cost, or a low carbon strategy at a somewhat higher short-term financial cost. This latter course offers local communities greater potential to become technological innovators, demonstrating a model of low carbon cities for the future.

A number of obstacles to this model arise due to the organisation of energy policy at the national level. Deregulation of energy in the interest of introducing competition into the energy market has handed many key decisions to the private sector. However, the inevitable business imperative of profits and market share requires that government must take a lead on regulation, especially in respect of environmental issues. The recent Energy White Paper sets a national policy framework, but does not suggest any major directional shift in policy. Australia's position as one of only three energy exporting countries in the OECD and the largest coal exporter globally poses special problems in respect of national policy which could be seen as inhibiting the development of renewable energy.

The discussion of local experience in South Australia and Victoria suggests that the enthusiasm and ability of Local government to push the renewable energy agenda is variable. Of the three cases discussed, Manningham and Onkaparinga are probably leaders in the field, so the majority of local councils in Victoria and South Australia may be viewed as having only a partial engagement with the issue as yet. The application of clear rules on the installation of solar PV panels is the most pervasive example of local policy initiatives, but was largely driven by State government in the initial phase. The need for leadership and a clear commitment from Local government to follow through is well illustrated by the initiatives taken by Onkaparinga council in developing its renewable energy strategy and plans for its Futurtec demonstration renewable energy park. But it is also clear that not all available initiatives emanate from Local government. Target setting appears to play an important part in progressing renewable energy development but has yet to be undertaken at the local level. South Australia's successful wind energy development programme may be viewed as a positive outcome of target setting. So too has the FiT, which is probably the most effective mechanism for solar PV uptake but which has been largely driven by State and Federal governments. While it could be argued that ambitious targets are more likely to stimulate development than easily achievable ones, over ambition could also be seen as possibly leading to disappointment and disillusion.

A key question concerns whether the current, centrally imposed energy policy framework, is best placed to drive change. Grassroots action by consumers, Local and State governments may offer greater potential, and give suitable incentives. The nature of such incentives and how far localised, bottom-up initiatives might achieve a broad uptake in renewable energy, are all significant questions raised by the Australian experience. *A further point for discussion is the extent to which it is appropriate for Local government to set clear mandates for renewable energy development rather than relying on markets to operate with a degree offreedom.* Is it better to steer the market in particular directions using outcomes such as carbon reduction targets and star ratings for buildings or to specifically mandate renewable energy contributions along the lines of the UK's Merton Rule? To return to our original question regarding the most effective level of government for promoting renewable energy technologies, it would seem that there is no clear answer as the levers of change vary at different levels and in different locations but there remains scope for further action. A significant change of political mindset is required to move from a high to a low carbon future. Whether this occurs as a top-down process, perhaps driven by international pressure, or as a result of consumer and voter opinion buying into a renewable future, each level of government will need to support its policy with appropriate governance arrangements at the local level to ensure that planning and development decisions result in appropriate on-ground action. Bottom-up action by Local government is considered to be essential for the uptake of renewable energy and the hesitation of Local government to act in this space, revealed by the analysis of capacity to act, should be explored further.

References

ABARE and Geoscience Australia (2010) Australian energy resource assessment. Government of Australia, Australia

ABS (2007) 2006 Census Quickstats. Australian Bureau of Statistics, Canberra

ABS (2012) 2011 Census Quickstats. Australian Bureau of Statistics, Canberra

Australian Government (2012) Energy white paper: Australia's energy transformation. Commonwealth of Australia, Australia

Bai X (2007) Integrating global environmental concerns into urban management: the scale and readiness arguments. J Ind Ecol 11(2):15–29

Beatley T (2000) Green urbanism: learning from European cities. Island Press, Washington

Betsill M, Bulkeley H (2004) Transnational networks and global environmental governance: the cities for climate protection program. Int Stud Quart 48:471–493

Bulkeley H, Betsill M (2003) Cities and climate change: urban sustainability and global environmental governance. Routledge, New York

Burch S (2010) Transforming barriers into enablers of action on climate change: insights from three municipal case studies in British Columbia. Glob Environ Change 20:287–297

Burton D (2007) Evaluating climate change mitigation strategies in South East Queensland. Research Paper 11, Urban Research Program, Griffith University, Brisbane

City of Onkaparinga (2008) Climate change—climate change strategy: a community plan 2028 initiative—(2008–2013). City of Onkaparinga, Noarlunga

City of Onkaparinga (2009) Community owned renewable energy project, Core Options Paper. Tabled at Council Meeting 15 December 2009, City of Onkaparinga, Noarlunga

City of Playford (2012) Environmental Sustainability. Picture Playford 2043, Strategic Discussion Paper 4, Elizabeth

Coenen F, Menkveld M (2002) The role of local authorities in a transition towards a climate-neutral society. In: Kok M, Vermeulen W, Faaij A and de Jager D (eds) Global warming and social innovation: the challenge of a climate-neutral society, Earthscan, London

Diesendorf M (2007) Greenhouse solutions with sustainable energy. UNSW Press, Sydney

Dollery B, Byrnes J, Crase L (2008) Australian local government amalgamation: a conceptual analysis of population size and scale economies in municipal service provision. Australas J Reg Stud 14(2):167–175

Droege P (2006) Renewable city: a comprehensive guide to an urban revolution. Wiley, Chichester

EU (2009) Progress towards achieving the Kyoto objectives. COM (2009) 630 final. Report from European Commission to the European Parliament and the Council, Brussels

Giddens A (1993) New rules of sociological method: a positive critique of interpretative sociologies, 2nd edn. Polity, Cambridge

Giddens A (2009) The politics of climate change. Polity, Cambridge

Girardet H (2003) Creating a sustainable adelaide. Adelaide thinker in residence program, Government of South Australia, Australia

Government of South Australia (2012a) Development Plan - Onkaparinga (City), Sept 2012

Government of South Australia (2012b) Development Plan - Playford Council, Sept 2012

Government of Victoria (2008) State planning policy framework–Clause 15–environment. Victorian Department of Sustainability and Environment, Melbourne

Gurran N, Phibbs P (2008) Planning for sustainable change: a review of Australian local planning schemes. In: Gilmour T, Blakely E, Pizarro R (eds) Dialogues in urban planning: towards sustainable regions. Sydney University Press, Sydney

Hammer S (2009) Capacity to act: the critical determinant of local energy planning and program implementation. In: Proceedings of the Urban Research Symposium, World Bank, Marseille

Hamilton C, Kellett J, Yuan X (2008a) Carbon profiling: an analysis of methods for establishing the local emissions baseline. In: Proceedings of the 3rd international solar energy society conference, Adelaide

Hamilton C, Kellett J, Moore T (2008b) Resourcing a low carbon future. In: Proceedings of the international solar energy society conference–Asia Pacific Region (ISES-AP-08), incorporating the 46th ANZSES conference, Australia and New Zealand Solar Energy Society, Sydney

Harrington L, Brown J, Ryan P (2006) Quantification of standby in Australia and trends in standby for new products. In: Bertholdi P, Kiss B Anastasiu B (eds) Proceedings of energy efficiency in domestic appliances and lighting, Proceedings of the 4th international conference, 985−997, London

Hatfield-Dodds S, Denniss R (2008) Energy affordability, living standards and emissions trading: assessing the social impacts of achieving deep cuts in Australian greenhouse emissions. CSIRO Publishing, Canberra

Kellett J (2003) Renewable energy and the UK planning system. Plann Pract Res 18(4):307–315

Kellett J (2011) More than a roof over our head: can planning safeguard rooftop resources? Urban Policy Res 29(1):23–36

Kellett J, Hamilton C (2009) Decarbonising the local economy: planning for renewable energy in urban areas. In: Proceedings of the State of Australian Cities conference (SOAC 2009), Perth, Western Australia

Lowe I (2005) The big fix. Black Inc, Melbourne

New Rules (2010). Merton rule–United Kingdom. http://www.ilsr.org/rule/climate-change/2543-2/. Accessed 14 June 2013

Pears A (2007) Imagining Australia's energy services futures. Futures 39:253–271
Rec Agents Association (2012) Geographical analysis of solar systems under the renewable energy target. http://www.recagents.asn.au/wp-content/uploads/2012/09/Research-note-3-Geographical-Summary-Sep-2012-Final.pdf. Accessed 5 Dec 2012
Sunter P, Hamilton C, Kellett J (2010) Renewable energy resource assessment for the City of Onkaparinga. Report to City of Onkaparinga, University of South Australia, Adelaide
Wong P (2009) Australia's contribution to a global agreement on climate change. Address to the lowy institute for international policy 20th April, Sydney

Endogenous Tourism Development Through Renewable Energy Governance: A Questionable Challenge

Dimitrios A. Parpairis and Dimitrios Lagos

Abstract Is Renewable Energy Governance (REG) finally capable to be used as a tool to achieve Endogenous Development within the Tourism Sector? Here, our effort is to challenge this question by critically assessing specific cases located within the Tourism industry, where various actors interact under specific conditions when the energy game comes to the forefront. Our final goal will be to question several REG schemes and explore which are the drivers affecting the specific relationship between sustainable tourism development and renewable energy (RE). This approach will help to identify processes that could facilitate the transformation of saturated tourism models to niche tourism markets and of exogenous threats to endogenous sustainability and development under certain conditions. We also question whether REG leads to endogenous tourism development or it is the endogenous tourism development that leads to RE exploitation. We conclude that things happen the second way.

1 Introduction

Today, within the frame of the world economic crisis and witnessing the results of the fiscal austerity and socio-economic conflicts, questions arise about the future of the world economies and their sustainability. Within the path of recovery and looking to create future growth, legislators are trying to create a framework of endogenous development that integrates traditional pillars of their economies. Within this context special attention has shifted towards two of the most significant

D. A. Parpairis (✉) · D. Lagos
Department of Business Administration, Univeristy of the Aegean, Chios, Greece
e-mail: dparpairis@ba.aegean.gr

D. Lagos
e-mail: d.lagos@aegean.gr

E. Michalena and J. M. Hills (eds.), *Renewable Energy Governance*,
Lecture Notes in Energy 23, DOI: 10.1007/978-1-4471-5595-9_5,

and promising pillars for economic growth and social recovery; those are Tourism and Renewable Energy Sources (RES).

Researchers have been keen to exploit in recent years, if these two pillars blend and could create fusion towards macroeconomic growth and a more sustainable future (Michalena et al. 2009; Michalena et al. 2010; Parpairis et al. 2011). Indeed, one of the most significant questions arisen is on what Tourism needs today and what can Renewable Energy (RE) offer. In combination, can there be leverage between these two economic sectors on the long-term?

Endogenous development is the ability to innovate at local level, by (*a*) *transforming the socio-economic system,* (*b*) *responding to external challenges,* (*c*) *promoting social information and* (*d*) *introducing specific forms of social regulation* (Garafoli 1992). Considering that tourism is a field where various actors (i.e. authorities, institutions, local governments, society, enterprises and people), function, interact and integrate with each other, the introduction of tools to boost endogenous development and create conditions towards sustainability is not an easy matter. The tourism field, being dominated by the mass tourism model but currently under the impacts of climate change, strives towards introducing new techniques to respond to contemporary challenges, something which intrigue tourism actors.

Certainly, significant exogenous threats to the sustainability of the tourism sector, as well as uncertainties on shaping its future policies and strategies, create a fluid and fragile socio-economic environment in which actors who operate within the tourism "industry" must find ways to ameliorate risks. Certain areas, for example islands and coastal areas, considered as the backbone of the tourist product, are most vulnerable to any rapid environmental and socio-economic variation, and thus, they present challenges that need to be addressed through specific tools and incentives that promote growth at regional and local level. On the contrary, within the era of globalisation, vulnerable yet attractive regions must find effective ways to achieve endogenous development, by strengthening their economic abilities, by building capacity for future growth and by becoming more competitive and innovative at a local level.

Already, in the sphere of macroeconomics, the idea of endogenous (bottom-up) development (Romer 1994) took root in the early 1980s, partly as a response to criticism of the exogenous (top-down) growth theory (Solow and Ramsey Models). The endogenous development theory embraces that economic growth is primarily the result of internal (i.e. investment in human capital, innovation and knowledge being significant contributors to economic growth) and not external forces. Moreover, the endogenous development due to its peripheral and local characteristics (i.e. local actors) plays an even more important role within the tourism socio-economic environment which is exercised mostly at local and regional levels (Lagos 2007).

It is therefore essential to investigate on how tourism actors can mobilise and effectively use their local resources (financial, social capital, natural and human resources) and which policies/strategies (i.e. Governance) are needed in order to achieve endogenous development. Tourism (frameworks) and the actors need to be coordinated and governed through special tools. Good governance is critical for success, since it involves authority, decision making and most of all accountability.

Within this context, specific types of energy governance (here being used as a "tool") can affect local growth and endogenous development by means of creating niche tourism or special tourism markets. Towards this direction, this chapter tests renewable energy governance (REG) on a specific sector of economy and reviews specific case studies and scientific literature on the use of REG as a macroeconomic tool for recovery and growth. Within this outline, an interesting question probes for an answer: *Is REG finally capable to be used as a tool to achieve endogenous development? And the most important: Is endogenous development capable to lead to a RE development?*

Our effort is to attempt to provide some answers to this question by critically assessing specific cases located within the tourism sector, where various actors are setting-up specific conditions in order to challenge endogenous development through REG.

2 Tourism Dynamics and Vulnerability: A Question of "Governance"

International tourism remained strong in 2012 (estimated to reach 1 billion tourists) and grew (4 %) compared with 2011, despite the uncertain global economic environment. The latest indexes and figures released from the UNWTO Tourism Barometer consolidate a strong international tourism mosaic. It is worth underlining that international tourism remains resilient within a fast changing and uncertain economic environment and this is further confirmed by the positive data on tourism earnings and expenditure. In the long-term, the UNWTO forecasts that Tourism activity, i.e. international tourism arrivals will increase by 3.3 % per year on average till 2030, where an estimated 1.8 billion people will travel. Still, the recent trend noticed a sharper tourism increase within the emerging countries versus the developed ones, which may well continue in the long-term.

This forecasted increase in tourism mobility within the global stage will impact positively the economic growth of the emerging and developed countries, but will have at the same time, an environmental impact, enhanced further by climate change, that needs to be controlled and sustained. This impact raises questions on the sustainability of fragile tourist areas, such as coastal zones and islands, were their Carrying Capacity limits their ability to absorb the expected tourism and economic growth as well as the direct effects from the mass tourism model and the climate change (Parpairis 1998, 1999). It is therefore important to investigate tourism's vulnerability within the future forecasted framework and trends, as well as the tools-policies that are required to be utilised from actors to leverage the expected impact.

In this respect, and in recent years, more and more of researchers and scholars have analysed in depth the vulnerability of the Tourism Sector in relation to the main "threats", such as climate change, obsolete or saturated tourism models (e.g.

mass tourism), and questionable "Governance". Furthermore, tourism is directly connected with climate change, since it is a truly global economic activity (WTTC 2012) and for many years was considered as the first industry at a global level in terms of expenses and employees. Tourism, today, is employing globally 255 million people and transports nearly 700 million international travellers per year—a figure that is expected to double by 2020 (WTTC 2012).

Apart from the dynamics (threat, risks, etc.) generated from climate change on Tourism and vice versa, the increasing pressure from mass tourism and the overexploitation of natural resources might endanger the very existence of the tourist industry itself. Mass tourism, dominant in the majority of cases, often leads to severe degradation of natural landscapes, a lack of water supply, pollution of coastal zones, and the construction of massive transport and building infrastructure. The answer for mitigating the exogenous and centralised forces, analysed above, existing within the Tourism macro-environment, must arise from "good governance" on a central, regional and local level. Governance, i.e. policy creation, strategy shaping, decision making and accountability, is a critical concept influencing all aspects within the tourism socio-economic framework and is therefore considered as critical for success.

The below Fig. 1, attempts to map the REG and the nexus/positioning of actors integration within the tourism idea/framework.

It is evident that REG is in the centre of policy integration, which can be led either by Enterprises, Governmental Authorities, Tourists or Society, depending on the interests of each actor and the direction of the wished outcome for each category. This figure could be also interpreted as frame for targeting the use of SWOT analysis of the interactions, i.e. using REG as a method of enhancing interaction between tourism actors, on how to achieve the transition from "old" saturated tourism models to niche markets and from today's exogenous threats to endogenous sustainability.

In this respect, it is worth trying to identify the conditions needed to achieve sustainability through the interaction of the actors within the Tourism framework.

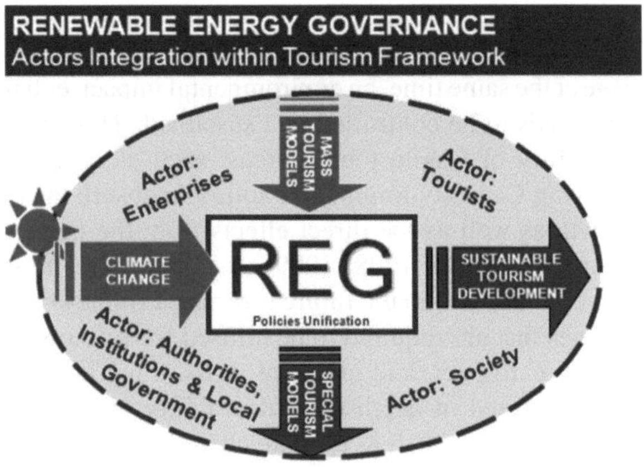

Fig. 1 REG and actors integration within tourism framework

3 Sustainable Tourism: Conditions Needed to Achieve Endogenous Development and the Role of RES

It is an academic axiom today that the realisation of the damaging effects of mass tourism, led to the concept of sustainable tourism. Indeed, there is a plethora of definitions and interpretations of the concept of "Sustainable Tourism Development" (Coccossis 1995; Hall and Lew 1998; Wearing and Neil 1999). Several authors believe that the sustainability of tourism development relies on the creation of particular characteristics of a tourism product in line with present and future tourists' needs (Burns and Montalvillo 2003).

Moreover, for the majority of scientists (see for example; Coccossis and Tsartas 2001), the concept of "Sustainable Tourism Development" concerns an economic, social and environmental tourism development which aims at the continuous improvement of tourists' experiences. For others, this kind of development is an additional opportunity for the local communities to benefit from the products of their own local identity and local natural resources (Burns and Holden 1995).

The need for sustainable tourism development within the exercised dynamics of climate change and mass tourism, leads to the necessity of creation and utilisation, from tourism actors, of "green" methods and tools (marketing plans, strategies, policies, sources, technologies, etc.). Within this context, the role of Renewable Energy Sources (RES) and Technologies (RET) is underlined: Especially for the tourism sector, scientific literature and practical case studies suggest that the role of renewable energy should not be underestimated for tourist destinations which suffer from environmental pressures (Government of Barbados 2001; Best 2002; Fitzerald 2005).

Furthermore, tourism vulnerability provoked by mass tourism, together with increased energy challenges generated from climate change, are now calling for new solutions based on endogenous development, which will strengthen local economic ability and local resources. Still, many intergovernmental organisations, such as the EU, seems to ignore that what works well in the political documentation might not apply at a local level, as local needs and aspirations may well be different (Michalena 2009; Michalena and Hills 2012).

For a local population what seems to be of a high importance is the endogenous development, i.e. the ability to innovate at local level (Garafoli 1992), by transforming: (*a*) *the ability of the local socio-economic system,* (*b*) *the ability to respond to external challenges,* (*c*) *the promotion of social information and* (*d*) *the ability to import specific forms of social regulation at local level to promote the above points.*

A question therefore arises on whether sustainable tools, such as RES and on a larger scale the concept of REG, can be utilised from tourism actors at local level and work towards boosting endogenous development, sustainable growth and building local capacity. Furthermore, it is challenging to investigate the necessary conditions, tools and incentives in order to make RES work within the tourism "micro" environment, i.e. achieve leverage. Garofoli (1992) in his effort to distinguish local from endogenous development, considers that local development is characterised from both endogenous processes and exogenous parameters, i.e. *"the basic element of the endogenous growth models lies in the formalisation of the accumulation process of input(s) that can be accumulated"*. This means that endogenous growth will have a structured multi-actor system at the local scale which catches either local initiatives or entrepreneurship, and/or grabs those that come from provincial or national authorities. It is this "sticky" structure which is the functional aspect that makes endogenous growth.

The question here is: *can benefits of REG and climate change be accumulated at a local scale or REG (mainly driven through European policies) will just be able to benefit the 20–20–20 targets?* In that sense, we wonder: *Where is the "battery limit" between local and European level?* We can consider this limit if we make the distinction between the concepts "mitigation" and "adaptation".

Indeed, whereas, at a global scale, "mitigation" is the only practical climate change response, at a local scale adaptation is what is mostly needed to cope with local-specifics in the way that vulnerability from climate change is expressed. To give a simple example that puts this questioning into practice: to promote a win–win situation between endogenous development and environmental enhancement through RE, requires public "productive" spending. Thus, the State should play a crucial role and impose taxes so as to increase the Feed-In-Tariffs to make RE more attractive to investors. Therefore, citizens finance RE and then RE reduces the carbon. Still, in some national situations (which are currently under financial problems) this win–win situation cannot be achieved because financial capability is reduced and therefore taxes cannot be collected, i.e. what is being today characterised as the "vicious cycle of recession".

Within this context, we have identified here below, specific case studies of several REG schemes, exploring *whether RE leads to endogenous development or whether endogenous development leads to RE exploitation.*

Our methodology will be based on critical assessment of specific case studies (see below), where various actors are setting-up specific conditions in order to challenge endogenous development through REG. Our final goal will be to attempt to verify the relationship and the drivers that coexist between REG and Sustainable Tourism Development.

4 Renewable Energy Governance: Case Studies

4.1 The Case of Barbados: Governance Drives the Relationship Between RE and Tourism

The case study of the island of Barbados, Michelle (2006), attempts to establish if sustainable tourism policy existed, and if so, how and at what level (local or central) was it being implemented, and if not, what were the identified main areas of weakness.

Barbados, one of the biggest tourist destinations in the Caribbean region, an island that has to put up with pressures for "green products", e.g. RES, demonstrates a centrally driven governmental influence on sustainable tourism development through regulatory mechanisms: several years after the Global Conference on the Sustainable Development of Small Island Developing States convened in 1994, sustainable tourism remains a priority for Caribbean countries. The main actors, responsible for sustainable tourism in Barbados, according to the author's argumentation, are the government and industry.

The conclusions raised for the case study of Michelle (2006) were that there was a vacuum in sustainable tourism plans and policies, that the government was indirectly influencing sustainable tourism through regulatory mechanisms (such as land use and infrastructure planning and market instruments), and that the industry was participating (in part) through voluntary compliance with green certification.

In fact, in the Barbados case study, there was a clear indication of government-industry cooperation at a central level. However, and despite all efforts of cooperation, interactions between actors were weak, as key infrastructure was inadequate, environmental impact assessments for potentially damaging tourism projects were not required by law and the public participation was weak.

The driver linking RE and Endogenous Tourism Development, in the case study of Barbados, was the holding of REG at central level avoiding it to share it with the local level.

4.2 The Case of Fiji: Adaptation to Climate Change Through REG is Driven by Reducing Costs at a Local Level

In the case of the Fiji islands (Pacific Ocean) Becken (2005) attempted to analyse the potential of tourist resorts towards reducing the effects of climate change, through reductions in carbon dioxide emissions. From the performed research it became evident that sustainable tourism was mainly led by the need to mitigate the climate change impact, since potentially devastating impacts, such as extreme events, sea level rise, transport and communication interruption, needed to be alleviated.

In Becken's study, the main actors involved were the hotel operators who were already prepared and had set-up actions for realising climate-related events and therefore were trying to adapt to potential impacts resulting from climate change. However, their "adaptation approach" was not really directed towards reducing emissions but on the opposite direction, i.e. towards decreasing energy costs for economic reasons.

In this case it has become evident that an exogenous condition (energy costs) has led to endogenous development of RE which has accrued climate change mitigation benefits. The "sustainable tourism" label is thus a passive reflection of fuel/energy/fuel tax system and has nothing to do with the tourism industry "sustainable" intentions. It looks like the whole approach is sustainability motivated, but, in the reality, is profit motivated!

4.3 A Case from Australia: How Actors Think

Research executed in the state of Queensland Australia (Dalton et al. 2007), investigated how tourist operators react to RE supply. Results indicated that the majority of operators were interested in RES as-such, but they were not sure about the positive marketing impact of RES in their operations.

On the contrary, tourist operators believed that their industry was threatened from RES implementation due to RES power supply limitations, reliability and economic viability. The responses provided in the study of Dalton et al. (2007) have varied depending on accommodation type offered from each tourism operator and his location (coastal or interior). For example, large type accommodation operators and self-catering accommodation showed more confidence in RES use and marketing capabilities.

This case-study makes us consider that major actors, e.g. large hotels need exogenous conditions to grow, while minor actors', e.g. small hotels need local tools.

4.4 A Study on the Cases of Calvià, Mallorca and Malta: Placing Climate Change into a Wider Context Driving Sustainable Tourism Development

Shifting our attention to the Mediterranean region and to the islands of Calvià, Mallorca and Malta, researchers Dodds and Kelman (2009) attempted to examine how sustainable tourism policies do and should function in order to reduce the associated vulnerabilities coming from climate change on the tourism sector.

Substantial data were collected from key actors responsible for policy implementation as well as tourism policy and planning documents from Malta's and Calvià's tourism industries. It has resulted that, although climate change could form important vulnerabilities to the tourism sector, climate change was rarely stated as being an important tourism issue! Even in the case where measures were implemented which would contribute to climate change adaptation, it was discovered in the end that those policies were implemented for reasons other than climate change! Some of those policies included: (1) enacting effective control systems to ensure that policies are implemented and monitored; (2) improving education and awareness on climate change and its potential impacts; (3) placing sustainable tourism and climate change within broader policy frameworks;

(4) implementing economic incentives to encourage adaptation strategies; (5) using accountable, flexible and participatory approaches for addressing climate change in sustainable tourism policies; and (6) filling in policy gaps while further integrating policies.

The research concluded that placing climate change into wider contexts reveals that some aspects of tourism might not be sustainable for small islands. Climate change should therefore be one dimension among many topics within sustainable tourism policies that are implemented both at central (governance) and local level (awareness), refer also to Fig. 1.

4.5 A Case-Study from Canada: How Endogenous Development Leads to REG (and not the Opposite)

During the past decade an innovative and emerging trend in Canada has been the creation of community energy plans (FRAME), where decisions that used to be left to regional level energy agencies or private individuals are now being considered at the community level. The driver behind this initiative is a local desire for change by reducing greenhouse gas emissions and by becoming more energy self-sufficient, e.g. a kind of "communal social responsibility". In theory, management at local level is desirable because it achieves these goals through improvements in the three (3) areas of energy efficiency, energy conservation and switching to REG.

Within this context, Denis and Parker (2009) attempted to analyse ten (10) of the first community energy plans in Canadian communities, ranging in population size from 500 to 1 million. The researchers have found that communities are choosing policies and programmes centred on increasing energy efficiency and conservation while RE receives much less attention.

This research provided insights on how communities recognised the substantial potential of RE, e.g. often focusing on technologies that the municipal sector could implement, such as biofuels for their transportation fleet. Wind, passive solar design, photovoltaic and solar thermal options were only recommended in a few cases. The research concluded that overall, only one of the five larger communities (Calgary) recommended implementing multiple RETs (i.e. increasing energy efficiency and conservation) while three of the five smaller communities proposed multiple RES (i.e. wind, photovoltaic, etc.).

Authors have finally suggested that smaller and more remote communities (actors) might most willing to lead in the planned introduction of renewable energy systems. *It is therefore evident from this case study that the associated driver for sustainability is generated at local level (endogenous) by using RE tools and not the other way around.*

4.6 A Case-Study from UK Which Demonstrates that Renewable Energy Cannot Lead To Endogenous Development Because Numerous Actors and Numerous Interests Exist

Focusing his research on the United Kingdom, Devine-Wright (2010), draw on evidence from a series of interviews with key actors involved in RE policy and development (including developers, local publics, politicians, activists and consultants) to examine their viewpoints.

Within this research, it has been noticed that the provision of community benefits had become a more common component of RE project proposals in the UK. The questions were focusing as to (a) the purposes RE benefits are fulfilling, and (b) if those benefits were an effective strategic element in negotiations around RE planning consents.

A variation was discovered in the extent and type of RE benefits offered, reflecting the maturity of different technologies and other factors. The normative case for providing community benefits appeared to be accepted by all involved, but the exact mechanisms for doing so remained problematic.

This research has actually challenged the perception that RE could lead to endogenous growth, since numerous actors and numerous interests exist; on the contrary *it seems that endogenous growth through actors' interaction and coordination may drive RE implementation.*

4.7 A Case-Study from Africa that Illustrates that Renewable Energy Governance is Central and not Following Endogenous Procedures: Initiatives

Pegels (2010) conducted a research in order to identify the relevant barriers to RE investments. In his research, the challenge of transforming entire economies was considered as an enormous task; even more so if a region was as fossil fuel based and emission intensive as South Africa. The South African electricity sector is a vital part of the economy and at the same time contributes most to the emissions problem.

The major barrier identified in Pegel's research was based on the economics of RETs, i.e. their cost and risk structures, two main factors in investment planning. First steps have been taken by the South African government to enhance energy efficiency and promote RE–through RE support measures, such as a feed-in tariff; however, they fail to show large-scale effects.

The paper leads to the conclusion that RE development is centrally driven and does not follow endogenous procedures.

4.8 In Maldives, Renewable Energy Should also be Governmental-Driven and not Following Endogenous Procedures

A team of researchers (Van Alpen et al. 2008) adopted a stakeholder-based approach to understand the size and nature of the market and financial issues to the widespread utilisation of RETs in the Maldives islands, located within the Indian Ocean.

Evaluating the different policy instruments, the research has found that pricing laws which regulate market access are more suitable than quota systems, provided that the current subsidies for conventional energy are removed or shifted towards RETs.

According to authors' views, and regarding alternative financial incentives, rebates on investments and low-interest loans should be provided to RET project developers to overcome the high initial project costs. In addition to these measures, the limited institutional capacity suggested that policy support should also focus on enabling the steady development of regional institutions rather than on supporting individual RET projects. These institutions should enable islands in each atoll to take advantage of efficient management and access to financial capital. This would allow the development of sustained RE projects that, in turn, would result in increased renewable capacity and decreased costs.

This research suggests for once more that RE is governmental-driven, based on financial measures and institutional capacity and not following endogenous procedures.

4.9 A Case-Study from Kenya, Africa, Which Relates to a Boost of Endogenous Tourism Development Through Governmentally Driven Renewable Energy Governance

Back in 2002, significant research (Mariita 2002) assessed the local environmental and socio-economic impact of geothermal power plant on poor rural community in Kenya. The author focused on the country's geothermal resources which were located in the Rift Valley region, in the middle of one of Rift Valley's major wildlife parks which are at the same time a major tourist attraction. Over the past two decades, the surrounding area has also become a major centre for Kenya's flourishing commercial flower farming, which is today partially powered by geothermal energy.

Mariita's research examined the environmental and socio-economic impacts on the nomadic low-income rural Maasai community from the simultaneous development of geothermal energy, flower farming and wildlife/tourism industry. While the near-term environmental impacts have been minimal, the research has warned of significant adverse impacts in the future if the competing demands of the fast growing geothermal energy, flower farming as well as wildlife/tourism sector are not adequately addressed. In the short term, however, the socio-economic impact of geothermal energy development was likely to be a main source of conflict, i.e. necessity to develop ways of extracting the geothermal resources without adding pressure on the remaining land.

To Mariita's opinion, the only solution that could be seen as viable in this case, would be that there should be appropriate policy and institutional measures in place to ensure that local communities will enjoy socio-economic benefits related to geothermal energy development.

In this case, it becomes once more evident that RE can be a driver for improving life and services quality for actors at local level, but this is possible if plans for endogenous tourism development are instigated at Governmental level where the associated policies are decided and implemented using REG as a tool.

4.10 Contribution of Renewable Energy Sources in Spain is Significant to the Creation of Endogenous Value (Job-Posts)

It has been argued by various scholars, that RES have a large potential to contribute to the sustainable development of specific territories by providing them with a wide variety of socio-economic benefits, including diversification of energy supply, enhanced regional and rural development opportunities, creation of a domestic industry and employment opportunities.

Within this context valuable research conducted by scientists (Del Rio and Burguillo 2009), have empirically analysed those benefits by applying a conceptual and methodological framework previously developed by the same authors to three RETs in three different places in Spain. Through case studies, this research shows that the contribution of RES to the economic and social dimensions of sustainable development is, indeed, significant. Particularly important is employment creation in these areas.

The specific socio-economic features of the territories, including the productive structure of the area, the relationships between the actors and the involvement of the local actors in the RE project may play a relevant role in this regard. Furthermore, other local (socio-economic) sustainability aspects beyond employment creation should be considered.

This research reinforces the conclusion that the contribution of RES to the economic and social dimensions of sustainable development might be significant; especially at local level with the creation of endogenous value, i.e. jobs.

4.11 A Case-Study from the Island of Crete in Greece Which Demonstrates the Interaction Between Actors, Using Renewable Energy as a Tool

Focusing their research on local challenges in the promotion of RES, two scholars (Michalena and Angeon 2009) dealt with the issue of RES acceptance in Crete, by designing some strong arguments for effective promotion of RE through local municipalities (actors).

Within the context of the research, it was considered that Regional Municipalities (actors) of Crete were playing the role of authorities of proximity (able to identify local needs and best organise local societies). Moreover, the cooperation of the local/regional level with the Hellenic Government (actor) at central level, and with the rest of regional municipalities of the island, could lead to fruitful results towards the adoption of a better sustainable future for Crete. In this respect, it was acknowledged that internal factors (such as local acceptance) and external elements (such as macrostructure) were playing a core role in the promotion of RES.

This research has concluded that, in a dynamic perspective, the combined analysis of these internal and external factors could provide a consistent framework to understand how small islands can reduce their economic and environmental vulnerability through use of RES. *This constituted a good starting point for analysing the island resilience, especially in relation to tourism sustainability generated endogenously.*

5 Discussion: Conclusions—Best Renewable Energy Governance for Specific Areas

It is commonly argued that areas that attract tourism interest are particularly vulnerable and exposed to the threats posed by climate change and mass tourism. At the same time, those areas could benefit from sustainable tourism development with lots of benefits for local communities; many of those through a properly designed REG . In their majority, studies have approached the development of integrated theoretical frameworks which allows a comprehensive analysis of the impact of RE on local sustainability in different territories. The question, thought, which is posed here is whether RE leads to endogenous development or the contrary.

Table 1 below, provides in a structured manner, a concrete consolidation of the critical points located within the case studies, i.e. actors, drivers, challenges and results. The results could provide a framework towards creating basic guidelines to be applied towards enhancing the role of REG within Tourism and at the same time boost the research on creating specific REG rules which could be applied for

Table 1 Renewable energy governance case studies critical points consolidation

Renewable energy governance case studies challenging sustainable tourism development (STD) through renewable energy sources

	Author	Location	Actors	Drivers	Challenges	Results
1	Michelle 2006	Barbados	• Government • Industry	(1) Vacuum in STD policies (2) Government indirectly influencing STD through regulatory mechanisms (3) Industrial "green" certification	(A) Government-Industry cooperation strong at a central level (B) Week interaction at local level	REG is central and not following endogenous procedures
2	Becken 2005	Fiji	• Hotel operators	(1) Climate change impact (2) Mass tourism	(A) Adaptation approach towards decreasing energy costs for economic reasons versus reducing emissions	Exogenous conditions (mitigation) leads to endogenous development and not in reverse i.e. endogenous development triggers adaptation through use of RE
3	Dalton et al. 2007	Australia	• Hotel operators	(1) RES use and marketing Impact	(A) RES power supply limitations (B) RES reliability and economic viability	Large tourism operators need exogenous conditions to grow. Minor actors (small hotels) need local tools
4	Dodds and Kelman 2009	Calvia, Mallorca and Malta	• Tourism industry	(1) Climate change impact (STD policies and functionality) (2) Mass tourism	(A) Adaptation approach towards climate change (B) Climate change rarely stated as an important tourism issue	Some aspects of tourism might not be sustainable for small islands. Climate change treated as a one dimension topic within tourism STD policies
5	Denis and Parker 2009	Canada	• Local communities	(1) Climate change impact (reducing greenhouse gas emissions) (2) Becoming more energy self-sufficient	(A) Policies centred on RES while RET (increasing energy efficiency and conservation) receives much less	Sustainability is generated at local level (endogenous) by using RES and not the other way around

(continued)

Table 1 (continued)

Renewable energy governance case studies challenging sustainable tourism development (STD) through renewable energy sources

	Author	Location	Actors	Drivers	Challenges	Results
6	Wright 2010	UK	• Developers • Local publics • Politicians • Activists • Consultants	(1) Community benefits had become a more common component of RES project proposals	(A) Exact mechanisms for achieving community benefits i.e. SD, remained problematic	RE cannot lead to endogenous development since numerous stakeholders/interests exist
7	Pegels 2010	South Africa	• Government • Industry	(1) Transforming "fossil fuel" to green economics	(A) Economics of RETs (cost and risk structures)	REG is central and not following endogenous procedures - initiatives
8	Van Alpen et al. 2008	Maldives	• Project developers	(1) Understanding the size and nature of the RES market	Provide: (A) Financial incentives (B) Rebates on investments	RE is governmenta-driven institutional capacity). RE not following endogenous procedures
			• Institutions	(2) Financial issues related to RE technologies (RETs)	(A) Low-interest loans to overcome the high initial RET project	
9	Mariita 2002	Kenya	• Communities	(1) Socio-economic impact (2) Environmental impact	(A) Short-term conflict: socio-economic impact of RE development	RES can be a driver for improving life at local level. Endogenous STD is instigated at Governmental level. Associated policies are decided and implemented using REG as a tool
				(3) From simultaneous development of RES and tourism	(B) Long-term impact: fast growing RE and tourism sectors competing demands are not adequately addressed (i.e. environmental and mass tourism)	

(continued)

Table 1 (continued)

Renewable energy governance case studies challenging sustainable tourism development (STD) through renewable energy sources

	Author	Location	Actors	Drivers	Challenges	Results
10	Del Rio and Burguillo 2009	Spain	• Stakeholders	(1) Diversification of energy supply (2) Enhanced regional and rural development opportunities (3) Creation of a domestic industry and employment opportunities	(A) Socio-economic features of the territories (B) Relationships (C) Between the actors (D) Involvement of the local actors in the RE projects	Contribution of RES to the socio-economic dimensions of SD is significant by creation of endogenous value (jobs)
11	Michalena and Angeon 2009	Crete	• Government • Municipalities	(1) Local Challenges in promoting RES	(A) RES promotion and acceptance (B) Macrostructure	Islands economic and environmental vulnerability mitigated through use of RES

specific choices of areas were tourism is present, emphasising in the endogenous dimension of tourism sustainability.

The combined critical approach attempts to place in perspective the identified results, thus facilitating on the identification of potential drivers that explain the relationship between RES and Tourism within the context of REG. This approach will help to identify processes that could facilitate the transformation of saturated tourism models to niche tourism markets and of exogenous threats to endogenous sustainability and development.

In our treatment of REG systems in the tourism sector we have tried to investigate *whether RE leads to endogenous development or whether endogenous development leads to RE*. We have approached our research question using the definition of sustainable tourism development which is directly related to the economic, social and environmental tourism development, aiming at the continuous improvement of tourists' experiences.

The sustainability of tourism is directly connected to the "local" factor. The importance of locality is underlined in all case studies. Indeed, according to the previously noted scholar's definition, endogenous development is the ability of actors to innovate at local level, by (*a*) *transforming the socio-economic system*, (*b*) *responding to external challenges*, (*c*) *promoting social information and* (*d*) *introducing specific forms of social regulation*. The transformation, response, promotion and introduction parameters are core parameters, established within the context of "Governance". It is therefore applicable to question whether REG can be finally capable to be used as a tool to achieve endogenous development.

In our case studies (of REG) we have seen **actors** of the tourism industry who interact in the energy market, having their own characteristics such as statutory instruments to wield, financial resources for investments, expertise in construction or generation, lobbying power, regulatory power, capacity and capability, flexibility and response time, etc. and demonstrating particular interests out of the use of RES.

The **"Frame of the System"** relates to the contextual features which shape the domain of REG systems. This could be for example, the present technical limits on RE generation types, obligatory national or international policy or political zeal. The way that the actors integrate with each other within the framed context is partly akin to "power" in the above definition and related to the process of governance.

This **interaction** of actors, within the "Frame of the System", has been led from drivers (process, competence, technology and institutional drivers) and ended in specific results. Consolidating the main **drivers** challenging todays tourism framework, in terms of the actors interaction, it has become evident that the issues of: (1) *climate change impact*, (2) *mass tourism pressure*, (3) *exploitation of resources* (4) *energy efficiency and finally* (5) *"green" economies*, sound critical for addressing the concept of REG, but they are centrally driven and have nothing to do with endogenous development.

Moreover, it is demonstrated, that the **challenges** faced today are multidimensional. It is worth underlining that in the case studies of islands and underdeveloped territories (Barbados, Fiji, Calvià, Mallorca and Malta, Malvides, Kenya and Crete), with their associated carrying capacities and fragile ecosystems, challenges are more positioned towards: (1) *weak governance at local level,* (2) *exogenous threats are mitigated rather than addressed i.e. one should expect adaptation at local level,* (3) *energy security and decreasing energy costs for economic reasons versus reducing emissions* (4) *climate change "unimportance" within tourism,* (5) *Missing incentives to overcome initial RET project cost and* (6) *sensitive actors' relationships, such as issues of promotion and acceptance conflicts.*

On the contrary, the case studies of the developed mainland territories (Australia, Canada, UK, Spain and South Africa), provided challenges related to: (1) *Policies centred on RES while RET (increasing energy efficiency and conservation) receives much less attention,* (2) *energy transformation and RES power supply limitations,* (3) *governance effectiveness, i.e. top-down mechanism for "diffusion" of policies to local level remained problematic,* (4) *understanding actors relationships and the socio-economic features at local level and* (5) *energy cost and risk structures, i.e. RES reliability and economic viability.*

The main **results** generated by analysing the case studies are quite substantial and seem to have a common platform for all locations (islands, underdeveloped or developed mainland locations). Clearly the most important are:

(1) *REG is central and governmental-driven and not following endogenous procedures (Barbados and South Africa),*
(2) *Exogenous conditions mitigation (e.g. climate change or mass tourism) leads to endogenous development, i.e. endogenous development triggers adaptation through use of RE (Fiji, Crete, Spain, Calvià, Mallorca and Malta),*
(3) *Sustainability is generated at local level by using RES as a tool and not the other way around (Australia, Canada and Kenya).*
(4) *RE is driven by Governance (Barbados, South Africa, Maldives and Kenya) and RE policies are influenced by contemporary challenges (energy security, diversification, economic and environmental vulnerability), and*
(5) *RE cannot lead to endogenous development since numerous stakeholders/ interests exist (Barbados, Spain and UK).*

It is evident from the above critical assessment that the questionable approach of achieving Endogenous Tourism Development through REG is not an easy matter, but would rather require the application of certain conditions.

The first question posed in our introduction was: *is REG finally capable to be used as a tool to achieve endogenous development?* From the above discussion it may be concluded that this possibility is excluded since endogenous sustainability is promoted at local level while REG is functioning mainly at central level. An important conclusion could be that centralised RE approaches need further local mobilisation (such as in the tourist industry) to help accumulate gains at the scale at which RE is actually implemented otherwise central RE targets or desires will not penetrate down.

The second and most important question posed was: *Is endogenous development capable to lead to a RE development?* From the discussion and analysis it is evident that this is a possibility.

Finally, from the results generated from the case studies analysed in this chapter, it could be conclusive as a central assumption that **REG** appears not to lead to the path of endogenous tourism development, but the other way around may be assumed, i.e. Endogenous Tourism Development leads to RES exploitation.

6 To Sum Up

In this chapter we attempted to challenge if REG is finally capable to be used as a tool to achieve Endogenous Development within the Tourism Sector. Our method was to question several REG schemes and explore the results. We have concluded that:

(1) REG appears not to lead to endogenous tourism development even if RES is one of a set of measures that move tourism towards a path to sustainability. It is Endogenous Tourism Development that leads to RE exploitation, and
(2) Endogenous growth through REG is a marketistic effect and looks like an illusion since local capabilities and tools are not exploited, but more centralised policies are in place to promote RE growth on a top-down axis.

Finally, it appears evident that specific conditions need to be assessed in order to make REG finally work within the tourism industry towards identifying processes that could facilitate the transformation of saturated tourism models to niche tourism markets and of exogenous threats to endogenous sustainability and development.

References

Becken S (2005) Harmonising climate change adaptation and amelioration: the case of tourist resorts in Fiji. Glob Environ Change 15(4):381–393

Best M (2002) The greening of Barbados' tourism industry: the private sector role. In: Eleventh annual caribbean water and wastewater conference in conjunction with the first caribbean environmental forum and exhibition, Pigeon Island Causeway, St. Lucia

Burns P, Holden A (1995) Tourism, a new perspective. Prentice-Hall, London

Burns P, Montalvillo S (2003) Local perceptions of tourism planning: the case of Cuellar, Spain. Tourism Manage 24:331–339

Coccosis H, Nijkamp P (1995) Sustainable tourism development. Avebury Press, England

Coccosis H, Tsartas P (2001) Sustainable tourism development and environment. Kritiki Publications, Athens (in Greek)

Dalton GJ, Lockington DA, Baldock TE (2007) A survey of tourist operator attitudes to renewable energy supply in Queensland, Australia. School of engineering, university of Queensland, Brisbane, Queensland 4072, Australia. Renewable Energy 32(2007):567–586

Del Rio P, Burguillo M (2009) An empirical analysis of the impact of renewable energy deployment on local sustainability. Renew Sustain Energy Rev 13(6–7):1314–1325

St. Denis G, Parker P (2009) Community energy planning in Canada: the role of renewable energy. Renew Sustain Energy Rev 13(8):2088–2095

Devine-Wright P (2010) Good neighbours, public relations and bribes: the politics and perceptions of community benefit provision in renewable energy development in the UK. J Environ Planning Policy Manage 12(3):255–275

Dodds R, Kelman I (2009) How climate change is considered in sustainable tourism policies: a case of the mediterranean islands of Malta and Mallorca. Tourism Rev Int 12(1):57–70

Fitzerald Y Jr (2005) Cleaner technologies for sustainable tourism: caribbean case studies. J Cleaner Prod 13:117–134

Garafoli G (1992) Endogenous development and Southern Europe. Avebury, Aldershot

Government of Barbados (2001) Barbados' first national communications under the United Nations framework convention on climate change (UNFCCC). In: St. Michael, Barbados: Ministry of Physical Development and Environment

Hall CM, Lew A (eds) (1998) Sustainable tourism development: a geographical perspective. Addison Wesley Longman, Harlow

Lagos D (2007) Regional economic development theories. Editions Kritiki, Athens (in Greek)

Mariita NO (2002) The impact of large-scale renewable energy development on the poor: environmental and socio-economic impact of a geothermal power plant on a poor rural community in Kenya. Energy Policy 30(11–12):1119–1128

Michalena E (2009) PhD Thesis on renewable energy policy and sustainable tourism development—geographic scales and complexities of sustainable development

Michalena E, Angeon V (2009) Local challenges in the promotion of renewable energy sources: the case of Crete. Energy Policy 37:2018–2026

Michalena E, Hills J (2012) Renewable energy issues and implementation of European energy policy: the missing generation? Energy Policy 45:201–216

Michalena E, Hills J, Amat JP (2009) Developing sustainable tourism, using a multicriteria analysis on renewable energy in Mediterranean islands. Energy Sustain Dev 13:129–136

Michalena E, Parpairis D, Lagos D (2010) Assessing the contribution of renewable energy in sustainable tourism development: the case-study of Mediterranean islands. In: The proceedings of the Conference: "Mass tourism vs. Niche Tourism", Limassol, Cyprus, 3–5 Nov 2010

Michelle M (2006) Sustainable tourism using regulations, market mechanisms and green certification: a case study of Barbados. J Sustain Tourism 14(5):489–511

Parpairis A (1998) The concept of carrying capacity of tourism development in coastal ecosystems. In: International Symposium, Bogazici University, Antalylia Turkey

Parpairis A (1999) Integrated coastal area management planning study: Fuka Matrouh, Egypt, In: UNEP/PAP/MAP, Regional Activity Center, Split, Croatia

Parpairis D, Michalena E, Lagos D (2011) Planning for the Hellenic tourism development in the era of climate change through the use of renewable energy sources. In: The proceedings of the conference: the 2011 Athens tourism symposium, international scientific congress on current trends in tourism management and policy, Athens, Greece, 2–3 Feb 2011

Pegels A (2010) Renewable energy in South Africa: potentials, barriers and options for support. Energy Policy 38(9):4945–4954

Romer MP (1994) The origins of endogenous growth. The J Econ Perspect 8(1):3–22

Van Alphen K, Kunz HS, Hekkert MP (2008) Policy measures to promote the widespread utilization of renewable energy technologies for electricity generation in the Maldives. Renew Sustain Energy Rev 12(7):1959–1973

Wearing S, Neil J (1999) Ecotourism. Butterworth-Heinemann, London

World Travel and Tourism Council (2012) In: Travel and Tourism Economic Impact 2012

Outliers or Frontrunners? Exploring the (Self-) Governance of Community-Owned Sustainable Energy in Scotland and the Netherlands

Niki Frantzeskaki, Flor Avelino and Derk Loorbach

Abstract Community owned renewable energy initiatives are the emergent and self-organized arrangements where communities become both producers and suppliers of energy. Cases of community energy developments form Texel (Netherlands), and Undy and Urgha (UK) are the empirical grounds that demonstrate this capacity. As highly desirable community owned renewable initiatives may seem, they face many enablers. However, they are also confronted with various tensions, as identified in this chapter. A closer look of the governance space in which these community initiatives operate, reveals that tensions and opportunities span from socio-cultural, political, and technological axes. These initiatives are both outliers and frontrunners of a sustainable energy transition: they create new forms of institutions, challenge even benefiting to them instruments, dare to uptake risks and seize opportunities, and operate outside demarcated institutional space. Community owned energy initiatives constitute a new form of local renewable energy governance that deserves to be explored.

1 Introduction

There have been numerous research papers and studies that advocate the benefits of renewable energy systems as the alternative—if not the substitute of current energy supply sources. The benefits of a shift to renewables include, among others,

---

N. Frantzeskaki (✉) · F. Avelino · D. Loorbach
Dutch Research Institute For Transitions, Erasmus University, Rotterdam, The Netherlands
e-mail: frantzeskaki@fsw.eur.nl

F. Avelino
e-mail: avelino@fsw.eur.nl

D. Loorbach
e-mail: loorbach@fsw.eur.nl

E. Michalena and J. M. Hills (eds.), *Renewable Energy Governance*,
Lecture Notes in Energy 23, DOI: 10.1007/978-1-4471-5595-9_6,

lowering overall greenhouse gas emissions, creating new energy markets, and when accounting all social-ecological costs and benefits, contributing to geopolitical stability (when oil and gas dependency are considered) (Verbong and Loorbach 2012). When this global perspective on a renewable future appears desirable, a local vision of a renewable-led future raises a number of questions based on our current experiences:

- Will local communities welcome or clash with the siting of renewables in their neighborhood? Regardless the spatial dynamics of the area in reference, (sprawling, densifying, or shrinking) regions will have to take more prominently into account perceptions and desires of local citizens that relate to their future (Wolsink 2012; Rygg 2012).
- Will a renewable energy future be governed by the state, by corporations, by hybrid arrangements such as public–private partnerships or, will new governance arrangements emerge? Current experience with the fossil fuel dominated system varies shows energy oligarchies around the world (Hoffman and High-Pippert 2010; Mumford and Gray 2010; Nikogosian and Veith 2012).
- Will a community that provided electricity from renewables rebound to more electricity intensive lifestyles, or is a renewable powered community possible to be energy responsible? (Darby 2006; Chitnis et al. 2013)

These critical points come to demarcate that beyond the technological advancements in the efficiency of renewable energy systems and the research on means to promote and better diffuse the use of renewables in the energy mix, *the role of communities in the energy system* has become more prominent, and should not be considered as negligible. This is further supported by the neoliberalism that steers energy infra-systems toward a more 'consumer'-responsible frontier.

Community owned energy infrastructures come to confront these three critical questions. Such initiatives have taken place sparsely in Europe over the past decade (Musall and Kuik 2011) and recently in Australia (Gross 2007) and in Canada (Corscadden et al. 2012). For a renewable-led energy future, community owned energy infrastructure appears as an ideal future micrograph that combines use of green energy technologies, governance by users, and fostering of energy responsibility.

Research on such initiatives has revealed that there are two socio-technological phenomena in synergy: the one of *experimentation* that the scholarships of sustainability transitions and energy policy studies have been investigating and one of the *self-organization* that numerous scholarships have been addressing, such as third sector studies, critical policy studies, infrastructure studies, and policy studies.

1.1 State of the Art: Experimentation and Self-Organization

The field of sustainability transitions studies has recently witnessed an increasing attention for self-organized innovation by citizens. Several authors have emphasized the need to pay more attention to the role of civil society in sustainability transitions (Seyfang and Smith 2007; Seyfang and Haxeltine 2012). Research on energy niches investigates the mechanisms and conditions that can nurture and grow promising innovations in order for them to offer viable alternatives to existing needs (Smith and Raven 2012). Community ownership of energy infrastructures is viewed as a new policy experiment for activating and establishing energy citizenry.

Energy policy studies explore the relations between community and renewable energy projects and a major focus has put on the resistance of community to neighborhood renewable installations. Energy policy studies have recently explored two types of facilitation mechanisms to fix community and renewable energy suppliers: (a) benefit sharing mechanisms that are market fixes to transfer benefits from energy projects to adjacent or nearby communities and the local administration (Corscadden et al. 2012; Cowell et al. 2011) and (b) enabling community owned renewable projects as niches of self-sustaining energy security.

A critical driver for community ownership of energy infrastructure includes policy incentives or an overall favorable policy regime that paves the ground for local leadership on taking-up energy initiatives (Saunders et al. 2012, p. 85). Full community ownership is often facilitated by community capacity building, via government-led initiatives supporting community energy (Walker et al. 2010, p. 2656); and via trust building between the community group that establishes and operates the energy facility (Walker et al. 2010, p. 2662). Shared ownership (between the local administration and community) is facilitated by profit expectations and a high degree of environmental awareness (Musall and Kuik 2011). Energy policy studies and cases conclude that community ownership is a policy-enabled or opportunity-enabled configuration.

In an overview of 'drivers and barriers' for Community-based Renewable Energy schemes, (Allen et al. 2012, p. 266) point out that drivers "span the entire sustainability spectrum"—including climate change, energy insecurity, job opportunities, lower energy prices, conservation of natural resources, social benefits (trust, cohesion)—while tensions can belong to three main categories: (1) institutional constraints (stakeholders, finance, dependency on expertise), (2) uncertainty and little knowledge, and (3) lack of time for actors and institutions to develop community renewable projects. These are particularly troublesome, because "energy consumers and producers are subject to complex, often confusing, and always changing set of policies" (Allen et al. 2012, p. 262).

A different focus on understanding infra-system governance by infrastructure studies proposes that instead of considering the demand side of infrastructures as a constant in the infra-system equation, one should consider the demand side as a dependent variable. Instead of users only representing the demand side, users can also adopt a dual role as infrastructure suppliers, supplying services that (better) meet their own demand. The concept of self-organization has received most attention in the recent work on *Inverse Infrastructures* (Egyedi and Mehos 2012). Egyedi and Mehos, describe inverse Infrastructures such as emergent, bottom-up, decentralized and citizen-owned infrastructures, and contrast these to the more traditional understanding of infrastructures as large technical systems, which are designed, top-down, centralized, and either publicly or commercially owned (Egyedi and Mehos 2012).

Self-organization is approached as "a mode of co-ordination in which control is dispersed and decentralized" (Egyedi and Mehos 2012). Institutional analysis of inverse infrastructures identifies that there is a considerable mismatch, i.e., policy gap, between emerging inverse infrastructures practices on the one hand, and the formal and informal institutions and policies that are oriented toward large technical systems on the other hand.

Within Third Sector studies, self-organization phenomena are primarily addressed in terms of 'co-operatives,' the 'social economy,' and/or 'social enterprises' (Birch and Whittam 2008; Moulaert and Ailenei 2005; Pestoff 1992). The *social economy* generally refers to the provision of services that satisfy needs that are not (or not adequately) fulfilled by the private and public sector. A *social enterprise* has "business orientation and innovative approach focused on the delivery of social benefits" (Birch and Whittam 2008, p. 440) and "an explicit aim to benefit the community, initiated by a group of citizens and in which the material interest of capital investors is subject to limits" (Moulaert and Ailenei 2005). Social enterprises can take various forms, including user or worker co-operatives, mutual aid associations, and even organizations with a commercial character. Several scholars have emphasized that the main contribution of the Third Sector is that it provides a diversity of alternatives to public or private sector (Scott-Cato and Hillier 2010).

Within the field of policy studies, self-organization has been typically addressed in approaches that are inspired by complexity theory and emphasize the limitations of governance in terms of steering and managing complex societal systems (Loorbach and Frantzeskaki 2012). Regarding the more specific issue of self-organization *by citizens*: this has been addressed (implicitly or explicitly) in terms of citizen participation and deliberative governance (Wesselink et al. 2011; Stirling 2008). Within that literature there is quite some attention for the ironies and dilemmas of the governmentality in civil society participation, such as the close relation between increasing calls for more citizen participation on one hand and the neoliberal political agenda on the other hand (see in particular recent discussions about UK government discourse on 'Big Society'). The literature also points out that there are different 'rationales' and 'design choices' that drive and shape public participation approaches; a distinction can be made between normative, substantive, instrumental, and legalistic rationales (Wesselink et al. 2011; Stirling 2008). These different rationales point to different motives, contexts, and problem definitions underlying public participation processes.

The presented approaches have different foci; some focus more on the 'civil society' dimension, while others are more centered on 'energy' or 'infrastructures'. These different foci also lead to different interpretations of what the main enablers and tensions, priorities and challenges are.

2 Research Approach

In order to place these academic discussions in a practical context, we have conducted a research project which aims to contribute to the state-of-the-art through: (1) a *cross-national* comparative analysis of empirical cases, (2) an *interdisciplinary* interpretation of empirical observations so as to bridge insights from different approaches, and (3) a *transdisciplinary* translation of research findings so as to allow for dialogue and communication between academics and non-academics. We argue that such cross-national, interdisciplinary, and transdisciplinary insight is still lacking in much of the literature. In order to realize these three aims, we start our research project with the empirical analysis of cross-national cases in which civil society is self-organizing energy infrastructures. Rather than attempting to 'integrate' or 'bridge' state-of-the-art research a priori, we take a more inductive approach in which we first use empirical observations to identify challenges and questions, which will then be viewed (in a later phase of the project) in relation to state-of-the-art research.

In our research we shed light on the *governance reality* in which these initiatives operate by exploring the policy-community interface and revealing institutional tensions, drivers, and controversies. Our focus is on capturing how participants in these empirical case studies *themselves* experience and explain the main drivers and tensions of their initiatives vis-à-vis the institutional.

We describe our cases at three different levels: (1) the actual organizations in which civil society self-organizes energy infrastructure (Sects. 3 and 4), (2) the socio-spatial context (Sects. 3 and 4), and (3) the national context (Netherlands and UK) (Sect. 1). By 'socio-spatial context' we refer to a geographical and/or social place that drives the project, in terms of providing it with a sense of community, identity, place, and motivation. This can refer to a geographically delineated community such as an urban district, an island, or a town, but it can also refer to a network or a social movement. As we will demonstrate in the case-study descriptions, these diverse types of 'socio-spatial contexts' play an important role in enabling the persistence of the individual projects and organizations.

Our empirical stories are based on research project in which we studied a total of eight case studies of community energy initiatives.

In this contribution, we only present three of the selected cases. The three nominated cases have in common the fact that: (1) they are practical examples of civil society self-organizing energy infrastructure, and (2) wind energy is used as the renewable energy technology by the communities.

A closer look at the case studies does not only reveal the role of the institutional context (Sect. 2), but also the institutional dynamics at hand (Sect. 3) in the form of similar or differing tensions and opportunities. We reflect on the case studies and their comparison in the form of governance implications and limitations (Sect. 6) and we conclude with a critical perspective on the role of community owned energy initiatives for a next-generation energy future (Sect. 7).

3 The Dutch Community Energy Case: Texel Energy

Texel Energie is one of the first and most famous energy co-operatives in the Netherlands. As the name indicates, it is geographically located on Texel, a Dutch island in the Wadden Sea (North Sea) that has 13,000 citizens spread over seven villages. Bureaucratically speaking the island Texel is a municipality within the province of South-Holland. *Texel Energy* is a co-operative with 3,000 members and 4,000 customer connections. One can become a member for 50 euros a year, for which one receives a share in the company, a discount on the energy price and a vote in the annual assembly. Each member gets one vote, independent of the amount of shares. *Texel Energie* was initiated by three islanders ("Texelaars") and formally founded in 2007. Initially, the main business was to buy and resell renewable energy, but in recent years it also started producing renewable energy through projects in solar energy, biomass and 'anaerobic digestion'. Our correspondent indicated that the organization is also working on investing in wind, geothermal, and tidal energy.

One of the main *drivers* for Texel Energy concerns *the local culture*; the island of Texel has a very strong local identity and an exceptionally strong cultural and historical strive for 'being independent'. An often referred to example thereof is the story of the company TESO—the ferry service that has connected the island to the mainland of the Netherlands for the past century. TESO is an abbreviation for *Texel's Eigen Stoomboot Onderneming*—literally translated as "Texel's Own Steam Ship Enterprise". Before TESO—until the end of the nineteenth century—the island was dependent on a commercial ferry company from the mainland. The story is that islanders were getting fed up with the increasing prices and decreasing quality of the ferry service. At the beginning of the twentieth century, a small group of respectable islanders started an initiative that would enable the islanders to buy 'their own steam ship'. By selling shares of 5–25 Dutch guilders, they managed to collect a total of 76,000 guilders (approximately 35,000 Euros).

In 1907, TESO commissioned the construction of its first steam ship. Today TESO still has over 3,000 shareholders, and the ferry ticket to go to Texel is still considerably cheaper (nearly 90 %) than those of the other Dutch islands (routes covered by commercial ferry services).[1] This historic tale illustrates the island culture of Texel, and TESO is very often celebrated and referred to as a model for island independence, also in discussion over renewable and sustainable energy. Our correspondent is not only a board member of Texel Energy but also born and raised on the island; he emphasized how important the local identity was, and that the need for independence was and is one of the strongest drivers for the success of *Texel Energie*, much more so than the environmental argument.

Still, there were issues arisen with Texel Energy, and those mainly concerned the newness of the business model, and related to that, the difficulty of getting financed:

---

[1] Historical story is taken from Texelse Courant, 7th of August 2007 + interviews.

It was new—we were one of the first of this type of initiatives in NL—we really had to invent everything ourselves. (...) [and another tension is] financing, especially for production—the banks are very hesitant. We need half million, that is so much money... you cannot finance that with 3,000 members. (...) It is especially the banks that create difficulties for us—because they don't know our model we have a very high-risk profile.

A related tension that was mentioned concerned the confusion over the differences between for-profit, non-profit, and not-for-profit:

Our goal is to provide reliable and sustainable energy for our members, our main goal is not to make profit. We fall in between profit and non-profit, that can be quite difficult: we have to explain and explain it all the time. Many people and government officials do not understand it. A few years ago the TESO also had many difficulties explaining their way of working to the EU and to the tax offices. It would be nice to have the social enterprise legally recognized... now it does not exist legally.

Our correspondent did however nuance that the main tensions were not necessarily legal or regulatory: "we were not that bothered by laws and regulations. Of course there were some obstacles when we made contracts and so on—since our concept was new—but one should not exaggerate the legal obstacles". Although "there is a lot of talk about getting rid of laws and regulations", our correspondent also warned against the tendency to get stuck in that legal focus.

When asked explicitly about the *interaction with government*, our correspondent answered that "there is hardly any interaction with government—we consciously choose not to involve them". Although the interest of government officials for local energy initiatives has considerably increased in recent years, *our correspondent doubts the usefulness of government involvement and facilitation.* Essentially, citizens and entrepreneurs do not necessarily 'need' to directly involve the government to start up an energy co-operative. As long as the basic legal regulations and institutional arrangements allow it, citizens and/or social entrepreneurs can start an energy co-operative, just like they can start a business enterprise or an association.

4 The Scottish Community Energy Cases: Urgha and Udny

Scotland is a breeding ground for community projects including a large number of community owned energy projects. We start our research for the UK energy context with the two frontrunners in community energy, Urgha and Udny Community Wind (Turbines). During the scoping of the UK case studies, representatives from the Scottish government and from Community Energy Scotland suggested Urgha and Udny as successful cases that "survived and succeeded in an unfriendly institutional landscape". As a result, the exploration of these two case studies reveals institutional controversies and opportunities faced by the two frontrunning communities when community energy in Scotland was in its infancy.

At present, the institutional context has changed with a Scottish Strategy about Community Energy explicitly stating conditions and targets for energy sufficiency and self-reliance.

4.1 Urgha Wind Project and the North Harris Trust

Urgha Wind is a community owned wind turbine by the North Harris Community Trust in UK. At a community-recycling site, a community group was established in 2003 so as to set up projects that benefit the larger community. North Harris is a sparsely populated area and the community consists of 700 inhabitants. At the beginning, the community-recycling site was contracted by the city council. The resources needed to support a healthy community economy however could not be covered by council's support. Hence, they erected a turbine (10 KW wind turbine) that generates electricity (that converts to heat and light) and the excess/surplus of electricity generated not used by the community is fed into the grid. The turbine generates 4,000 pounds/year of income. The primary objective of operating the wind turbine is to support job creation. Adjacent to the turbine's location is a small business district/area. There is a future plan to erect a second turbine.

A recognized and experienced tension for Urgha Wind is the *risk aversion of banks* when it concerns loans for communities rather than private energy investors. Though there are funds available and favorable conditions in loan packages for community owned energy projects, the majority of the banks is risk averse and avoids lending to community organizations. Given the changes in the funding schemes, grant funds are not anymore available for community energy projects; therefore if one wants to apply for feed-in-tariff, different sources of funds need to be explored such as commercial landing and private funding. This brings new tensions given that community projects are seen as of high-risk from banks making them reluctant to grant loans to community initiatives. Another risk to be considered is the *financial viability risk* of the Community's Trust. In the case of North Harris Community Trust, since the erection of the turbine, the company went bust so the wind turbine was not fully installed (monitoring equipment was not installed). The turbine was operated and generated electricity even without seizing its full design potential. The Trust had to find new sources of money so as to finalize the project.

Additionally, the community trust had *to compete for loans and grants with private energy operators in an open market*. A tough lesson taught was that in the beginning, commercially owned projects take over benefits due to their scale and better marketing-devised strategies. This however changed due to the support given by Community Energy Scotland (see following Sect. 4.3).

At the same time, *the enforced feed-in-tariff scheme creates extra complications* rather than an enabling environment for community energy. For the North Harris Community Trust, the grant fund that was awarded to cover the first three years excluded the community operators to apply for and as such, benefit from the feed-in-tariffs. Another complication concerns the benefit-holders of the community owned energy projects. The existing Planning Law does not specify about the beneficiary of community owned energy projects (who gets the benefit); a fact that creates accountability issues within the group from the community who operates the wind turbine and the community as a shareholder of it.

The *time* that is required from the proposal stage to the operation of the wind turbine creates additional hurdles due to the group stamina it requires to deal with the uncertainty and the ad hoc demands that were created given that the community groups undertake these activities at voluntary capacity.

4.2 Udny Community Wind Turbine in Aberdeenshire

The Udny wind turbine is owned by Udny Community Trust, which is a community founded and owned organization. It started with five members of a community (professional engineers and farmers) that showed interest in community energy in order to generate income for the local community. Udny Wind is a leading community project in Aberdeenshire. The installation of wind turbines by the community exemplified how to work toward the outcome for the follower communities.

In the case of Udny wind, a helpful condition was the fact that the *local council officers welcomed the idea of a community owned wind turbine* and were as helpful as they could to the community group. The council officers recognized that there is demand by community for facilitation and advice and respond to it by working in partnership with community to establish energy projects; they remained available and open to communicate and interact with the community throughout the project cycle.

A controversy faced by Udny community wind was the *incompliant funding possibilities*. Udny Community Trust were granted 400,000 pounds by the national development fund which had to be declined since it was considered as double funding after having been granted a bank loan with favorable conditions.

4.3 Energy Communities in Scotland

A common feature in the Scottish context is the willingness and tendency of communities to strive for self-sufficiency and independence, an aspect also present in the energy sector. The establishment of a mediating organization—Community Energy Scotland, with the task to enable communities to undertake initiatives, further reinforces this cultural aspect and aims at succeeding in having community owned energy projects. After the energy strategy being laid down by the Scottish government, there were 200 villages that subscribed for community owned projects. The Scottish government responded to this demand by forming a consultancy-support group to aid these villages to become more energy efficient. Community Energy Scotland started as a Highlands and Islands Enterprise and in 2002 changed into the structure that operates today. It shifted from a government-based organization to an economic aid-based organization. The different community groups elect directors and members of Community Energy Scotland.

Community Energy Scotland currently is a non-profit organization that helps communities to initiate proposals for community energy projects and to seek support from local authorities. Its role is to empower group initiatives and to respond to community requirements for initiating such projects. Development officers are now placed around Scotland to assist communities. It also functions as learning diffusion channels: (a) they transfer lessons learned from operating successful projects (b) good practices and lessons from the interaction between community and banks or other funding.

Community Energy Scotland helps communities during the first phase to conduct a report, a feasibility study, and to put together a planning application that complies with the Community Renewable Scheme (CARES) of the Scottish government. The process is transparent and adaptable to community requirements and capacity. Different types of support are available: Internet sources, publications, general information, RE-toolkit, community-energy toolkit, mentorship program (with one-to-one consultations) training events (on demand and regular), and a practitioners-community conference with organizations (e.g. corporate actors). Recently, the initiation grant for community energy projects is a loan scheme of 150,000 pounds that covers almost 90 % of the funding. Communities seek funding from a number of resources such as private and commercial sectors. The requirement is that approximately 20 % of the total cost has to be matched by community's resources. A way to ensure this and to succeed in matching resources is to establish a community share scheme.

Community Energy Scotland is involved in different types of community projects and initiatives such as: (a) community-buildings, where advice is provided in one-to-one basis (via phone talks and consultation) about energy efficiency in community-buildings and other broader needs, (b) communities generating profits, and the profit is given to the community for benefiting its welfare and wellbeing (not utilized by one person or limited few), (c) community paradigm program where 30 groups are involved in a networking project sharing an agenda for locally produced food so as to reduce carbon footprint of the food chain.

5 A Closer Look at the Cases

5.1 Institutional Context

The Netherlands: The Netherlands does not have a strong tradition in citizen collectives and co-operatives. Although it does have a history of collective citizen efforts to fight the threat of the sea, this has over centuries aggregated to formal organizations and resulted in the strong corporatist model that the Netherlands are renowned for. Political action and socio-economic co-operation is something that primarily happens between formal organizations, not between citizens or neighbors. In this context, the very concept of citizen co-operatives is quite a marginal phenomenon. Until recently, when one would mention the word 'coöperatie' (Dutch for co-operative), many Dutch citizens would associate and confuse this with the phenomenon of 'housing corporations'; large centralized organizations that have been entrusted by the government to organize social housing.

However, this relative marginal position of co-operatives and citizen collectives is currently changing in the Netherlands. Several energy co-operatives have started to emerge in recent years, and are receiving considerable attention from civil society organizations and governments. One of our correspondents emphasized that recently there has been a bewildering amount of departments and organizations enthusiastically 'jumping on top of citizen initiatives' and commented that all this attention is not always particularly helpful.

When asked to compare the Netherlands to the other national contexts (UK, Germany and Belgium) both Dutch as well as non-Dutch correspondents emphasized that in the Netherlands there seems to be (1) a lot of talk and debate and not so much on action, (2) a lack of collectivism and co-operative movement tradition, and (3) an unstable and unpredictable investment climate. The first point also relates to the elaborate public debates about legal and financial details. For instance, currently in the Netherlands there is a considerable debate going on regarding commercial tax laws related to renewable energy, in which one side argues that if a collective of citizens produces renewable energy (e.g. through a collective investment in solar panels) they should not pay value added tax when subsequently using that energy for their own use. One of our non-Dutch correspondents—although he acknowledged that there was something to be said about this argument—could not help but wonder why the Dutch would not just get on with solar energy panels rather than first endlessly debating tax law adaptations. Regarding the second point—a lack of collectivism and co-operative movement tradition—correspondents reported their impression that the Dutch field of energy co-operatives seemed to exist of "separate islands" of small organizations who all want "their own cooperative" rather than co-operating with one another.

Another more general legalistic discussion that dominates the Dutch public debate in recent years, concerns the idea that in the Netherlands there are 'too many laws and regulations' that hamper sustainability and innovation. On the one hand, most of our correspondents from all four Western-European countries agree that the overload of regulations and complex bureaucracies form considerable obstacles to their initiatives. On the other hand, one Dutch correspondent also pointed out how the focus on 'getting rid of laws' in itself can fall in the trap of a legalistic discussion.

Last but not least, one correspondent mentioned that the Dutch focus on its natural gas resources limits the opportunities for renewable energy technologies. It is indeed not a coincidence that these phenomena—i.e., the presence of a natural resource limiting investments in alternative economic developments—is widely known as 'the Dutch disease'.

Scotland and the UK: Scottish government and the UK government have positioned energy sufficiency high on the *political agenda*. This is a strong driver for all energy related projects. More specifically, there is a clear direction from the Scottish government to realize the energy ambition of the region to become energy independently mapped out in the Strategy for Energy Scotland ("Energy Roadmap 2020"). Clearly defined targets within the Energy Roadmap 2020 are considered useful by different stakeholders because they provide legitimacy, (constitutional) support, and an institutional stepping-stone for mechanisms and venues to gain support (financial, policy, and organizational). Wind-projects owned by communities were promoted and prioritized as action plans for communities to be energy self-sufficient and financially profitable. Economic development and benefits for communities were triggering motives also communicated by council and government.

In addition to this, the Scottish government provides *financial incentives* for promoting energy projects in Scotland. Existing financial motives include the feed-in-tariff and low interest rate loans and the national lottery fund. Grants from councils cannot be used; available grants include the lottery fund, LEADER EU Initiative (EU rural development funds), and loans to support community projects. An office and project for corporate investments on renewable energy is the Energy4All office.

An additional enabling factor is the *availability and mobilization of resources (time and personnel) for community capacity building*. At present, Aberdeenshire council has an office and appointed officers that provide advice to small businesses and households about energy installations and measures for energy efficiency. Resources were made available for having pilot projects with small-scale energy projects. The councils have seminars to disseminate information and in this way to create and educate the community with the goal to create and if possible, grow the demand side. Seminars targeted housing associations and neighborhoods for introducing district-heating installations.

A technological and financial burden common to all community energy initiatives in Scotland is the grid coverage and connection cost. Grid connection in remote locations is limited and when available, grid connection cost is a hurdle. After the mediation of Community Energy Scotland, grid operators are invited in community consultations to inform community groups about future (planned) grid operations and installations. Investments in the grid for grid expansion are however not planned and scarce. At the same time, for communities that want to have their energy project, they face the difficulty of grid disconnect. The grid connection is important to consider given the fact that these communities set the energy projects in agricultural or grazing land that is outside the existing grid coverage area. Therefore, the choice lies at the community's hands to upgrade the grid on their own cost (an amount of approximately 500,000 pounds, based on 2011 estimates and information) before installing any green energy technology. The grid operating company enjoys a monopoly and their interactions with community (when not mediated by Community Energy Scotland) are slow and not open to information sharing and to creation of informal routes for co-operation.

5.2 Institutional Dynamics

Community owned renewable energy initiatives and projects are confronted with a number of drivers and tensions as the three cases illustrate. We conceptually place them across the following axes: (a) socio-cultural axis, (b) policy interest and agenda axis, and (c) socio-technological axis. The main drivers are summarized as follows:

Drivers	Texel case (NL)	Urgha and Udny cases (UK)
(a) Socio-cultural axis	Local culture of 'being' independent	Local culture and legacy on collective arrangements
(b) Policy agenda axis		Political push and on the policy agenda Existence of mediating organization (Community Energy Scotland)
(c) Socio-technological axis	Existing technology at hand (wind energy)	Existing technology at hand (wind energy)

The main tensions of the three cases are summarized as follows:

Tensions	Texel case (NL)	Urgha and Udny cases (UK)
(a) Socio-cultural axis	Newness of the business model	Risk aversion of banks to grant loans
(b) Policy agenda axis	No interaction with government	Incompliant funding mechanisms
(c) Socio-technological axis		Time-management and project management risks Financial viability risks Grid coverage, connectivity, and cost

6 Lessons Learnt From the Cases

Though there are drivers in place that create an enabling context for community owned renewable systems, there are two prevailing governance controversies distinct to each case. In the context of Scotland, there is a political push for community owned energy projects and resources are mobilized to support communities in any possible way. The existing facilitation mechanisms and policies however are inconsistent and incompliant and as such create a 'valley of death' for initiatives with the future risk to discourage communities from taking-up such initiatives.

Based on our empirical exploration of community energy projects in two countries that differ significantly on the socio-cultural and political context, we can argue that community energy projects have features of both outliers and front-runners for a sustainable energy transition. Their *outlier* characteristics include and are distinctive since they:

- operate at a new community-policy territory questioning fixed arrangements such as not-for-profit or for-profit forms
- viewed as exceptional cases outside the norm or conventional energy supply–demand systems
- viewed as exceptional cases of communities' attitudes toward renewable energy systems.

At the same time, community energy initiatives and projects exemplify *front-runners*' characteristics when considering a renewable energy future since they:

- dare to own a renewable energy system and take on all the risks (financial, project-related, technological) that come with it
- create a new energy governance arrangement in the absence of a policy agenda or push toward it (Texel case) or prior to policy agendas (Udny and Urgha cases)
- tap into community's skills, expertise, and innovative potential for pursuing community needs such as security and energy independence.

The empirical cases show that self-organization of renewable energy projects by communities operate at the interface of community and policy and are hybrids that challenge institutions that are in place to benefit or enable them. Given that there are no perfect institutions, there is no flawless institutional space for renewables governance as well. The implication that community owned renewables bring forward that the governance focus has to shift from designing diffusion instruments (alone) toward examining which institutions fit the operation of community owned renewables.

Beyond controlling or beyond governing? Such a perspective asks for a new type of governance that focuses on providing space for initiatives to act (themselves) upon the tensions or paradoxes rather than intervening by controlling or formalizing. This type of governance may appear as an 'invisible governance'; we do not imply to undermine the role of government in guiding and regulating, we do however propose a new type of reflexive governance (Grin 2010) (Voss et al. 2009) (Voss J and Bornemann 2011) that diagnoses paradoxes and facilitates space for self-correction and action without neglecting government's roles and responsibilities. As such, a meta-governance approach of governing self-organized energy will need to ensure careful assessment and monitoring to safeguard that self-organization delivers services, maintains equity in access and service for all citizens (especially when public goods and services are operated by self-organized operators e.g. think about drinking water to be operated by self-organized initiatives).

7 Conclusions

Community owned renewable energy projects may appear as the ideal initiatives for a sustainable energy future. This however does not imply that they are confronted with less tensions or controversies or that the institutional space that they operate is smoothly paved. Our empirical research shows that community owned renewable initiatives operate as frontrunners of a sustainable energy transition: they experiment with current institutional arrangements, dare to take risks and seize opportunities, and create new institutions or, simply, dare to self-organize energy. From an energy governance perspective, such initiatives require new institutions to fit their operation and new modes of governance that can navigate energy transition processes in a valley of institutional plurality and emergence.

Acknowledgments This research has been funded by the ARVODI research contract with the Dutch Ministry of Infrastructures entitled "Governance of sustainable infrastructures" and by the Research Programme Next-Generation Infrastructures.

References

Allen J, William RS, Rocio Diaz-Chaveza (2012) Community-based renewable energy in the Lake District National Park—local drivers, enablers, barriers and solutions, local environment. Int J Justice Sustain, 17(3) DOI:10.1080/13549839.2012.665855

Birch K, Whittam G (2008) The third sector and the regional development of social capital. Reg Stud 42(3):437–450

Chitnis M, Sorrel S, Druckman A, Firth SK, and Jackson T (2013) Turning lights into flights: estimating direct and indirect rebound effects for UK households, Energy Policy, 55, 234–250

Corscadden K, Wile A, Yiridoe E (2012) Social license and consultation criteria for community wind projects. Renew Energy 44:392–397

Cowell R, Bristow G, Munday M (2011) Acceptance, acceptability and environmental justice: the role of community benefits in wind energy development. J Environ Plann Manage 54(4):539–557

Darby S (2006) Social learning and public policy: lessons from an energy-conscious village. Energy Policy 34:2929–2940

Egyedi T, Mehos D (eds) (2012) Inverse infrastructures. Disrupting networks from below. Edward Elgar, Cheltenham

Grin J (2010) Understanding transitions from a governance perspective, Part III in Grin J, Rotmans J, Schot J (eds) Transitions to sustainable development, New directions in the Study of long term transformative change, pp. 223–236, Routledge, USA

Gross C (2007) Community perspectives of wind energy in Australia: the application of a justice and community fairness framework to increase social acceptance. Energy Policy 35:2727–2736

Hoffman SM, High-Pippert A (2010) From private lives to collective action: recruitment and participation incentives for a community energy program. Energy Policy 38(12):7567–7574

Loorbach D, Frantzeskaki N (2012) Why taking complexity seriously implies a paradigm shift for policy studies, Chapter 15, In: Lasse Gerrits and Peter Marks (ed) COMPACT I: public administration in complexity, pp 327–345, Emergent Publications, USA

Moulaert F, Ailenei O (2005) Social economy, third sector and solidarity, relations: a conceptual synthesis from history to present. Urban Stud 42(11):2037–2053

Mumford J, Gray D (2010) Consumer engagement in alternative energy—can the regulators and suppliers be trusted? Energy Policy 38:2664–2671

Musall FD, Kuik O (2011) Local acceptance of renewable energy—a case study from southeast Germany. Energy Policy 39:3252–3260

Nikogosian V, Veith T (2012) The impact of ownership on price-setting in retail-energy markets—the German case. Energy Policy 41:161–172

Pestoff VA (1992) Third sector and co-operative Services—an alternative to privatization. J Consum Policy 15:21–45

Rygg BJ (2012) Wind power—an assault on local landscapes or an opportunity for modernization? Energy Policy 48:167–175

Saunders RW, Gross RJK, Wade J (2012) Can premium tariffs for micro-generation and small scale renewable heat help the fuel poor, and if so, how? Case studies of innovative finance for community energy schemes in the UK. Energy Policy 42:78–88

Scott-Cato M, Hillier J (2010) How could we study climate-related social innovation? Applying Deleuzean philosophy to transition towns. Environ Politics 19(6):869–887

Seyfang G, Haxeltine A (2012) Growing grassroots innovations: exploring the role of community-based initiatives in governing sustainable energy transitions. Environ Plann C Gov Policy 30:381–400

Seyfang J, Smith A (2007) Grassroots innovations for sustainable development: towards a new research and policy agenda. Environ Politics 16(4):584–603

Smith A, Raven R (2012) What is protective space? Reconsidering niches in transitions to sustainability. Res Policy 41:1025–1036

Stirling A (2008) "Opening up" and "closing down": power, participation, and pluralism in the social appraisal of technology. Sci Technol Hum Values 33(2):262–294

Verbong G, Loorbach D (eds) (2012) Governing the energy transition, reality, illusion or necessity?. Routledge, USA

Voss JP, Bornemann B (2011) The politics of reflexive governance: challenges for designing adaptive management and transition management. Ecol Soc 16(2):9 (www.ecologyandsociety.org/vol16/iss2/art9)

Voss JP, Smith A, Grin J (2009) Designing long-term policy: rethinking transition management. Policy Sci 42:275–302

Walker G, Devine-Wright P, Hunter S, High H, Evans B (2010) Trust and community: exploring the meanings, contexts and dynamics of community renewable energy. Energy Policy 38:2655–2663

Wesselink A, Paavola J, Fritsch O, Renn O (2011) Rationales for public participation in environmental policy and governance: practitioners' perspectives. Environ Plann A 43:2688–2704

Wolsink M (2012) Undesired reinforcement of harmful 'self-evident truths' concerning the implementation of wind power. Energy Policy 48:83–87

Part III
Hidden Problems Behind the "Re Miracle"

Renewable Energy Governance in Kenya: Plugging into the Grid ‘Plugging into Progress’

James Mwangi, Nicholas Kimani and Maina Muniafu

Abstract The dearth of commercially viable renewable energy (RE) operators in Kenya is telling. Despite a rising need for clean and reliable energy, a progressive institutional framework, and a new RE funding scheme, which should result in a genuinely competitive and a self-sustaining investment proposition, the poor results on the ground show that technological innovations, costs and prices, and policies have yet to be fully aligned to achieve full RE potentials. We argue that the resulting negative effects of such poorly administrated RE sector is best understood by considering the practical challenges faced by RE power generators and lost opportunities for ordinary consumers to enhance their socioeconomic well-being which hinges upon access to affordable and reliable energy, and for whom barriers impeding that growth must be removed.

1 Introduction

Renewable energy (RE) generation should currently be big business in Kenya. Take a rising need for clean, reliable energy, add a new RE funding scheme, and then consider erratic weather conditions that have affected over-relied upon hydropower generation in Kenya, ideally what you should get is a genuinely competitive and a self-sustaining investment proposition. This outcome, is yet to

J. Mwangi (✉)
GreenNext Sustainability Limited, Nairobi, Kenya
e-mail: jimmy_mwangi@yahoo.co.uk

N. Kimani · M. Muniafu
Sustainable Development Initiatives Center, United States International University, Nairobi, Kenya
e-mail: nkimani@usiu.ac.ke

M. Muniafu
e-mail: mmuniafu@usiu.ac.ke

E. Michalena and J. M. Hills (eds.), *Renewable Energy Governance*,
Lecture Notes in Energy 23, DOI: 10.1007/978-1-4471-5595-9_7,

eventuate, and to understand why, we highlight key developments in Kenya's RE sector from the perspective of two ordinary (and hypothetical consumers). Both share something in common: a desire to access affordable and RE supply to improve their livelihoods, while improving their quality of life by minimizing health problems while not feeling threatened by insecurity. For these people and potential consumers, plugging into a national grid that supplies affordable and reliable power is analogous to 'plugging into progress'.

We consider Jane Shiro, a typical 'Mama Mboga', or small-scale fruit and vegetable seller, who could be found in any of Kenya's small busy urban areas. She lives in the far end of Korogocho, a vast unplanned urban settlement in Nairobi County, which is teeming with wooden and rusted tin-roofed buildings. For her, having more street lighting on the road where her modest stall is situated, would enable her to stay open after dark, while not worrying about insecurity. The extra income would also enable her to pay for electric lighting, which her young children could use to do their homework, rather than the smoky and dim kerosene lamps. But the street lighting is at best sporadic, and the resulting insecurity along her road worries her. Meanwhile, her childrens' health is at risk from the dim lighting, and the kerosene lamps, whose fumes are creating a host of respiratory problems arising from indoor air pollution.

We also consider Nduku Musau, in King'oti shopping center situated about 60 km from her county headquarters, Machakos Town. A successful small-scale dairy farmer for several years, she recently got connected to the grid—in January 2013—courtesy of the government-funded rural electrification policy. She is mulling over whether to increase the size of her herd, and install an electric milk cooler so as to supply the rapidly expanding Machakos town. Already milk companies have expressed an interest in purchasing bulk supplies of raw milk at attractive prices. But she worries about the frequent blackouts and the ever-rising costs of electricity.

In viewing Kenya's RE governance framework through the eyes of both Jane and Nduku, we seek to answer the important policy question, "why is it that despite the fact RE is expected to be a significant energy supply option in Kenya," there is still a dearth of commercially viable RE operators, and what are the effects on ordinary consumers?

To do this, we produce an evidence base that describes key legal and institutional aspects of the RE governance framework, while identifying some key factors that affect the deployment of RE supplies in the country. We hypothesize that while renewables are intended to play a key role in driving Kenya's economic growth, the dearth of commercially viable RE operators points to the fact that technological innovation, the economy (costs and prices), and policies have yet to be fully aligned to achieve full RE potentials. For the sake of ordinary consumers, whose socioeconomic development depends on access to affordable and reliable energy, barriers impeding that growth must be removed.

2 Kenya: Good for Doing Business

Kenya is East Africa's largest economy, and has enjoyed stable economic growth since the early 2000 s, mainly due to the introduction of market-based reforms and providing more incentives for both local and foreign private investment, which have made Kenya a much better place for doing business (World Bank and IFC Doing Business 2013). In recent years there has been appreciation by the government that current supply, although increasing slowly, cannot keep up with demand. The oil import bill in 2008 consumed 55 % of the country's foreign exchange earnings from exports. On the other hand, the over reliance on biomass has exerted considerable pressure on the remaining forest and vegetation stocks, thereby accelerating the processes of land degradation (Kiplagat et al. 2011).

Moreover, despite the abundance of potential and a strong growth in demand for electricity, the country faces constraints in satisfying electricity demand. A substantial proportion of RE resources are unexploited. Of the potential renewable sources, Kenya has harnessed only about 30 % of its hydropower sources, approximately 4 % of the potential geothermal resources and much smaller proportions of proven wind and solar power potentials (Kiplagat et al. 2011). Furthermore, a large potential exists for the development of biomass-based energy such as biogas, biodiesel, and power generation from bagasse (Cameron et al. 2012).

Table 1 presents details of installed capacity and generation of electricity by different producers. Total installed capacity expanded by 4.7 % to 1606.1 MW in 2012. The increased installed capacity is attributable to commissioning of new power generation facilities, imports from Uganda and Tanzania (for hydro), increased generation on by independent power producers (IPPs) and emergency power producers (EPPs). Co-generation and wind energy jointly accounted for a paltry 1.5 % of the total generated.

Details of demand and supply balance of electricity are shown in Table 2. The total domestic demand for electricity expanded by 2.2 % to 6414.4 million kilowatt hour (kWh) in 2012. Small, medium, and large commercial and domestic sectors remained the main consumers of electricity, accounting for 93.2 % of the total domestic consumption. Domestic and small commercial consumption increased by 3.9 % to 2568.5 million kWh. Sales to the Large and Medium commercial consumers declined to 3409.2 kWh in 2012 but accounted for the largest share of the total demand.

3 The Governance Mix

3.1 Active Policymaking

Kenya's liberalized energy sector has made significant progress in the formulation of RE policies. Policymakers appreciate that the commercialization of RE—or energy derived from natural processes (e.g., sunlight, geothermal, hydro, and

Table 1 Installed capacity and generation of electricity, 2008–2012

	Installed capacity					Generation GWH								
	Hydro	Thermal Oil	Geothermal	Co-generation	Total	Hydro	Thermal oil				Geothermal	Co-generation	Wind	Total
							KenGen	IPP	EPP	Total				
2008	719.0	418.9	128.0	2.0	1267.9	3267.0	521.4	883.0	741.0	2145.4	1039.0	4.0	0.2	6455.6
2009	730.0	421.5	158.0	2.0	1311.5	2160.0	654.0	1208.0	1135.0	2997.0	1293.0	50.0	7.2	6507.2
2010	728.0	469.2	189.0	26.0	1412.2	3224.0	291.0	1307.0	540.0	2201.0	1442.0	92.0	16.8	6975.8
2011	735.0	582.7	190.6	26.0	1534.3	3217.2	903.0	1538.8	358.7	2800.5	1443.7	80.9	17.6	7559.9
2012	769.0	610.6	199.6	26.0	1606.1	4015.9	682.5	1208.9	309.0	2200.4	1515.9	104.7	14.4	7851.3

Source Kenya government, Kenya National Bureau of Statistic, economic survey 2013

Table 2 Electricity supply and demand balance, 2008–2012

	2008	2009	2010	2011	2012
Demand					
Domestic and small commercial	2030.8	2058.1	2200.3	2571.4	2568.5
Large and medium (commercial and industrial)	319.0	3058.1	3204.9	3440.3	3409.2
Off-peak	36.2	36.8	38.2	37.9	36.0
Street lighting	26.3	21.3	20.5	17.9	20.6
Rural electrification	239.1	254.4	290.8	306.1	380.1
Total domestic demand	5352.2	5428.7	5754.7	6273.6	6414.4
Exports to Uganda and Tanzania	41.0	27.0	29.6	37.3	32.7
Transmission losses and unallocated demand	1062.4	1051.5	1191.5	1248.9	1402.2
Total demand = total supply	6455.6	6507.2	6975.8	7559.8	7851.2
Of which imports from Uganda and Tanzania	25.0	39.0	30.0	33.9	39.1
Net generation	6430.6	6468.2	6945.8	7525.9	7812.1

Source Kenya National Bureau of Statistics, 2013

wind) that are replenished at a faster rate than they are consumed—is essential for a secure and sustainable energy supply. This appreciation is evident in several national policy documents, long-term plans, and strategic initiatives designed to promote and invest in RE for the country, including: Kenya's Scaling Up Renewable Energy Program (SREP) Investment Plan, Updated Least Cost Power Development Plan (ULCPDP), Rural Electrification Master Plan, The Energy Act of 2006, The Feed-in Tariff (FiT) and Revised FiT (ReFiT) Policies, The Kenya National Climate Change Action Plan, the Kenya Vision 2030, and the National Energy Policy. By way of illustration, we consider two of these instruments.

First, the National Climate Change Action Plan (NCCAP) of 2013, which actualizes the National Climate Change Response Strategy of 2010, which was adopted to guide integration of climate change adaptation and mitigation measures into government planning and budgeting. The NCCAP identifies energy, agriculture, and water as some of the most vulnerable sectors. It notes that RE will play an increasingly key role in climate change adaptation and mitigation measures. It recommends that the government should accelerate the development of green energy including wind, solar, and renewable biomass (GOK 2013).

Another instrument, the ULCPDP 2011–2031 envisions that Kenya's electricity peak demand will increase from the current 1,302–15,026 MW by 2030. This is in line with the Vision 2030 which envisages energy as a key enabler for economic growth across the country. To meet the increased electricity demand due to the enhanced economic activities, the ULCPDP has identified various generation sources targeting 5,110 MW from geothermal, 1,039 MW from hydro, 2,036 MW from wind, 3,615 MW from fossil thermal, 2,000 MW from imports, 2,420 MW from coal, and 3,000 MW from nuclear. The investments required for generation, transmission, and distribution to meet this demand are enormous (GOK 2011).

3.2 Multiple Instruments of Benefit

To facilitate the uptake of renewables, the government has developed three main policy tools: FiTs, 0 % import duties, and VAT exemption.

3.2.1 Feed-in-Tariffs

A FiT can be a very effective tool to promote the development of renewable electricity in a targeted, cost-effective and controlled manner. They are well suited to younger liberalized electricity markets in that they can provide investor certainty in such fledgling markets—by setting prices at a level, which is purposely designed to attract and stimulate new investment in renewables.

In 2008, the Ministry of Energy (MOE) introduced a FiT for RE sources. At present, it enables power producers to generate and sell Renewable Energy Sources generated Electricity (RES-E), to a distributor at a predetermined fixed tariff for a given period of time, in this case, 20 years). The FiT was applied to three technologies namely wind, small hydropower, and biomass (municipal waste and cane bagasse). In 2010, the FiT was revised (now known as the ReFiT) to accommodate additional RE sources and reviewed the tariffs (which now includes geothermal, solar, and biogas sources). By 2011, the FiT Policy had elicited a total of 49 expressions of interest from potential investors to develop RE sources. Table 3 illustrates the proposals received, with most being for wind projects.

In January 2013, Kenya revised its power pricing policy for RE to allow the sale of solar generated power to the national grid. According to the new policy, suppliers of solar power will earn US$0.12 per kWh, but the investors must be running projects generating a minimum of 0.5–40 MW. Those selling solar power directly to consumers are currently allowed to charge up to US$0.20 per kWh, a move aimed at encouraging more investors to develop such projects, especially in remote areas of the country where the national grid has not reached.

Table 3 Proposals for power purchase agreements (PPA's)

Received proposals				Approved proposals		
Technology type		No.	Capacity (MW)	No.	Capacity (MW)	% of total approved capacity
1	Wind	23	1,118	20	1,008	74
2	Biomass	4	164	4	164	11
3	Hydro	19	111	16	81	7
4	Geothermal	1	70	0	0	5
5	Biogas	1	40	1	40	3
6	Co-generation	1	18	1	18	1
Total		49	1,521	42	1,311	100

Source SREP investment plan, 2011

The new policy also introduces geothermal FiTs for generation of 35–70 MW. This gives IPPs an opportunity to sell power to the national grid. The previous policy recognized geothermal projects of up to 35 MW. The tariffs have been raised from the previous US$0.85 per kWh to US$ 0.88. Unlike the previous tariff that recognized biogas projects of up to 40 MW, the revised tariff has roped in smaller producers of up to 10 MW. The tariffs that suppliers of biogas power will earn have also been increased from a maximum of US$ 0.08 per kWh to US$0.22.

Kenya's draft Energy Policy and Bill 2012 allows for the biennial reform of the ReFiT policy. As such, it is expected that the recent review of the current FiT will lead to the creation of a standard, non-negotiable FiT (i.e., one with a fixed price), and to the revision of the price per kWh for each renewable technology type. Second, the draft Energy Policy and Bill 2012 will if passed allow for the expansion of the number of off-takers beyond just Kenya Power (KP), allowing off-grid IPPs to sell to independent off-takers (ECA 2012).

3.2.2 Tax Breaks

The SREP introduced a zero-rated (0 %) import duty on RE equipment and accessories in 2011. This is unusual as all goods in Kenya have an import duty unless explicitly exempted by an Act of Parliament. The same program removes VAT (value-added tax) on RE materials, equipment, and accessories. Prior to 2011, there had been a 16 % VAT on RE materials.

The Energy Act 2006, draft Energy Policy of 2012 and the Energy Bill 2012 all contain fiscal incentives such as subsidies and tax holidays, but, with a small number of exceptions, they have not been put into practice by the relevant authorities.

4 Institutional Framework: Unpacking Responsibilities

The reforms in the energy sector pursuant to the Energy Act of 2006 have seen a complete reorganization of functions hitherto concentrated in the MOE and the Kenya Power and Lighting Company (KPLC) limited. This was a result of the need to place responsibilities with specific institutions that would specialize in the mandates vested in them under the Energy Act to enhance efficiency.

Accordingly these were unbundled into generation, transmission, distribution, oversight, and policy functions. The institutional structure in the electricity sub-sector in Kenya comprise the MOE, Energy Regulatory Commission (ERC), Kenya Electricity Generating Company (KenGen), KPLC, the Rural Electrification Authority (REA), Kenya Electricity Transmission Company (KETRACO), Geothermal Development Company (GDC), IPPs, and EPPs.

5 Kenya's Renewable Energy Sources: FiT for the Challenge?

5.1 Renewable Energy Sources Covered by FiT Policy

5.1.1 Big Potential of Wind Energy

The fact that wind energy is free, non-polluting, and virtually inexhaustible makes it an attractive means of meeting Jane's and Nduku's need for cheap power to realize their development needs.[1] Kenya has immense wind energy potential, particularly in the northwest of the country in Marsabit district where average wind speeds are 10–11 m/s in the early morning hours (Kamau et al. 2010). The MOE developed a Wind Atlas in 2008 with indicative data. To augment the information contained in the Wind Atlas, the ministry, with the assistance of Development Partners, installed 55 wind masts and data loggers (as at 2011) to collect site-specific data. A wind resource map using satellite data has also been developed under the Solar and Wind Resource Assessment Project (SWERA) (Government of Kenya 2011).

Perhaps the most promising example of a wind energy project is the Lake Turkana Wind Project (LTWP), which is led by a consortium of investors who include Kemperman Paardekooper and Partners Africa (Netherlands), Aldwych International Ltd. (UK), the Investment Fund for Developing Countries of Denmark, Norfund, and Vestas Wind Systems AS (Denmark).

The LTWP aims to provide 300 MW of reliable, low cost wind power to the Kenya national grid, equivalent to approximately 20 % of the current installed electricity generating capacity. The Project is of significant strategic benefit to Kenya, and at Ksh70 billion (€600 million) will be the largest single private investment in Kenya's history. Covering 40,000 acres (162 km^2), and located approximately 700 km north of Nairobi, the project will replace the need for Kenya to spend approximately Ksh15.6 billion (€120 million) per year on importing fuel, and the tax contribution to Kenya will be approximately Ksh3 billion (€22.7 million) per year and Ksh58.6 billion (€450 million) over the life of the investment. During the 32-month construction period, up to approximately 2,500 jobs will be created followed by over 200 full time jobs throughout the period of operations. The first phase of the project is expected due for completion in 2014 (LTWP 2013).

[1] In our analysis we exclude clean coal and nuclear energy. Coal power is not currently a part of the generation mix in Kenya, but there are plans to generate 620 MW of electrical power by 2018 from coal, and anywhere from 2,400–4,490 MW in 2030 depending on figures in the ULCPDP or the National Energy Policy, respectively. We also exclude nuclear energy because it is yet to feature prominently in international discussions on support for mitigation actions. We do acknowledge, however, that to meet Kenya's the increased electricity demand by 2013, the ULCPDP has targeted 3,000 MW from nuclear.

Fig. 1 Wind turbines overlooking Kenya's picturesque Rift Valley, which also holds vast geothermal generation potential

Another example is the Ngong wind farm, which is operated by KenGen, which presently produces 5.1 MW, but plans to boost this to 25.5 MW in the near future (KenGen 2013). What follows are photographs of some of the wind turbines situated at the site (Figs. 1, 2).

5.1.2 Unstable Hydropower Potential. Small Hydroplants Might be the Solution?

Historically, Kenya has been dependent on hydroelectric sources of energy. Currently, over 750 MW is exploited, mainly in existing hydropower plants owned by KenGen. Since the late 1990s, however, hydroelectric production has been unstable and even declined, leading to reduced generation capacity across the country, and greater reliance on thermal electric power. Stakeholders have, as a result, often cited the potential impact of climate change on hydropower as a reason to reduce its future role in the country's electricity mix (Government of Kenya 2012).

Hydropower potential is presently estimated in the range of 3,000–6,000 MW. Of this, just over 750 MW is currently exploited, mainly in large installations owned by the national power generation utility, KenGen. According to the

Fig. 2 Wind turbines facing the top of Ngong Hills. What is not visible is the new power substation, which is situated after the third wind turbine

ULCPDP at least half of this overall potential is from small hydroplants and that 1,249 MW of projects of roughly 30 MW or larger remain. Presently, the installed grid-connected small-scale hydroelectric projects contribute about 15.3 MW, though there are several other small hydro schemes under private and community generation especially in several tea estates across the country, which are not grid-connected. One such example is the 'Greening the Tea Industry in East Africa' (GTIEA) project, which has developed several small hydropower projects intended for internal consumption and to supply excess electric power to the national grid.[2]

[2] GTIEA Project, is a small hydropower initiative, that is co-implemented by UNEP and the African Development Bank (AfDB) and executed by East African Tea Trade Association (EATTA) to develop Hydro power sites in tea growing areas to decrease the tea factories overdependence to the main grid.

5.1.3 Biomass (Municipal Solid Waste and Cane Bagasse): A RE Source Close to the Hearts of People but Away from a Regulatory Framework

The term biomass refers to organic matter that can be used to provide heat, make fuel, and generate electricity. The two key types of biomass under consideration here are municipal waste and cane bagasse.[3]

Municipal Solid Waste

The municipal solid waste (MSW) created by a population is a significant management problem for any country. MSW consists of solid waste including durable and nondurable goods, containers, food scraps, yard waste, and inorganic waste from homes, institutions and businesses, wastes generated by manufacturing, agriculture, mining and construction, and demolition debris, as well as sludge and liquid waste from water and wastewater treatment facilities, septic tanks, sewerage systems, slaughter houses.

With appropriate waste-to-energy technologies, MSW can be used to provide energy while helping to clean the environment (Cameron et al.). Unfortunately, although the generation of MSW has grown rapidly, the capacity to collect and dispose the residues in Kenya has declined for several reasons. In Kenya there is no statute or national policy or organization established to regulate the management of solid waste. In one study on Nairobi, the solid waste management situation was found to be characterized by low rates of collection of waste due to inadequate fleet of municipal vehicles to transport waste, encroachment on the city's designated dumpsite by cartels of waste pickers and dealers and the growth of unauthorized dumpsites in various parts of the city (Government of Kenya 2008). Similar weaknesses in solid waste management systems and practices have been found in other major urban centers in Kenya (Munala and Moirongo 2011).

Cane Bagasse (Biomass Co-generation)

Co-generation refers to the simultaneous production of heat and power from one single fuel source. It is common where plant processes require both heat and power such as sugar processing and offers opportunity for improved plant energy

[3] We did not consider woody biomass (or commercially grown biomass) as a source of bioenergy for electricity because deforestation through burning of woody biomass is already a concern for the Government of Kenya (wood and charcoal provide roughly 70–75 % of total energy consumption in Kenya), which has an aggressive reforestation target to achieve 10 % forest cover. This will prove a significant challenge, even without finding additional land for biomass feedstock. Woody biomass or commercial biomass, if produced in large quantities for electricity, could also compete for scarce arable land with food.

efficiency besides reducing energy costs and providing additional revenue stream through surplus power export to the national grid. Mumias Sugar Company, Kenya's largest sugar miller, has taken advantage of its co-generation potential from sugarcane bagasse by generating 38 MW, out of which 26 MW is exported to the national grid (Government of Kenya 2012).

6 Renewable Energy Sources Covered by ReFiT Policy

In this section, we highlight key RE sources covered under the ReFiT policies already highlighted previously in Sect. 2.1.

6.1 *Sympraxis Between Kenya and Germany*

Large-scale electricity generation potential in Kenya, using waste from slaughterhouses, agroprocessing from sisal, flower, and coffee production, has been identified. In one study, it has been estimated that the combined gross annual electrical energy potential of selected biowaste residues (residues of barley, maize, tea, sugarcane, and seedcotton) amounting to 3.92 TWh is equivalent to 73 % of the total annual electricity generation in the country, and that even half of this potential still exceeds the combined country's current fossil fuel dominated IPPs annual generation capacity of 1.77 TWh (Nzila et al. 2010).

The private sector has rolled out some small commercial biogas facilities for heat and electricity generation. One example is found in Kilifi, Mombasa County, where the company, Biogas Power Holdings, owns and operates which is currently the biggest biogas plant in Kenya. The plant transforms agricultural wastes—cow manure and sisal—into energy in the form of electricity and heat. They have a close working relationship with German Technical Co-operation(GIZ), an organization which has operated in Kenya since 1975, and during this time, have been involved in providing technical skills assistance in various RE technologies, including energy saving cook stoves, solar PV, and biogas.

6.2 *Geothermal Energy: Huge Potential that Requires Huge Investments*

With probably the world's largest single source for geothermal in the Rift Valley, a continental scale volcano-tectonic feature that stretches from northern to southern Africa, Kenya's geothermal industry has the capacity to produce enough energy to power its own and its neighbor's needs. Being a non-polluting and virtually inexhaustible source of cheap power, one would naturally expect it to help transform Jane's and Nduku's lives.

All of the geothermal-based electricity generation constructed in Kenya to date has been in the Olkaria geothermal field near Naivasha. In 2010/2011, Kenya had 198 MW of installed geothermal capacity. Surface studies suggest that 5,000–10,000 MW could be generated, clustered into three regions namely the Central Rift (1,800 MW), South Rift (2,450 MW), and North Rift (3,450 MW) (Ngugi 2012).

Since geothermal power projects require extensive preliminary test drilling, they are more capital intensive than many other RE projects. The experience in Kenya has been that significant investment is required to ascertain whether a site has the potential to recover the costs. It has also been observed that even when viable sites are identified, there is a relatively long lead time from the start of exploration to power plant commissioning and the first revenues.[4]

6.3 Solar Energy: Consequences of the Lack of "Net-Metering System"

Kenya's geographical location astride the equator gives it an excellent opportunity for electricity generation using photovoltaic (PV) systems, and for solar water heating systems, which are mainly used in homes, hotels, hospitals, and learning institutions.[5] These observations, notwithstanding, two considerations come to mind as regards why solar power (in general) is less attractive for commercial power generation in Kenya.

First, despite the ReFiT policy that allows the sale to the national grid of electricity generated by photovoltaic solar panels, investors have raised concerns that the prices are not attractive, as compared to countries like Tanzania, Uganda, and South Africa, which offer higher FiTs for solar. Second, the FiT policy does not yet allow for 'net-metering', a practice that allows ordinary small-scale RE power producers to 'bank' or 'store' their electricity in times of over-production (for example, for solar energy, during peak production in the day) in the national grid, and to balance out their grid consumption with this 'banked' or 'stored' electricity during other times (for example during night, morning and evening hours) (GIZ 2012). This is despite the potential for the introduction of net-metering in Kenya

---

[4] It should be noted that there are also direct-use, or using the heat directly rather than generating electricity, options for geothermal. This is already done in at least one tourist facility in Kenya and for greenhouse heating in the flower industry.

[5] The Energy (Solar Water Heating) Regulations, 2009 require all premises within the jurisdiction of a local authority with hot water requirements of a capacity exceeding 100 l per day to install solar water heating equipment. Both commercial and residential premises will be required to install solar equipment. The policy is intended to achieve a 60 % contribution of solar energy to hot water demand in a building in order to reduce demand from the national grid.

with more than 300,000 potential net-metering clients. A policy framework for the connection of such development under net-metering basis is therefore necessary if it is to be legally and economically feasible.[6]

7 Practical Challenges Coming Out from Already Existing Experiences in RE Governance

We acknowledge that organizing the energy transition from non-sustainable to RE is challenging owing to barriers that stand in the way of aligning the technological innovations, costs, prices, and policies required to achieve full RE potentials (Verbruggen et al. 2010). We explore this argument by highlighting certain practical challenges arising under the FiT and ReFit policies, and, financial challenges, as experienced by an independent project developer (also a close associate with one of the authors), with interest in two 'undersubscribed' RE technologies. From these, we glean insights on certain cross-cutting challenges affecting commercial RE power generation in Kenya.

7.1 Technical Challenges

Among the key concerns arising from the GTIEA project was that power wheeling should be introduced to small hydro producers who wish to wheel energy to their customers (e.g., tea factories for tea companies) through KP distribution lines. GTIEA project management, who had already done a case study and found it viable, also considered that the measure would avoid the construction of new lines parallel to the existing KP lines.

They also felt that further reforms to the RE governance structure were necessary so that a third party, other than the SHP generator, were allowed to act as a private distributor. Such private distributor, from where energy is being wheeled from, should be able to deposit surplus energy, withdraw deposited energy when required, and get energy in credit. Following from this, a yearly balance sheet should be prepared by the end of every fiscal year and the amount of energy wheeled and credited or deposited worked out so as to come up with the final energy billing.

[6] We note that the new office facilities at the Nairobi-based United Nations Environment Program (UNEP), constructed in 2011, is capable of contributing to Kenya's net-metering program, once rolled out. With 515 kilowatt-peak (kWp) rooftop solar plant, which comprises 6,000 square meters of solar panels, energy saving lighting, natural ventilation systems and other green features, the office facility has been designed to generate as much electricity as its 1,200 occupants consume (http://www.unep.org/gc/gc26/Building-for-the-future.pdf).

As regards environmental issues, the experience from GTIEA was that it would be desirable to simplify and standardize the environmental study, through conducting a standardized initial environmental examination (IEE), which is normally practiced up to 25 MW hydropower projects in other countries, rather than a full fledged Environmental Impact Assessment (EIA), given the relatively modest dimensions associated with small hydropower projects.

7.2 *Financial and Regulatory Challenges*

Local financing conditions present hurdles to developers seeking upfront capital for RE projects. We consider experiences of a private developer with keen interest in developing certain 'under-subscribed' RE technologies, namely solar and wind. He has struggled to show potential investors that it is possible to attain a reasonable return on investment on the long-term power purchase agreements with KP, which are based on the FiT rates currently in force. He has even tried to rely on revenues from carbon financing, which are unpredictable at best. That said, we acknowledge that high profile project developers have faced similar challenges, most notably the LTWP,[7] and that it is possible to draw upon a small measure of financial assistance by a small—but growing—pool of commercial and concessional lenders both locally and abroad.

We note that a related challenge arises in respect to carbon financing. In principle, RE energy projects should find it a relatively straightforward process when seeking carbon financing. However, uncertainties abound regarding compliance market, given the questions hanging over the future of the Kyoto Protocol, particularly for a non-LDC like Kenya, and the resulting low prices for certified emission reductions (CERs). While accessing the voluntary market appears to be a viable alternative, the fact is that there is an unclear regulatory framework surrounding carbon trading in Kenya. The National Policy on Carbon Finance and Emissions Trading has remained in draft form since 2010. Furthermore, while the Africa Carbon Exchange (ACX) was launched in Nairobi in 2010 with a view to serving as a primary trading platform for carbon credits generated in Kenya and in the East African region, to date the exchange has not traded any carbon credits. This is because most developers of carbon projects in Kenya have opted to lock in their carbon credits with prequalified buyers to reduce the risks associated with price fluctuations of carbon credits.

[7] As we have pointed out, the Consolidated Energy Fund proposed in the draft National Energy Policy of 2012, once operationalized, could be used to assist developers in securing loans or providing lower interest rates. However, its scope and capitalization is unclear at this stage.

8 Conclusions: Both Bitter and Sweet!

There can be no doubt that Kenya government recognizes that RE sources hold massive potential for cost-effective electricity generation. Nevertheless, the dearth of commercially viable RE operators suggests that this potential is yet to be realized. What then does this imply for the efficacy of the country's RE governance framework? Two points stand out.

The first is obvious. An understanding of the challenges facing private investors is a prerequisite for developing sound RE strategies, particularly given circumstances where private investment in the sector has been identified as the only means of meeting projected future energy demand. An appropriate combination of public instruments are required to overcome the challenges identified, including making greater use of economic incentives within such cornerstone instruments as the FiT and tax rebates.

Following from this, the ultimate success of these measures should, however, be viewed from the eyes of ordinary consumers like Jane Shiro and Nduku Musyoka. It is unlikely that both individuals would have a conceptual understanding of the *esoterica* that is the policy, strategy, legal frameworks, and assessment documents that reflect the Government's position on RE. Rather more likely is that the fact they are concerned primarily with and exhausted, in every sense, by the struggle for simple survival. That said, it would be facile to suggest they are generally devoid of any hope of bettering their lots due to inadequate access to reliable and affordable electricity. In all likelihood, their businesses will still operate, even as we remain hopeful that Jane does not fall victim to the rampant night-time insecurity in her area, and that Nduku's dairy business will somehow manage to benefit from the economic boom currently experienced in Machakos town.

This bitter-sweet picture of blighted economic and social progress juxtaposed with resilience and untapped RE potential should remind policymakers everywhere that in the final analysis, the true mark of an effective RE governance framework lies, not so much in the design of its institutional arrangements (bodies and agencies), or in the array of its institutional policies, instruments and laws. Maybe what is needed is for the voices of ordinary consumers to be heard more clearly affirming that given access to clean, reliable, and affordable energy, their lives can be transformed, and by extension, those of their communities, and the nation as a whole.

References

Cameron L, Wurtenberger L, Steibert S (2012) Kenya's climate change action plan: mitigation. Chapter 5: electricity generation. Report prepared for the Ministry of Environment and Natural Resources, Nairobi

Economic Consulting Associates and Ramboll (2012) Technical and economic study for development of small scale grid connected renewable energy in Kenya. Submitted to the Ministry of Energy through the Energy Regulatory Commission

GIZ (Deutsche Gesellschaft für Internationale Zusammenarbeit) (2012) Grid connection of solar PV: technical and economical assessment of net-metering in Kenya. Business Daily, 28 Nov 2012
Government of Kenya, Kenya National Audit Office (2008) Managing solid waste in Nairobi City
Government of Kenya, Ministry of Energy (2011) Least cost power development plan (LCPDP), 2011–2031, Nairobi
Government of Kenya (2011) Scaling up renewable energy program, investment plan for Kenya. Ministry of Finance, Nairobi
Government of Kenya, Kenya National Bureau of Statistics (2012) Kenya facts and figures 2012
Government of Kenya, Kenya National Bureau of Statistics (2013) Economic survey 2013
Government of Kenya, Ministry of Energy (2012) National energy policy, third draft, 11 May 2012, Nairobi
Government of Kenya Ministry of Environment and Mineral Resources (2013) National climate change action plan 2013–2017, Nairobi
Kamau JN, Kinyua R, Gathua JK (2010) An analysis of the wind energy potential in Kenya. Renewable Energy 35:1298–1302
Kenya Electricity Generating Company (KenGen) (2013) The generator. Nairobi
Kiplagat JK, Wang RZ, Li TX (2011) Renewable energy in Kenya: resource potential and status of exploitation. Renew Sustain Energy Rev 15:2960–2973
Lake Turkana wind project (LTWP) (2013) Website: http://ltwp.co.ke/home. Accessed 25 May 2013
Munala G, Moirongo BO (2011) The need for an integrated solid waste management in Kisumu. Kenya JAGST 13(1):65–78
Ngugi P (2012) Financing the Kenya geothermal vision, short course on geothermal development and geothermal wells. United Nations University – Geo thermal Training Programme and LaGeo, El Salvador
Nzila C, Dewulf J, Spanjers H, Kiriamiti H, van Langenhove H (2010) Biowaste energy potential in Kenya. Renew Energy 35:2698–2704
Verbruggen AN, Fischedick MB, Moomaw WC, Weir TD, Nadai AE, Nilsson LJ, Nyboer JG, Sathave JH (2010) Renewable energy costs, potentials, barriers: conceptual issues. Energy Policy 38:850–861
World Bank and IFC (2013) Doing business in Kenya report 2013. Washington DC

Renewable Energy in New Zealand: The Reluctance for Resilience

Hugh Byrd and Steve Matthewman

Abstract This chapter explores renewable energy governance in the context of New Zealand's "energy culture". New Zealand enjoys an international reputation as being a clean and green country. Yet surface appearances can be deceptive. Image frequently trumps reality. The green label is largely an exercise in branding (the country is the latest recipient of the "Fossil Award"), although energy is one of the areas where this might not hold. New Zealand's energy supply mix is impressive, the majority of it being drawn from renewable sources. However, global warming will severely impact upon our ability to generate adequate amounts of electrical power in a sustainable manner, and our centralised corporate-dominated supply system is poorly placed to deal with the challenges that lie ahead. These issues are compounded by various political problems such as ownership of resources and access to the grid. Numerous questions arise: can water be commodified or is it held in common? Does it properly belong to the indigenous people of this country? Why is there no feed-in tariff and why are smart meters not being installed? To explore the topic of renewable energy governance we examine various components of the national energy culture, energy policies and resources. We then look at the likely impacts of climate change, the current state of the deregulated electricity supply industry and why the "business as usual" model is set to prevail. This is illustrated by reference to two case studies—of the potential for distributed generation to contribute towards future electricity demands in Auckland and the proposed district energy system in Christchurch—in both cases we identify a worrying reluctance for resilience.

H. Byrd (✉)
Faculty of Art Architecture and Design, University of Lincoln, Lincoln, UK
e-mail: hbyrd@lincoln.ac.uk

S. Matthewman
Faculty of Arts, University of Auckland, Grafton, New Zealand
e-mail: s.matthewman@auckland.ac.nz

E. Michalena and J. M. Hills (eds.), *Renewable Energy Governance*,
Lecture Notes in Energy 23, DOI: 10.1007/978-1-4471-5595-9_8,

1 Introduction

New Zealand has an international reputation as being 'clean and green' as well as '100 % pure'. For the most part these brand images are empty signifiers (Grinlinton 2009; Pearce 2009), they do, however, have some basis of truth in terms of renewable energy. New Zealand's electricity supply is largely generated by renewable sources such as hydro, geothermal and wind. The proportions vary from year to year but they typically supply about 75 % of total electricity demand. This is an enviable energy supply mix that many other countries would covet. However, what at first sight may appear to be a fuel's paradise, disguises unsettled issues of the governance of energy supply and the maintenance of adequate renewable resources in the face of climate change.

Harnessing energy from the wind, rain and sun raises key governance issues regarding ownership of these resources and the distribution of the energy gained from them. For example, some of the key energy questions in 2013 in New Zealand are: who owns the water in the rivers (in anticipation of selling off the country's rivers, see Bosselmann and Round 2012), why is there no feed-in tariff, why has the subsidy for solar hot water heating been removed and why are authentic smart meters not being installed? These questions increase in significance as the price of electricity rises. To understand such questions, and related ones, we begin by outlining the contours of New Zealand's energy culture. This will provide the context for the subsequent discussion.

2 The Forming of New Zealand's Energy Culture

New Zealand's energy policies and resources form an energy culture that differs significantly from other OECD countries. The concept of an 'energy culture' in New Zealand has been previously proposed as a micro-sociological conceptual framework to trace interactions between household energy technologies, energy norms and behaviours (Stephenson et al. 2009). However, this paper will look at the pragmatics of the national culture, its implications on governance and its consequences on renewable energy.

At the heart of the nation's energy culture is the pioneering settler spirit. Dakin (2007) portrays the immigrants to New Zealand throughout its history as being people that were *escaping* both church and state in Europe to a place where they were free from a class structured society and where the pioneering spirit was unhindered by regulation (see also Phillips 1987). This individualistic spirit has endured and manifests as a resistance to being told what to do by an interfering "nanny state". A recent example of this, discussed further below, was when a political leader was obliged to apologise to an outraged public for attempting to regulate for energy efficient light bulbs in 2008.

This energy culture is also a culture of complacency in as much as it adheres to the notion of 'infinite energy' (Fanning 2012); a mind-set that assumes that demand will always be satisfied by supply. This mind-set was formed by the historical low price of electricity and the seemingly endless supply of natural resources to generate it. In consequence, the energy culture is one that fixates on the present, paying little attention to future energy security. We discuss the implications of this reluctance for resilience in the following section.

The energy culture is also linked to politics in two important senses: grassroots politics concerning public protest and government politics concerning privatization. Let us briefly discuss each in turn. 'Green politics' are an important part of the country's political landscape. One of the world's first green parties was in New Zealand. The 'Values Party' was set up in 1972, almost a decade before European green parties, and was spawned from protests against the destruction of the environment caused by hydroelectric development (Rainbow 1992). These types of protests have carried on through anti-nuclear protests in the mid-1970s (Priestly 2010) to the present day and are currently focussed on wind farms. For example, the successful campaign, from the protestor's point of view, against a large wind farm in Turitea (New Zealand Board of Enquiry 2011).

The nation's energy culture has been decisively shaped by the privatisation of the electrical supply industry. This industry has a vested interest in a centralised supply of energy and a financial disincentive to practice energy conservation. This state of affairs is best described by the chairperson of the Australian Photovoltaics (PVs) Association, Muriel Watt: "They don't want you to use less (power) because they make less money when you use less. The retailers make money so they don't want you to use less. The generators make money (by selling more)... and the networks make money from every kilowatt hour that goes through their network, so they don't want you to use less either" (Fanning 2011).

It has also been suggested that the electrical supply industry has presented barriers to the integration of distributed energy generation. A report on these barriers concluded that a major barrier was the unwillingness of the electrical supply industry to change the 'status quo' of centralised power generation (Stevenson 2010).

3 Energy Policies and Resources

These characteristics of the energy culture lead to vulnerability and a lack of energy security for the end-users economically, with precious little control of energy prices. Additionally, they face the risk of power failure that may be caused by both natural events or by inadequate supply systems, including generation and infrastructure capacity (Matthewman and Byrd 2011). As Nair and Zhang (2009, p. 3419) write: 'One current challenge facing the New Zealand power system infrastructure is the transmission capability of the national grid system. The current backbone of the national grid was built in the 1950s and 1960s, and there is a need for upgrading capacity to meet future requirements. The electricity demand in NZ is estimated to grow at around 1.3 % per annum which puts stress on both transmission grid and distribution networks'. These systems lack resilience. In this section, we explain the sequence of changes in energy supply policy that have taken place over the last half a century and *have led to the governance failure which we identify as a reluctance for resilience.*

3.1 Hydroelectricity: The Backbone of New Zealand's Electricity Supply

Hydroelectricity has long been the backbone of the energy supply mix in New Zealand. Small scale private power stations started in the late nineteenth century but it was not until 1896 that the state gained control over the water supply for hydroelectricity (Martin 2011). Power stations were gradually developed on both the North and South Islands but by the onset of the Second World War there were serious power shortages. New dams and power stations were rapidly developed in the 1950–1960s, but by the late 1960s a strong environmental movement had been mobilised that was to impact on the development of power industry to the present day.

3.2 Strong Opposition to Nuclear Power

Such were the electricity shortages in the 1950s that alternative means of generation were considered and nuclear power was seen as the viable means of overcoming New Zealand's power problems. The first governmental report on nuclear power was produced in 1957 but it was not until 1968 that the Planning Committee on Electric Power Development in New Zealand recommended a reactor turbine to be in operation by 1977 with three others to follow on soon after (Priestly 2010). Plans were drawn up and sites north of Auckland were selected.

New Zealand could so easily have introduced nuclear power into its generation mix and, although the New Zealand Nuclear Free Zone, Disarmament and Arms Control Act 1987 prohibits nuclear power for military purposes, it is entirely possible that the country could once again consider nuclear power for its electricity generation.

Public opinion against nuclear power grew during the mid-1970s and the Royal Commission Enquiry into Nuclear Power Generation, set up in 1975, was confronted with strong opinion against its implementation on the basis of environmental safety. It also met with criticism that the total electricity consumption of New Zealand could be reduced rather than supply increased. The Royal Commission eventually concluded that, "nuclear power is not justified for New Zealand until about the turn of the century, or even perhaps later" (Priestly 2010).

3.3 The Rise and Fall of Natural Gas

However, there was another reason behind the decision to delay the adoption of nuclear power. By 1970, it had been firmly established that there were substantial quantities of natural gas in the offshore Maui gas field that could be economically recovered. Gas-fired electricity generation stations are substantially less expensive than nuclear powered and with relatively less public opinion against their construction, they proliferated. Annual production of gas increased rapidly over the 30 years, peaked in 2002 and has been in decline in the decade since. With few new significant finds, the economic extraction of natural gas is predicted to dramatically fall by the end of this decade.

3.4 Coal: Focus of Strong Public Opinion

The Huntley power station is central to the story of gas-fired electricity generation. Constructed in stages between 1973 and 1985 and further upgraded in 2004, it produces about 17 % of the country's total power demand. In the 1990s, with the realisation that gas reserves were dwindling, coal partially substituted gas and now the power station contributes about 50 % of the total greenhouse gases produced by electricity generation in New Zealand (Gluckman 2009). The power station has been the focus of strong public opinion and protests (New Zealand Herald 2007) and has come to symbolise the ire of environmentalists. In response to this, a government report suggested that the plant may be closed down in 2015. However, following a change in government, the station has been given consent to operate for another 25 years (Genesis Energy 2012).

3.5 Focus on Price Rather than Efficient Use of Resources

The Maui gas field successfully halted nuclear power in New Zealand and gave about 30 years of energy to New Zealand. The gas field is now nearly depleted. The search for more gas fields continues, even into 'deep-water'. However, in November 2012, one of the largest international oil companies, Petrobras, withdrew its search on the grounds that there were insufficient signs of gas and oil to justify further exploration (Energy News 2012). During those 30 years, there has been some growth in geothermal power generation and, more recently, wind power with each contributing about 20 and 7 %, respectively, to total electricity demand (Ministry of Economic Development 2012). But these forms of power generation are unlikely to fill the energy gap that is emerging unless demand is reduced. An adequate supply of resources to generate energy is, once again, facing the country as the demand for electricity increases and fossil fuels deplete. Only this time the country is also faced by the impacts of climate change that conspire to both increase electricity demand and reduce the supply from hydro. It also faces the problem of the impacts of 'peak oil'. While New Zealand has a relatively 'clean' electricity supply, about half the country's total energy demand comes from oil that is used principally for transport, leaving New Zealand the obvious choice to adopt electric vehicles.

The depletion of fossil fuels, the assumption of 'infinite energy' and climate change challenge the energy culture of New Zealand. The deregulated electricity supply industry has shaped public opinion to focus on price as opposed to the efficient use of resources. The following sections analyse how these factors will impact on New Zealand.

4 The Heavy Impact of Climate Change

Climate change will impact on energy supply and demand in several ways in New Zealand. One of the biggest problems with New Zealand's existing hydro schemes is the lack of water storage capacity. New Zealand's hydro schemes do not benefit from large reservoir capacity; most have just a few months of storage capacity (International Waterpower 2006). They are therefore more vulnerable to annual or even seasonal fluctuations in precipitation and snow melt. New Zealand Glaciers have an estimated volume of about 53 Km^3 and have been exponentially decreasing in volume (World Glacier Monitoring Service 2009).More than half the water entering hydroelectric lakes and rivers comes from glacial water (Fitzharris and Hay 1989). However, global warming will have an impact on this. Predictions of a 3 °C temperature rise and 15 % increase in precipitation indicated a significant decrease in snow accumulation resulting in increased flows of 40 % in the winter and a 13 % decrease in the summer (NIWA 2012). Other recent research has shown that snow accumulation, at 1,000 m elevation, may reduce by up to 44 % by 2040, and 79 % by 2090 (Poyck et al. 2011).

Short periods of dry weather have significant consequences on water storage for hydro in NZ. For example, in 2012 the South Island experienced several weeks without rain with the result that large hydro lakes were some 64 % below levels usual for that time of year. Retail prices increased fourfold (Evans 2012). Dry years such as this combined with reduced glacial melt will have an even greater impact.

Also of relevance, although still inadequately researched, is the energy-water nexus in New Zealand. With increased temperatures, the peak demand for electricity will shift towards summer rather than winter. There will also be an increased demand for water for irrigation during the summer. This will reduce the ability of the hydroelectric power sector to provide an unfluctuating supply and could result in significant reduction in the hydroelectricity supply in the future.

A further impact that climate change will have on energy security is the risk of infrastructure failure. An extreme, though sobering, example is that of the cable failures that fed the Central Business District of Auckland in 1998. High summer temperatures increased the air-conditioning load as well as ground temperatures until cables melted. After several attempts to mend the cables, the CBD eventually went back to work after 2 weeks of closure costing billions of dollars (BBC 1998).

The importance of this example for New Zealand is that while many blackouts are caused by systems failures, there is an increasing risk of failures due to inadequate energy; whether due to depletion of resources such as oil and gas or due to the vagaries of the climate in the supply of renewable energy. As we enter the period of peak oil, peak gas and climate change the security of energy supply for electricity generation is under threat. Understanding the nature of blackouts is more than just a record of past systems failures; blackouts are dress rehearsals for the future in which they will appear with greater severity and frequency (Matthewman and Byrd 2011).

5 Cheap Electricity by Ignoring External Costs

The New Zealand Government's monopoly on generation was removed by the State Owned Enterprises Act (1986). Over the following 10 years, various legal changes took place in the structure and ownership of the generation and transmission of the electrical supply industry until in 1996 the wholesale electricity market was in full operation and consumers could choose their suppliers. The deregulation of the New Zealand electricity supply was a copy of that in the UK, driven by the need to reduce government budget deficits rather than a concern about electricity prices (Evans and Mead 2005).

Ownership and management of the electricity supply industry is a key issue for renewable energy as it dictates policy, investment, prices and the stewardship of resources. In general, monopolies are not good at controlling prices but can be good at protecting resources while a competitive market does the reverse. The purpose of those in a competitive market is to sell as much of a product as possible at a rate that will eliminate the competition. While resources are abundant, the consumer may benefit from this; the problem arises when resources are constrained. Policies that encourage consumers to use electricity efficiently or use an alternative source are not in the interests of a competitive electrical supply industry. Hence, energy efficiency or self-supply of electricity is a potential economic threat to be resisted. As a consequence of this, the energy culture in New Zealand focuses on the price of electricity. The success of the electricity industry is measured in 'power switches', the number of people who change back and forth between companies. The media reinforce this agenda set by the electricity supply industry (on this see Jones 2012).

Reducing electricity costs of consumers by reducing demand or the introduction of requirements that aim to conserve resources is a sensitive topic. It was an important factor in the 2008 general election. The government of the time proposed that shower heads should be 'low-flow' and that light bulbs should be 'energy efficient'. There was public outrage at this and accusations of the government becoming a "nanny state" (New Zealand Herald 2008). Consequently, the party leader apologised (South 2008) and the party subsequently lost the election. While the loss of the election may have been due to several factors, the issue of trying to impose energy efficiency requirements on a public accustomed to an unfettered electricity supply should be acknowledged.

While cheap electricity is desirable it should not be at the cost of appropriate investment in the supply industry in response to forthcoming resource constraints. To respond to the issues of climate change, depletion of natural gas or maintenance of fragile infrastructure requires mitigation strategies that come at a price. Investment for preparedness combined with public awareness would justify additional expenditure and consequent price rises. However, the supply industry and its governance have set the agenda and everyone else is dancing to their tune, inadvertently strengthening a very unhelpful energy culture; that energy should be cheap, or at least not pay for its true external costs.

The problem of not accounting for externalities is that low prices eventually come back and hit in some form or another. Eventually they are paid for. This was the basis of the 'boomerang paradox' in the electricity industry in Australia (Simshauser et al. 2010). Years of cheap electricity had lured customers into habits and appliances that consumed electricity excessively. Most of the electricity in Australia is generated by burning coal. As coal prices escalated and the price of carbon increased, electricity bills also rose. The growth in air-conditioning also accelerated and burdened the capacity of the electrical infrastructure. In order to respond to this demand, new cables and new power generation had to be installed. It was estimated that for every AUD$1,000 spent on air-conditioning, AUD$7,000 had to be spent on upgrading the infrastructure (Fanning 2011). Ultimately the consumer has to pay for this. While the same externalities may not apply to New Zealand, climate change and resource depletion demand an energy policy that looks beyond 'power switches' as a measure of the electrical supply industry's effectiveness. Clearly the status quo needs to be changed. We need an energy culture based on resilience and sustainability. In the next section, we look at reasons why the barriers to distributed generation of electricity remain.

6 Distributed Generation

6.1 Disincentive for Distributed Generation

Distributed generation (DG), in particular by PVs, has been victim of cheap energy. Unlike almost every country in the OECD and in many developing countries, New Zealand has not introduced a subsidised feed-in tariff. *There is no obligation for the New Zealand Government to commit to either reducing carbon emissions or increasing renewable energy generation as there is in Europe.*

Indeed, New Zealand recently had the dubious distinction of being given 2012's "Fossil Award" by the UN's civil society conference for the worst performance on climate change (New Zealand Celsias 2012). There is also no financial incentive for the electricity supply industry to engage with DG and this has created obstacles to this area of the renewable energy sector. In 2010, a report was produced (Stevenson 2010) for the Government's Energy Efficiency and Conservation Authority (EECA) entitled "Analysis of Barriers to Distributed Generation". The primary obstacle in the report was described as 'the electricity industry status quo prevails'. *The industry is based on central generation and is unwilling to move from this model.* As previously noted, they have no financial incentive to do so.

In response to this report a 'Retail Advisory Group' was set up by the Electricity Authority to examine whether there were any obstacles to DG (Retail Advisory Group 2012). The report found that there were no obstacles of any significance. However, of greater significance was the proportion of the Group with a vested interest in centralised generation and the overwhelming submissions to the report by the centralised electricity supply industry. The recommendation of the group was that electricity retailers should not be required to purchase electricity from small scale DG or offer any rate for exported electricity to the grid. In other words, there is no incentive for small scale renewable energy to be connected to the grid in New Zealand.

Due to electricity price increases in New Zealand combined with the reduced capital cost of photovoltaics, DG electricity produced by PVs has now reached grid-parity (Byrd and Ho 2012). Distributed generation has the potential to reduce transmission losses from centralised generation plants, to be able to offset electricity demand from the grid and to be able to share surplus electricity with the grid. In so doing, it necessarily alters the energy markets that it enters, for this reason it is often perceived negatively as a 'disruptive technology' (Sustainable Electricity Association New Zealand 2012). Significant changes have been observed in countries where feed-in tariffs have been introduced, for example Germany and Japan.

6.2 Solar Hot Water Heating Subsidy Removed: An Even Greater Dependency on the Grid

A further issue that has raised concern in New Zealand is the removal in 2012 of a government subsidy for solar hot water heating. The cheap price of electricity has historically made investment in solar water heating unattractive. With the majority of houses in the country using electricity for water heating, which accounts for about 30 % of household electricity use (Isaacs et al. 2010), the removal of the subsidy was a surprise on energy efficiency grounds.

A report by the Parliamentary Commissioner for the Environment put forward the case that the problem in New Zealand is not so much adequate energy but more the problem of peak demand (Parliamentary Commissioner for the Environment 2012) and that solar energy is not available during times of peak hot water demand. While this is true, it is a curious argument because there are simple and proven technologies that can delay the electric heating of hot water until times outside peak demand. 'Ripple control' has been in operation since the 1950s in New Zealand. Even more curious was the recommendation in the report that heat pumps should be used to heat hot water rather than solar hot water heaters. Such a recommendation results in an even greater dependence on the grid for electricity and, even though heat pumps are more efficient than electrical resistant heaters, solar water heating in New Zealand is estimated to provide up to 75 % of household hot water needs (EECA 2012).

6.3 Increased Demand Due to Buildings and Vehicles

Elsewhere in the world countries have attempted to reduce energy demand by introducing requirements that increase the energy efficiency of buildings and appliances. With the risk of being called a nanny state, politicians in New Zealand have steered clear of using legislation for reducing energy demand. For example, the first Building Code (minimum legal standards) requiring houses to be insulated were implemented in 1978. It remained unchanged for 30 years and even then required only a modest change in standards in 2008 (Byrd 2012).Energy performance standards for commercial buildings are also low compared to other OECD countries. Large areas of unshaded single glazing and a lack of control of artificial lighting are not only common but also compliant with Code. The New Zealand Green Building Council has introduced a 'green' accreditation scheme for voluntary improvements above Code compliance. However, 'best practice' (4-star) accreditation can be achieved by buildings that meet, but do not have to exceed, Building Code standard for energy use. In other words, 'best practice' can be achieved even when a building is on the threshold of breaking the law (Byrd and Leardini 2011).

Apart from a general increase in energy use as the built environment grows, there are two areas of growth that are predicted to rapidly increase. The first is the increase in electricity use by domestic heat pumps for both cooling and heating. The Government has been offering a subsidy for heat pumps with the original intention of reducing the amount of heating by electrical resistance. The subsidy has had several unintended consequences that have resulted in an increase in energy consumption (Byrd and Matthewman 2012). The most significant of these consequences has been the extent that the heat pumps have been used for cooling. New Zealand is a temperate climate and cooling is not necessary in appropriately designed buildings. However, aggressive marketing, subsidies and a perceived improvement in lifestyles have led to a rapid increase in the uptake of heat pumps for cooling in New Zealand, a relatively new pattern of energy consumption to New Zealand. Research has indicated that, in Auckland, energy consumption of houses could increase by 180–250 % in real terms by the year 2041 due to increased cooling loads (Page 2009).

The second significant anticipated area of growth in electricity demand is electric vehicles. Almost half the total energy use in New Zealand is for transportation and the majority of this is imported oil which makes New Zealand very vulnerable to the global oil market and the consequences of 'peak oil'. A logical response to this would be to shift to electric vehicle use, not only because the electricity produced is from renewable sources but also because about 85 % of the population are urbanised and 90 % of daily travel by private vehicles is less than 60 km; which is well within the capabilities of electric vehicles. The electricity Authority (Hemery and Smith 2008) has predicted an exponential increase in the use of electric vehicles into the next decade. And although the additional energy required could be provided from the proposed wind power developments (assuming they will obtain consent to be constructed) (Duke et al. 2009) the problem with electric vehicles is not their average energy use but their impact on peak electricity demand in the evenings when vehicles return home and recharge. As one electricity market commentator has observed (Scott 2012): "the cables will melt".

6.4 *Slow Movement Towards Smart Meters and a Smart Grid*

Prior to the report on solar water heating, the Parliamentary Commissioner for the Environment produced a report on smart metering (Wright 2011). It was a strongly worded report, targeting the electricity supply industry that challenged their interpretation of 'smart'. Electricity meters were being installed that were intended only for remote reading and not for 'home area network' (HAN) functionality. Therefore, the meters were 'smart' for the supplier but not for the customer and, for only a few dollars more could have included HAN technology. This technology is required for DG so that electricity can be directed from the home to the grid and would allow DG, such as photovoltaics, to feed into the grid.

A report on smart grids (Strbac et al. 2012) in New Zealand was commissioned by a company within the electricity supply industry. The report anticipated the growth in electricity demand due to electric vehicles and heat pumps and identified that managing peak loads was of greater importance than overall energy demand. It concluded that investment in a smart grid on a national level would become feasible in about 2030. However, of greater interest was what was omitted from the report. No mention was made of the potential impact of distributed generation and its impacts on a smart grid. The smart technology that is required to make the maximum potential of distributed electricity generation is in the hands of an industry that will not significantly benefit from it financially. The status quo of the electrical industry and its focus on its own bottom line is at odds with a resilient electricity network.

6.5 Case Studies: Opportunities Coming from Disasters and Depletion

The analysis above has outlined the conflict that exists within New Zealand's energy culture. The desire not to be controlled by the state has resulted in a country where there has been little regulation to reduce energy demand. One would expect that the same culture would embrace distributed generation of electricity as a means of self-determination and resilience. However, the state has sold much of its control of the electricity sector to private industry that now controls the network and pricing to the extent that distributed generation, with the exception of remote areas, can be effectively excluded from contributing to the network. It is in the interests of the electricity supply industry to centralise rather than decentralise and this conflicts with the idea of resilience.

These centralised systems lack resilience which, within the context of energy, is defined as *the ability to respond to a disturbance by resisting damage and recovering quickly* (Folke et al. 2002). Resilience of the electricity supply in New Zealand is of increasing importance as changes take place both suddenly and gradually. Sudden changes in New Zealand include such things as earthquakes and dry weather. Gradual changes include climate change and the shift towards greater electricity use for transport as a response to oil insecurity.

To illustrate these conflicts, two case studies will be briefly described. Both are recent proposals that have not been implemented. The first concerns the potential for using distributed generation to contribute towards the future electricity demands for transport in Auckland. The second concerns the proposals for a DES in Christchurch.

6.5.1 Powering Electric Vehicles in Auckland

While the historical focus of the 'green' movement in New Zealand has been on the environmental and aesthetic implications of electricity generation by hydro and wind, the internal combustion engine vehicle, with all its negative environmental implications, has been embraced by society. Auckland is a car-dependent city with an average of one car for every 1.5 people. Only 5.5 % of those travelling to work use public transport (Huang et al. 2010). As a consequence, New Zealand also depends on imported oil with almost half of its total energy consumption being for transport. A report by the consultancy firm ARUPs (Auckland Council 2011) showed that the single largest contribution to reducing carbon in Auckland would be by a transition to electric vehicles (EVs).

One of the main problems of transitioning to EVs in Auckland is the fragility of the electrical distribution system. Without a smart grid, the only means of attempting to control the electrical demand for recharging in the evening would be by increased pricing. An alternative to this is to charge electric vehicle by PVs on residential roofs. Research was carried out into the feasibility of this (Byrd and Ho 2012) and it was found that the electricity generated on household roofs could potentially supply an all-electric vehicle fleet and also generate excess electricity that could offset other forms of generation. EVs and PVs have a synergy that could significantly reduce carbon outputs but could also reduce electricity demand and peak demand on the grid as well as reducing the need to upgrade electrical infrastructure.

6.5.2 Energy for Buildings in Christchurch

The devastation caused by the earthquakes that hit Christchurch in February and June 2011 is such that much of the city needs to be rebuilt. This gives an opportunity to review the best methods of generating and transmitting energy in a resilient manner so that there is greater preparedness should there be similar events in the future. Such a system would not only need to be robust but also be able to operate so that there was some form of energy-autonomy in the event of infrastructure failure.

An early decision was made to use a DES that would pump hot and chilled water around a circuit in the City (BECA 2011). The circuit is made up of large underground pipelines that are heated by boilers that can use a mixture of fuels such as agricultural waste or timber; both in abundance in the area. The chilled water is to be derived from tri-generation plants.

It is not clear what the main motivation is behind decision for a centralised DES; the feasibility study appears to have been drawn up after the decision was made and there is little evidence of any alternatives being considered.

The questions that arise from this decision are: why place heavyweight pipes below ground in an earthquake prone area where liquefaction of the ground is an historical problem? Furthermore, if the system fails in one place it is all likely to fail leaving the City without this energy system. The other issue that never appeared in the feasibility study is the question of reducing demand. The opportunity to design new buildings with low energy usage has not been adequately considered. Indeed, there is no need for cooling in buildings in Christchurch if they are appropriately designed (with perhaps the exception of a hospital). Energy demands could be brought near to zero and those that remain produced by distributed generation through smart meters.

7 Discussion of Case Studies

The two case studies described illustrate the dichotomy in the energy culture of New Zealand. Both studies propose systems that are based on renewable energy but their governance, ownership and control are at opposite ends of the energy-system spectrum. On the one hand, we have a system for Auckland that responds to the need for reducing carbon emissions and gives building occupants both autonomy and choice. On the other hand, we have a system in Christchurch that is centralised and has little resilience in the event of breakdown. The former has all the issues of the supply industry, discussed above, to overcome (such as an inadequate feed-in tariff and lack of smart meters), were it to be adopted. The latter is likely to proceed as it is centralised in both its decision making and its technology. The lack of governance to support the distributed generation of renewable energy for transport, in the former case study, and the apparent support for centralised generation, in the latter case study, might be judged the wrong one if the ultimate goal is to create a resilient and sustainable energy culture.

8 Conclusions

While the shift away from fossil fuels to renewable energy is unquestionably a positive step towards resilience, it introduces its own set of issues relating to the governance of energy supply. New Zealand claims to be ‘clean and green’ and much of this claim is based on a high proportion of renewable energy in the electricity supply mix. However, RE has brought with it a conflict within the energy-culture of the country that has been masked by an historical abundance of energy resources that has allowed prices to remain cheap. This has created the illusion that we possess ‘infinite energy’. As these resources deplete, in particular in the case of gas, or become less dependable because of climate change, in particular in the case of hydro, the ownership and management of these resources becomes significantly more important (and arguably more politicised).

In addition to being informed by the beliefs that we have infinite energy and that we are clean and green, the energy culture has also been framed by consumers who are adverse to regulation and by government who are committed to deregulation. The consequence of this is a non-resilient electrical supply industry based on centralised generation. It is actively antagonistic towards more flexible practices of distributed generation. It will take a significant political turnaround to change this energy culture, including legislation aimed at reducing electrical demand (new building codes, energy efficient lighting and solar hot water heating subsidies). Unfortunately, as our two case studies demonstrated, we seem to be heading in the wrong direction. Our earnest hope is that New Zealand does not continue to win the Fossil Award in the future.

References

Auckland Council (2011) Auckland city council carbon strategy technical report. Unpublished report carried out by ARUPs

BBC (1998) Auckland faces at least six weeks without power. http://news.bbc.co.uk/2/hi/asia-pacific/61828.stm. Accessed 19 Oct 2012

Beca (2011) Christchurch central city district energy system information report. Prepared for Christchurch Agency for Energy Trust. http://www.cafe.gen.nz/sites/default/files/NZ1-

4474091-ChristChurch%20City%20District%20Heating%20and%20Cooling%20Scheme%20Information%20Report.pdf. Accessed 10 Dec 2012

Bosselmann K, Round D (2012) Debate: who owns the water? New Zealand Herald, 26 September. http://www.nzherald.co.nz/nz/news/article.cfm?c_id=1&objectid=10836420. Accessed 9 Jan 2013

Byrd H (2012) The case for policy changes in New Zealand housing standards due to climate change. J Environ Planning Policy Manag 14(4):360–370

Byrd H, Ho A (2012) Exergy and the city, solar progress, Spring: 13–14. http://issuu.com/commstrat/docs/solar_spring12_issuu

Byrd H, Leardini PM (2011) Green buildings: issues for New Zealand. Procedia Eng 21:481–488

Byrd H, Matthewman S (2012) Warm up New Zealand: heat smart – a critical overview. New Zealand Sociol 27(2):53–74

Dakin J (2007) The secular trend in New Zealand. NZARH/Humanist Trust, Wellington

Duke M, Andrews D, Anderson T (2009) The feasibility of long range battery electric cars in New Zealand. Energy Policy 37(9):3455–3462

EECA (2012) Solar water heating. http://www.energywise.govt.nz/how-to-be-energy-efficient/your-house/hot-water/solar-water-heating. Accessed 10 Dec 2012

Energy News (2012) Signs of oil and gas off East Cape insufficient to continue search – Petrobras. http://www.energynews.co.nz/news-story/10120/signs-oil-and-gas-east-cape-insufficient-continue-search-petrobras. Accessed 10 Dec 2012

Evans G (2012) Hydro storage slides. Energy News, 16 January. http://www.energynews.co.nz/node/8195. Accessed 14 March 2012

Evans L, Mead R (2005) Alternating currents or counter-revolution? Contemporary electricity reform in New Zealand. Victoria University Press, Wellington

Fanning E (2011) The hidden cost of infinite energy. The Global Mail, 6 February. http://www.theglobalmail.org/feature/the-hidden-cost-of-infinite-energy-part-1/19/. Accessed 8 Dec 2012

Fanning E. (2012) The hidden cost of infinite energy. The Global Mail retrieved from http://www.theglobalmail.org/feature/the-hidden-cost-of-infinite-energy-part-1/19

Fitzharris J, Hay J (1989) Glaciers: can they weather the storm of climate change? In: Welch R (ed) Proceedings of the 15th New Zealand geography conference, New Zealand geographical society, pp 284–291

Folke C, Carpenter S, Elmqvist T, Gunderson L, Holling CS, Walker B (2002) Resilience and sustainable development: building adaptive capacity in a world of transformations. Ambio 31(5):437–440

Gluckman P (2009) 'Climate change', office of the prime minister's science advisory committee. http://www.pmcsa.org.nz/climate-change/. Accessed 10 Dec 2012

Genesis Energy (2012) Resource consents granted for huntly power station. http://www.genesisenergy.co.nz/genesis/about-us/media-centre/news/resource-consents-granted-for-huntly-power-station.cfm. Accessed 10 Dec 2012

Grinlinton D (2009) Clean green New Zealand-reality or myth? Georgetown University, Washington DC

Hemery E, Smith B (2008) Electric vehicles demand forecast. www.ea.govt.nz/dmsdocument/7962. Accessed 10 Dec 2012

Huang N, Vale B, Vale R (2010) Car or public transport? some lessons from policy-making for Auckland's transport. Int J Arts Sci 3(13):1–19

International Waterpower and dam construction (2006) Hydro in the mix in New Zealand. http://www.waterpowermagazine.com/story.asp?sectionCode=166&storyCode=2039414. Accessed 10 Dec 2012

Isaacs N, Camilleri M, Burrough L, Pollard A, Saville-Smith K, Fraser R, Rossouw P, Jowett J (2010) Energy use in New Zealand households: final report on the household energy end-use project [HEEP] (BRANZ study report 221). BRANZ, Porirua

Jones N (2012) Power switches show market works. New Zealand Herald, A6, June 19

Martin J (2011) Hydroelectricity – Hydro, 19th and early 20th centuries. The Encyclopaedia of New Zealand. http://www.teara.govt.nz/en/hydroelectricity/2. Accessed 2 Nov 12

Matthewman S, Byrd H (2011) Blackouts: a sociology of power (failure), looking forward: trends, horizons and utopias. In: Sociological association of Aotearoa (New Zealand) annual conference, Victoria University of Wellington, 7–9 Dec 2011

Ministry of Economic Development (2012) NZ energy quarterly. Issue 19, September. http://www.med.govt.nz/sectors-industries/energy/energy-modelling/publications/new-zealand-energy-quarterly. Accessed 10 Dec 2012

Nair NK, Zhang L (2009) SmartGrid: future networks for New Zealand power systems incorporating distributed generation. Energy Policy 37:3418–3427

New Zealand Board of Enquiry (2011) Final report and decision of the board of inquiry into the Turitea wind farm proposal. New Zealand Environmental Protection Authority, Wellington

New Zealand Celsias (2012) New Zealand takes colossal fossil award for worst performance on climate change. December 9. http://www.celsias.co.nz/article/nz-takes-award-worst-performance-doha/?utm_source=CelsiasWeekly&utm_medium=Email&utm_campaign=20121212. Accessed 18 Dec 2012

New Zealand Herald (2007) Protesters could be charged over power station climb. February 26. http://www.nzherald.co.nz/nz/news/article.cfm?c_id=1&objectid=10425858. Accessed 10 Dec 2012

New Zealand Herald (2008) Are you in favour of shower pressure limitations to conserve water use? October 10. http://www.nzherald.co.nz/nz/news/article.cfm?c_id=1&objectid=10536784

NIWA (2012) Climate change scenarios for New Zealand. http://www.niwa.co.nz/our-science/climate/information-and-resources/clivar/scenarios. Accessed 10 Dec 2012

Page I (2009) Regional heat pump energy loads. BRANZ Report No. E528. Building Research Association of New Zealand: Wellington

Parliamentary Commissioner for the Environment (2012) Evaluating solar water heating: sun, renewable energy and climate change. http://www.pce.parliament.nz/assets/Uploads/Hot-Waterweb.pdf. Accessed 10 Dec 2012

Pearce F (2009) New Zealand was a friend to middle earth, but it's no friend of the earth. The Guardian, November 12. http://www.guardian.co.uk/environment/cif-green/2009/nov/12/new-zealand-greenwash. Accessed 1 Jan 2013

Phillips J (1987) A man's country? the image of the pakeha male, a history. Penguin, Auckland

Poyck S, Hendrikx J, McMillan H, Hreinsson EÖ, Woods R (2011) Combined snow – and stream flow modelling to estimate impacts of climate change on water resources in the clutha, New Zealand. J Hydrol 50(2):293–312

Priestly R (2010) Nuclear power: how close we came. Energy NZ, Vol.4 No.5 September–October. http://www.contrafedpublishing.co.nz/Energy+NZ/Vol.4+No.5+September-October+2010/Nuclear+power+How+close+we+came.html. Accessed 3 Oct 12

Rainbow S (1992) Why did New Zealand and Tasmania spawn the world's first green parties? Environ Politics 1(3):321–346

Retail Advisory Group (2012) Review of arrangements for purchase of power by retailers from small-scale distributed generation. June 29. http://www.ea.govt.nz/our-work/programmes/market/distributed-generation/. Accessed 2 Dec 12

Scott J (2012) Maintaining voltage with extensive distributed generation and charging electric vehicles without melting the wires. Podcast. http://www.energynews.co.nz/blog/john-hancock/7700/podcast-maintaining-voltage-extensive-distributed-generation-and-charging. Accessed 2 Dec 12

Simshauser P, Nelson T, Doan T (2010) The boomerang paradox: how a nation's wealth is creating fuel poverty – and how to defuse the cycle. AGL Economic and Policy research. Working Paper No. 17. http://ideas.repec.org/a/eee/jelect/v24yi1p72-91.html. Accessed 10 Dec 2012

South G (2008) Don't say that 'sustainability' word. New Zealand Herald, October 31. http://www.nzherald.co.nz/business/news/article.cfm?c_id=3&objectid=10606288. Accessed 7 July 2010

Stephenson J, Lawson R, Barton B, Carrington G, Thornes P, Ford R, Williams J (2009) Energy cultures – energy efficiency in households, centre for sustainability: university of Otago. http://www.csafe.org.nz/energy_cultures_2. Accessed 9 Jan 2013

Stevenson T (2010) Analysis of barriers to distributed generation (dg): a report prepared for the energy efficiency and conservation authority (EECA). http://www.eeca.govt.nz/distributed-generation. Accessed 10 Dec 2012

Strbac G, Pudjianto D, Djapic P, Aunedi M, Stanojevic V, Castro M, Ortega E, Mancarella P, Telfar G, Tucker B, Corney S, McDonald J (2012) Smart New Zealand energy futures: a feasibility study. http://www.meridianenergy.co.nz/assets/Uploads/Smart-Enegy-Futures-Summary-Report-Jan2012.pdf. Accessed 17 Nov 2012

Sustainable Electricity Association New Zealand (2012) Solar photovoltaics. http://www.seanz.org.nz/Resources/The-Technologies. Accessed 1 Dec 2012

World Glacier Monitoring Service (2009) Glacier mass balance data 2008 & 2009. http://www.grid.unep.ch/glaciers/pdfs/6_3.pdf. Accessed 10 Dec 2012

Wright J (2011) Advanced metering infrastructure: nomination of the MEP and access to data. http://www.pce.parliament.nz/assets/Uploads/PCE-EA-Submission.pdf. Accessed 10 Dec 2012

The Development of Renewable Energy Governance in Greece. Examples of a Failed (?) Policy

Antonis Metaxas and Michael Tsinisizelis

Abstract Renewable energy sources (RES) implementation has been crucial for countries throughout Europe, and this led to several legislative efforts to enhance RES development. However, the persisting financial crisis frequently proved to constitute an obstacle in taking correct and efficient measures. The need to reduce fiscal market specific deficits seemed to overshadow RES policies' efficient implementation, leading to situations that drive away investments rather than facilitate them. In some cases, the governments, in their effort to mitigate the existing fiscal deficits and adapt to the financial crisis persisting in Europe, often seem to take a wrong turn. The measures they have adopted seem to be in the opposite direction, since they only consider short-term financial results, ignoring the medium- and long-term negative effects in the overall economy. Moreover, regulatory issues that remain unsolved (i.e., unforeseen delays in the licensing procedure) also discourage investments in the renewable energy sector, leading to a recession rather than economic growth. This deficient national governance fails to get the "broader picture" and assess correctly the many advantages of an increased, but also well-structured, RES penetration.

For the purpose of drafting this chapter, legislative developments until the 31st December 2012 have been taken into account.

A. Metaxas (✉)
School of Law, Economics and Political Science, University of Athens,
154 Asklipiou Street 114 71 Athens, Greece
e-mail: ametaxas@pspa.uoa.gr

M. Tsinisizelis
School of Law, Economics and Political Science, University of Athens,
57 Solonos Street 106 79 Athens, Greece
e-mail: deannope@uoa.gr

E. Michalena and J. M. Hills (eds.), *Renewable Energy Governance*,
Lecture Notes in Energy 23, DOI: 10.1007/978-1-4471-5595-9_9,

1 Importance of the RES Expansion in Today's Financial and Social Environment

Fiscal austerity and economic growth seem to be the most significant challenges which Europe faces today. Many European governments aim at reducing spending, following their extensive bailouts, and stimulus packages, but at the same time they need to be aware of the impact their policies will have on prospects for growth and social cohesion. As the trend shows a clear shift to a resource-efficient and low-carbon economy to address the rising consumption of energy, most European governments are keen on adopting national renewable and low carbon strategies enhancing their competitiveness and growth, since such policies ensure jobs creation, enable achieving energy security, and at the same time combat environmental degradation. While each country is following a different path to renewable energy deployment, governments' strategies and their impact on business investment will play central roles in both the energy sector and the broader economic development. But is it always easy for a government to impose/introduce the appropriate measures that balance rightfully between the need to mitigate the existing fiscal deficits or structural deficiencies and economic growth especially in the energy sector? Do short-term positive financial results also go hand in hand with long-term development and energy efficiency? What are the consequences of inconsistencies observed in such policies, especially in countries facing the most severe fiscal problems, such as Greece, which are gifted with an abundance of renewable resources?

In order to detect the possible inefficiencies of policies adopted by the Greek State, it is necessary to first get acquainted with the basic cornerstones of the existing relevant legal framework in the renewable energy sector.

2 The European Legislative Framework[1]

In Europe, renewable energies are continuing to expand both in terms of investment and geographical spread. In doing so, they increasingly contribute to combating climate change, countering energy poverty, and energy insecurity. The Directive 2009/28/EC on the promotion of the use of renewable energy sources (RES) sets the overall target to reach 20 % of renewable energy in gross final energy consumption in 2020. Reaching this target will require a huge mobilization of investments in renewable energies in the coming decade.

More specifically, Directive 2009/28/EC, adopted on April 23, 2009 (repealing Directives 2001/77/EC and 2003/30/EC), establishes a common framework for the promotion of energy from RES and sets mandatory national targets for the overall share of energy from RES in gross final consumption of energy. It also lays down rules relating to statistical transfers between Member States, to joint projects between Member States and with third countries, as well as to joint support

[1] Metaxas (2010)

schemes and their effects. Additionally, it regulates issues regarding guarantees of origin, administrative procedures, information and training, and access to the electricity grid for energy from RES. The Directive also provides for the adoption of national renewable energy action plans, whereas it also includes sustainability criteria for biofuels and bioliquids. Moreover, it provides for the obligation of the Member States to report on the progress made in the promotion and use of energy from RES and for the formation of a transparency platform. It also includes provisions relating to energy from RES in the transport sector. Member States were obliged to bring into force the laws, regulations, and administrative provisions necessary to comply with this Directive by 5 December 2010.

Other basic European regulation regarding the energy market includes Directive 2009/72/EC of 13 July 2009 concerning common rules for the internal market in electricity and repealing Directive 2003/54/EC. This Directive establishes the common rules for the generation, transmission, distribution, and supply of electricity markets in the European Union. It also lays down the rules relating to the organization and functioning of the electricity sector, open access to the market, the criteria and procedures applicable to calls for tenders, and the granting of authorizations and the operation of systems. Last, it also stipulates universal service obligations and rights of electricity consumers and clarifies competition requirements.

3 Basic Pillars of the Greek Legislative Framework on RES

Although, as already mentioned, Greece is gifted with an abundance of renewable resources, it lies far from fully exploiting them. The liberalization of the Greek energy market is still an ongoing process, which commenced with the implementation of Directive 96/92/EC through law 2773/1999. Since then, there has been considerable regulatory activity aiming at the improvement of the legal framework, the elimination of existing obstacles and the provision of investment incentives mainly through subsidies. Even though many unresolved issues still exist, the opening up of the Greek energy market intended to create significant new investment chances for enterprises, especially those active in the field of RES. Despite the attractiveness of the sector, foreign companies are reluctant to enter the Greek energy market due to the complexity of the pertinent legal framework, the long duration of the procedure for the issuance of the necessary licenses as well as the inconsistency of RES policy, as described in detail below.

3.1 RES: Licensing Procedure (Laws No. 3851/2010 and 4001/2011)

Due to the fact that, under the previous licensing regime (provided for in Law No. 3468/2006), the market uptake of RES in Greece was lower than expected, mainly due to long licensing and administrative procedures and grid-related issues, the Greek Government revised the RES policy framework partly to reduce administrative burdens on the renewable energy sector. The legislation currently in force on Renewable Energy Sources adopted by the Greek Parliament in May 2010 (L. 3851/2010, ΦΕΚ. 85A, 4-6-2010, partially transposing Directive 2009/28/EC), simplifies some of the procedures of the previously existing licensing regime. The law aspires to enable Greece to be provided nation-wide with electricity stemming from RES (up to 2020) up to a percentage of 40 %.

Certain provisions of the abovementioned law regarding the licensing procedure were amended by L. 4001/2011, in force as of 22nd August 2011 (transposing Directives 2009/72/EC and 2009/73/EC). This law also provides for, among others, specific extensive competences of the Regulatory Authority for Energy (RAE) and the complete unbundling of the electricity transmission and distribution systems.

Additionally, provisions included in Law No. 3851/2010 were also supplemented or amended by Article 39 of Law No. 4062/2012 (in force as of 30th March 2012) which fully transposed Directive 2009/28/EC). This Law also provided for an additional bonus of up to 10 % in the Feed-in-Tariff (FIT), based on a Local Content Requirement clause for PV plants using equipment of EU origin.

Apart from these, more general provisions, specific procedures are provided for in the Licensing Regulation by the National Regulatory Authority (RAE) and also in certain Ministerial Decisions (such as Ministerial Decision No. 9154/14.04.2011 on the prerequisites to implement a photovoltaic station on land and buildings) as well as important internal circulars (such as Y.A.Π.E./Φ1/*οικ.* 26928).

The licensing steps for a photovoltaic station of more than 1 MW include application to RAE for a Production License, application to the competent Department of the Region for an Installation License, permission to proceed to minor-scale-interventions by the Department of Urban Planning, application to the System Operator (SO) for grid connection, signing of a contract with the System Operator for selling electricity to the grid and, finally, after the PV system has been installed, application to the competent Department of the Region for an Operation License.

The Production License is issued by RAE, which examines if the criteria provided for in the Law (article 2 of L. 3851/2010) are met and decides whether to issue or not a Production License within 2 months from the application submission date, provided that the application file is complete, otherwise, from its completion. The Production License is valid for up to 25 years and can be renewed for another 25 years. If the Installation License is not issued within 30 months of the issuance of the Production License, the latter is being automatically revoked.

Exceptions can be granted for photovoltaic stations up to 100 KW meant to be operated by farmers, as well as for stations which are being expressly exempt from the obligation to possess a Production License. Moreover, according to the Law (L. 3851/2010), a Production License or any other declaratory decision (known as "exception") is no longer required for photovoltaic systems up to 1 MW and for wind farms of up to 100 KW.

Regarding the pricing of electric energy produced by RES stations, it is calculated based on the provisions of L. 3851/2010 and Art. 27A of L. 3734/2009. The new FITs are guaranteed for 20 years and are adjusted annually by 25 % of the previous year's consumer price index.

However, FITs, as provided for in the above laws, have significantly decreased, through the Ministerial Decision No. Y.A.Π.E./Φ1/*οικ*.2262/31.1.2012, as amended by Ministerial Decision No. Y.A.Π.E./Φ1/2301/*οικ*.16933/09.08.2012 concerning energy produced by Photovoltaics. Furthermore, apart from the significant decrease of the FITs a new tax measure has been imposed on all RES producers, the detailed description of which follows in the next paragraph.

3.2 Article I Par. I.2 of Law No. 4093/2012

The latest legislation regarding energy production from Renewable Energy Sources, adopted by the Greek Government on 12.11.2012 (L. 4093/2012), imposes a retroactive tax levy (so-called "solidarity contribution") on RES producers' turnover for the years 2012–2014. As provided for in the provisions of the aforementioned Law (Art. I para I.2) this special retroactive tax is imposed on electricity producers from RES and C.H.P. (Cogeneration of Heat and Power of High Efficiency), which is calculated upon the price of sale of electricity, during the period 1.7.2012 to 30.6.2014, and refers to both the operating RES stations, as well as to those which will be connected thereafter. This retroactive taxation has raised a series of substantiated doubts, whether the relevant provisions include elements that constitute infringements of EU Law provisions and principles. Not only that, the de facto retroactive character of this measure collides with a basic principle of any good RES governance: The need for predictability and legal safety of the investment environment. (see below, under 4.II.).

4 Legal "Angles" of the RES Legal Framework in Greece and Their Consequences

Despite the fact that the extension of the use of RES in electricity generation is both desirable and necessary and that the Greek Government has made certain efforts to reduce the licensing bureaucracy and eliminate obstacles, mainly by rendering the Grid Connection procedure less complicated, however, the novelties introduced by Law No. 3851/2010 and following amending legislative provisions created a "congestion" of applications, thus rendering the holding of binding timeframes questionable. This is the main drawback RES investors are facing along with other issues not directly related to the grid connection procedure itself but to inefficiencies of the Network Operator.

Obligations of grid operators toward RES producers are binding only after the conclusion of a connection agreement contract. However, this can be signed only if the necessary grid infrastructure has been built, which in its turn can be fulfilled only if the relevant Grid Connection Quotation has been issued and approved, leading to a "chicken and egg" situation that does not enable RES producers to effectively pursue their rights.

4.1 Delay in Providing Grid Connection Quotation Within the Deadline of 4 Months and its Consequences

As already mentioned, pursuant to article 39 par. 6 of Law No. 4062/2012, amending article 187 par. 2 of Law No. 4001/2011, which amended in turn article 3 of Law No. 3851/2010, "the competent operator issues the Grid Connection Quotation within 4 months from the submission of the relevant application." However, what usually happens in practice is that due to (actually existing or alleged) work overload and lack of personnel, this deadline is rarely respected, whereas in many cases the sometimes selective delays extend to almost 2 years from the initial submission of the relevant application.

According to general Administrative Law Principles the use of a present tense or of the words "shall" or "must" shows a clear tendency for an obligation that needs to be fulfilled within the deadline provided.[2] However, according to article 10 par. 5 of Law No. 2690/1999 (Code of Administrative Procedure) the deadlines included in laws are generally indicative as regards the Administration, unless clearly stated as being exclusive. This means that no concrete consequences follow in case the Administration fails to respect these deadlines if no relevant provision exists. In other words, in case the legislator omits to declare that the deadline is exclusive and to provide for specific consequences, then the deadlines provided for are obligatory on one hand but their violation remains without concrete legal implications on the other hand.

This "loose" wording included in the Law regarding the Operator's obligation to issue the Grid Connection Quotation within the deadline of 4 months facilitates the de facto violation of the deadline, since the Operator in many cases delays the issuance of the Grid Connection Quotation. Thus, the investor remains unprotected against the Operator's arbitrariness, not being able to put a pressure on the Operator to respect the provided deadline of 4 months. This results in the investor suffering multiple damages, since he faces unforeseeable delays in the scheduled investment's implementation. Plus, given the fact that the Feed-in-Tariff for RES gradually reduces, the investor usually ends up with getting a much lower guaranteed price for the energy produced by RES than initially scheduled. This very negative result has a domino effect on other aspects as well, since the investor has

---

[2] Dagtoglou (2012)

developed the project's business plan based on different financial data (i.e., bank financing agreements based on different expected revenues, etc.).

What is more, the moratorium recently imposed on the PV licensing procedure, pursuant to the Ministerial Decision No. Y.A.Π.E./Φ1/2300/*οικ*.16932 (Official Government Gazette B' 2317) due to the "coverage" of the national targets set by the Ministerial Decision no A.Y./Φ1/*οικ*.19598/01.10.2010 has significantly affected many investors, whose investments were "trapped" by this moratorium. More specifically, investors who didn't have a binding Grid Connection Quotation from the Operator, despite the fact that they submitted the relevant application almost 2 years ago in some cases, can no longer proceed with implementing their investments. Therefore, the violation of a deadline, which—as already mentioned—is regarded as indicative, has lead to a semi-permanent situation where investors suffer huge financial losses and are unable to complete their investments, despite having already invested huge amounts of money in, i.e., buying or renting the land and possibly some equipment. At the same time, the only protection they can have against this situation they found themselves in is to pursue compensation for the losses suffered due to this delay through the time—and money consuming—legal path. However, this path is also not very easy to follow, since they have the burden of proving the extent of the damage suffered and of establishing the causal link between the Operator's behavior and the damage suffered.

It is also worth noting that this awkward situation has caused a difficult constellation for the Regulator. Many investors have filed complaints to RAE regarding the repeated delays as regards the issuance of the Grid Connection Quotation by the Operator. However, RAE has rejected[3] such complaints by stating that "the provided deadline of 4 months cannot be applied." Such a statement is problematic from a legal point of view, since it fails to grant the necessary normative value to the respective legislative provisions that are violated by the Operator. So the Operator remains de facto unpunished for the relevant breach and investors de facto unprotected.

4.2 Retroactive "Solidarity" Tax on RES[4]

Apart from this still unresolved issue in the licensing procedure, which creates uncertainty as regards the timeframe for the execution of RES investments, the recent imposition of the retroactive tax measure on RES producers' turnover further destabilizes the investment environment. Moreover, the relevant legislative provision arguably infringes EU Law. More in particular, the measure in question is considered to constitute illegal State aid, as well as being incompatible with the EU Energy Law and, specifically, clear RES development Policy objectives and principles.

---

[3] See indicatively RAE's Decision no 750/2012.

[4] Kapros (2011)

4.2.1 Illegal State Aid Within the Meaning of Art. 107 I TFEU

Wholesale Energy Market[5]

In order to better understand why the imposed tax constitutes a state aid measure favoring fossil fuel energy producers and energy suppliers, a quick overview of the Greek electricity wholesale market and its basic elements is deemed necessary. The Greek wholesale electricity market is based on a day-ahead unit commitment market clearing and generation dispatch formulation. On the day ahead, injectors, i.e., producers, importers, hydro, and RES, are mandated to place their energy and reserve offers, their techno-economic declarations defining the features of the units they are using to offer energy or nonavailability declarations in case they are unable to offer energy for the dispatch day. At the same time, energy suppliers place their demand bids. The objective is to minimize the cost of balancing the energy to be absorbed with the energy to be injected in the system, while meeting the reserve requirements and the generation units' technical constraints.

The Day Ahead Scheduling (DAS) solution is provided by the Operator of the Electricity Market, which is responsible for clearing the transactions within the DAS (Article 118 of Law No. 4001/2011), determining the clearing prices of the energy, i.e., the System Marginal Price (SMP), and of the reserves. Thereafter the Operator of the Electricity Market ranks the offer bids in an offer curve by increasing price, while it ranks the demand bids in a demand curve by decreasing price. The quotations should reflect at least the unit's variable cost. The intersection of the demand and the offer curves sets the SMP—all generators that offered bids lower than the market price are scheduled to inject energy, while all buyers that offered demand bids higher than the market price are scheduled to be served.

Support Mechanisms

Apart from the mandatory wholesale pool, conventional electricity producers from natural gas are remunerated through additional mechanisms, i.e., the so-called "Variable Cost Recovery Mechanism" (VCRM), as well as a Capacity Payment Mechanism (CPM). The VCR mechanism was initially introduced as a temporary measure in order to compensate new private entrants in the energy market and provides for additional payments, so that the natural gas unit committing energy for the DAS ends up with a profit equal to 10 % of its variable cost, in case this profit is not reached through the market revenues for energy. In other words, the recovery mechanism provides for the natural gas fired units to always recover their commitment costs, thus eliminating a potential deficit and "neutralizing" financial dangers linked with possible distortions of the Greek electricity market. As far as the Capacity Payments Mechanism is concerned, it is based on the obligation of the suppliers to present sufficient guarantees. Each supplier, self-

[5] Andrianesis et al. (2012)

supplying customer and exporter (hereon: “load representatives”) is assigned a Capacity Adequacy Obligation, according to the energy consumed by each load representative during the periods of increased probability of loss of load, as calculated by the SO. Furthermore, each producer issues, for each of his units, Capacity Availability Tickets. The total number of Tickets issued for each unit equals its net capacity. The Tickets constitute a call to load representatives for the conclusion of Capacity Availability Contracts. Currently, only a Regulated Mechanism (RM) is offered to the market participants. The producers and the load representatives participate in the Capacity Availability Market by concluding Contracts with the SO. In this case, the producers receive a regulated price for the available capacity with which they participate in the RM.

RES Special Account

In Greece, renewable energy generation is mainly promoted through a guaranteed FIT system, the main rules for the application of which have been set by Article 13 par. 1 of Law No. 3468/2006, as amended by Article 5 par. 2 of Law No. 3851/2010. Moreover, the Operator of Electricity Market (LAGIE) and the Electricity Transmission Grid Operator (DEDDIE) collects the amounts paid to RES electricity producers from a Special Account (so-called “RES Special Account”), which is financed by:

(i) The amounts paid by the electricity producers and suppliers, corresponding to the renewable energy injected in the transmission system and distribution grid situated in mainland and on the interconnected islands,
(ii) The amounts paid by the Suppliers of the Noninterconnected Islands for the power absorbed in these islands’ systems,
(iii) The Special Levy for the Reduction of Greenhouse Gases (SLRGG), which varies among different Customers’ categories, and which is paid by every single electricity consumer in the country.

Moreover, the lignite fired units are imposed a special levy of 2 Euros per MWh of energy produced. This special levy is also included in the Special Account’s resources. This Account is exclusively intended to cover the cost difference between the higher FIT paid to RES producers and the SMP paid to LAGIE. This means that the receivables of the Special Account should equal the total RES costs that LAGIE must pay to all RES producers for the total amount of energy generated by their RES and cogeneration power plants for high-efficiency heat and energy in Greece (mainland and noninterconnected islands) at the FIT which applies under their power purchase agreement.

Wholesale Electricity Market Distortions

Due to the fact that the SLRGG was initially introduced 12 years ago and has been always calculated in the same way, it fails to reflect the real additional cost/benefit of meanwhile higher RES penetration: According to two scientific studies performed by the National Technical University of Athens and the Greek

Foundation for Economic and Industrial Research (IOBE), the fees collected through SLRGG correspond only up to a percentage of 40–50 % to compensations granted to the RES producers. The reason for that is that SMP (which is crucial for the fair pricing of the SLRGG) does not represent the costs avoided by the suppliers due to every RES electricity megawatt-hour substituting a conventional electricity megawatt-hour. This happens for a number of reasons. Indicatively:

(1) *The merit order effect*. As the RES produced energy contributed in the DAS increases, the SMP decreases. This is due to the fact that the more RES production increases, the less more expensive electricity producing units are needed to cover the demand in electricity. Since the SMP is determined by the most expensive unit to provide electricity every hour, the increase in the energy produced by RES results in the reduction of the SMP.
(2) *The RES capacity credit and the subsequent value due to RES*. RES provide capacity to the System and therefore, save capacity payments.
(3) *The distortion of the SMP due to Public Power Corporation's (PPC S.A.) dominant position*, which keeps it at an artificially low level. The fact that the SMP is distorted is largely due to the fact that PPC is currently the dominant electricity generator and supplier company in Greece. At the electricity generation level, PPC operates a quite diversified portfolio of power plants including lignite, natural gas, and large hydro, while its competitors have been permitted to invest only in gas power plants. At the retail level, PPC is virtually the only supplier since small alternative suppliers only hold an insignificant market share.

The motives for PPC retaining this artificially low SMP can be summarized in the following: (a) to contribute less, as a supplier, to the RES Special Account, since the suppliers only pay the SMP for each RES megawatt-hour fed into the system, and (b) to pay less, as a supplier, for electricity imports, since the latter are also paid only the SMP. If the VCR Mechanism would not exist, then PPC would also pay—as supplier—for the natural gas power units the SMP (and the Capacity Payment).

What actually happens is that the suppliers (currently mainly PPC) pay, through the wholesale mandatory pool, an unfairly low amount toward the entities which inject energy in the system (i.e., the conventional producers, the RES producers, and the importers) due to the distorted SMP, creating thus a deficit on the generation side as a whole. In order to partly compensate for this deficit, PPC pays an additional amount only to private natural gas fired power producers through the VCRM. By doing so, the whole deficit on the generation side (resulting from the distorted SMP) is transformed to a deficit in the RES Special Account only. On the contrary, this additional amount paid through VCRM should have been used proportionally to reduce the whole deficit existing on the generation side and not to protect only the conventional producers.

This results in the alleged need for the imposition of the tax levy in question "in order to limit the deficit of the RES Account."

Why the RES Tax Constitutes Illegal State Aid according to Art. 107 I TFEU[6]

The tax measure constitutes a state aid measure in the meaning of Article 107(1) TFEU, since it is found to meet all the 4 criteria included in that provision. More specifically:

1. *Transfer of state resources:* The measure is attributed to the State. The RES tax has been imposed by Law No. 4093/2012. Furthermore, the revenue from the RES tax comes under the control of the state, since it is collected by the Electricity Market Operator (LAGIE) and the Distribution Grid Operator (DEDDIE) and credited to the RES Special Account which is managed by LAGIE. LAGIE and DEDDIE are also controlled by the State (the latter as far as its duties regarding the Noninterconnected Grid Administration are concerned) and their duties with respect to the RES Special Account are mandated by Law No. 4001/2011 (articles 118 par. 2 and 129 par. 2). Additionally, the State forgoes potential revenue. The RES tax is not imposed on all energy producers, i.e., producers of electricity from fossil fuels, which is unfair. According to standard practice, the forgoing of tax revenue is equivalent to "consumption" of state resources (T-67/94, Ladbroke v Commission; C-66/02, Italy v Commission).
2. *Economic advantage:* It is well established in the case law that exemption from tax or nonliability for tax confers an advantage to undertakings which would otherwise have to bear the tax in their budget (C-6/97, Italy v Commission; C-487/06 P, British Aggregates v Commission). Producers of electricity from fossil fuels are not liable for the RES tax and therefore obtain an advantage in relation to their competitors (electricity producers from renewable sources).
3. *Effect on trade and distortion of competition:* Since electricity is traded across Member States (C-379/98, Preussen Elektra) it cannot be excluded that trade between Member States may be affected. Moreover, it cannot be excluded that there can be an indirect effect on trade because thousands of Greek undertakings use nontaxed electricity produced from fossil fuels as input into their operations. Electricity produced from fossil fuels accounts for more than 80 % of total electricity consumption in Greece (for a similar analysis of indirect effect on trade see Commission Decision SA.21918 on electricity tariffs in France, OJ C 398, 22/12/12). It follows that competition between producers of electricity which are not subject to the tax or undertakings which use nontaxed electricity and their competitors in intra-EU trade is distorted.
4. *Selective measure:* The beneficiaries of this solidarity are not specified, nor is the need for it explained. However, Law No. 4093/2012 stipulates that the proceeds from the tax will be credited to the "RES Special Account." It can be inferred, therefore, that the "solidarity" intended by the RES tax is solidarity with the RES Special Account, and it is needed in order to reduce the budgetary deficit of the RES Special Account. Although, as already mentioned, terminology issues ("solidarity contribution") and intentions linked with the

[6] Hellenic State Aid Institute (2011)

adoption of a state measure are absolutely irrelevant as regards its legal categorization as a state aid for which only its effects are crucial, given the fact that the remuneration mechanisms of all electricity producers and the interdependence of various structural distortions of the Greek electricity market contribute jointly to the formation of the existing deficit in the RES Special Account, it must be concluded that if the aim of the tax is to strengthen solidarity with the RES Special Account, then the tax should apply to all those operators.

In any case, since the scope of the RES tax should extend to all electricity operators, then it must be inferred that the system differentiates between undertakings which are in the same legal or factual situation. This differentiation is arbitrary because it is not based on any objective distinction between electricity undertakings. The RES Special Account was established in the context of the Greek policy of encouraging production of electricity from renewable resources and reducing greenhouse gas emissions. Since producers of electricity from conventional fossil sources also contribute to greenhouse gas emissions, they are in a comparable legal or factual situation and should have also been subject to the RES tax. There is no objective reason for their exclusion. Even if it cannot be established a priori which undertakings should contribute to the solidarity aims of the RES tax (and therefore it cannot be defined which undertakings should be subject to the tax and which should be excluded), it can still be shown that the tax in its application differentiates between undertakings which are in comparable factual situation because they are competitively interlinked.

In addition, this differentiation cannot be justified from the nature of the contested RES tax. On the contrary, the aim of strengthening solidarity with the RES Special Account would require that the tax is levied on all producers who impact on the budgetary deficit of the RES Special Account. Moreover, since the very purpose of the RES Special Account is to internalize the environmental costs of conventionally produced electricity, it would follow that it must be the producers of electricity from fossil fuels that should be making additional contributions to cover the deficit.

4.2.2 Incompatibility of the Measure with the EU Normative Framework on RES as Well as EU Energy Policy Goals

Besides constituting a selective state aid measure, as established above, the tax measure in question also infringes provisions of the EU Energy Law and is found to be incompatible with the EU Energy Policy. The European Commission has repeatedly criticized the application of retroactive measures and has highlighted the negative effects that such measures may have in promoting investment in RES in the context of reaching EU Energy and Environment Policy targets in 2020. The imposition of this tax measure on the gross revenues of all already operating RES projects in Greece (apart from those for which the reference value of energy sold taken into account for the Power Purchase Agreement is being calculated according to the one in force after 09.08.2012) is moving in the opposite direction from the aforementioned EU Energy policy. Moreover, the European Parliament and Council have repeatedly stressed (both in the Directive 2009/28/EC and the relevant Position Paper) the importance of retaining a stable investing environment for RES and the need the energy prices to reflect the external costs of energy production and consumption, "including, as appropriate, environmental, social and health care costs." Additionally, the Commission, in its Communication to the European Parliament, the Council, the European Economic and Social Committee, and the Committee of the Regions, titled "Renewable Energy: a major player in the European Energy Market" of 6th June 2012, renounces policies that hinder investment in renewables and in particular, policies that continue to subsidize fossil fuels, which, according to this document, should be phased out. In view of the complementarity of climate and renewable energy policies, a well functioning carbon market is deemed to be necessary, together with "properly designed energy taxes to give investors clear and strong incentives in low carbon technologies and their development." Further on, the Commission considers that the retroactive changes suddenly imposed on support schemes, despite the fact that they are often triggered by unexpectedly high growth, which rapidly increases expenditure rendering them not sustainable in the short term, undermine investor confidence in the sector.

4.2.3 Effects of the Tax Measure on the RES Producers

The adopted tax measure has a retroactive effect in the sense that it did not exist, nor was it known at the time of the conclusion of the Power Purchase Agreements between the RES producers and the competent Operator and of the launching and financing of the relevant investments. Thus, the producers find their investments "captured" in a scheme that significantly reduces their income, whereas at the same time they have organized their business plans and scheduled their obligations according to the predicted income, which of course did not consider the later imposed tax.

This situation, coupled with the fact that the Market Operator pays the producers with a significant delay, today reaching a period of more than 5 months since the relevant invoices become due, worsens the situation even more, since many RES producers face serious problems of viability, with the banks on the one hand asking for default interest on top of the monthly installments and the tax further diminishing their income and hindering their financial situation.

5 Discussion–Conclusions

It becomes clear from all the above that the current legal framework regarding RES is suffering from over regulation and complexity, leading to a lack of investment safety, irrational bureaucracy, high implementation costs, and unforeseeable delays. It is lacking strategic structure, rather forming a patchwork consisting of provisions from previous not so successful laws, therefore not leading to the covetable growth.

The national policies for RES in Greece seem to have been set rather superficially, without properly evaluating their consequences on the market operation, the economy, and the society. The RES regulatory framework seems to be moving in the wrong direction and despite the natural and geographical advantages Greece presents as far as RES development is concerned, growth and competitiveness in the economy that could very well arise out of a well-structured energy policy are still lagging behind.

It is expected that in the near future the photovoltaic production cost will equate the grid parity in other European countries, therefore leading to the abolition of the feed-in-tariff system with RES energy production entering the competitive energy market. It is essential, therefore, to simplify the RES legal framework, in order to adapt to the market rules for the economy and consumers' benefit.

Acknowledgments Many thanks go to my colleague at Law Firm "Metaxas and Associates" Ms. Chr. Sarantidou, Attorney At Law, LL.M. (NYU) for her assistance in drafting this chapter as well as to Ms. S. Katemoglou for her editorial support.

References

Andrianesis P, Liberopoulos G, Kozanidis G, Papalexopoulos AD (2010) Recovery mechanisms in a joint energy/reserve day-ahead electricity market with non-convexities. In: Proceedings of EEM10 conference, Madrid, Spain, 23–25 June 2010

Andrianesis P, Liberopoulos G (2012) The "Hidden Cost" of renewable energy sources in electricity pool markets. In: 9th international conference, 10–12 May 2012

Andrianesis PE, Liberopoulos G, Papalexopoulos AD (2012) Greek wholesale electricity market: forthcoming market changes and bid/cost recovery. In: 8th Mediterranean conference on power generation, transmission, distribution and energy conversion, MEDPOWER 2012

Dagtoglou PD, General Administrative Law, 6th Edition 2012, p. 158.

Hellenic State Aid Institute (2011) State aid. Basic normative texts. Nomiki Vivliothiki, Athens

Kapros P (2011) The special RES levy: analysis and predictions. http://www.eletaen.gr/drupal/sites/default/files/meleti_emp.pdf

Metaxas A, EU Energy Law in G. Calavros—Th. Georgopoulos (2010) European Union Law (II). Nomiki Vivliothiki, Athens

Maroulis G, Piria R, Zane EB (eclareon), Frank R, Bauknecht D (Öko-Institut) (2011) Integration of electricity from renewables to the electricity grid and to the electricity market—RES-INTEGRATION. http://www.oeko.de/oekodoc/1378/2012-012-en.pdf

Sakellaris K (2010) The greek capacity adequacy mechanism: design, incentives, strategic behavior and regulatory remedies. http://ideas.repec.org/p/pra/mprapa/24642.html

Lost in the National Labyrinths of Bureaucracy: The Case of Renewable Energy Governance in Cyprus

Paris A. Fokaides, Andreas Poullikkas and Constantinos Christofides

Abstract In July 2011, Cyprus experienced an abrupt disruption of power supply as a result of an explosion that happened near the biggest conventional power station of the island, which destroyed about 60 % of the its power generating capacity. This was a major drawback for the Cypriot economy that could turn into an opportunity to restructure Cyprus's energy mix towards a more sustainable one by further promoting renewable energy sources. However, due to the time-consuming licensing procedures which exists in Cyprus nothing has materialised. This chapter searches for the reasons for this situation and examines the potential of the single stop principle for national renewable energy governance. Within this context, a simple numerical model is considered; what-if scenarios employing the single stop rationale by means of time and cost savings are quantified and examined, and specific suggestions for the improvement of the licensing procedures are performed.

1 Introduction

The promotion of renewable energy technologies (RET) is a policy that has been established in recent years by the vast majority of developed countries. As every other construction, the overall design and installation of a RET development must

P. A. Fokaides (✉)
Visiting Lecturer in the Frederick University, 7, Y. Frederickou street 1036 Nicosia, Cyprus
e-mail: p.fokaides@frederick.ac.cy

A. Poullikkas
American University of Sharjah, P.O. Box 26666 Sharjah, United Arab Emirates
e-mail: apoullikkas@aus.edu

C. Christofides
University of Cyprus, P.O. Box 20537 1678 Nicosia, Cyprus
e-mail: ccc@ucy.ac.cy

E. Michalena and J. M. Hills (eds.), *Renewable Energy Governance*, Lecture Notes in Energy 23, DOI: 10.1007/978-1-4471-5595-9_10,

follow some rules which will ensure that this will be constructed in a legitimate and sustainable way. The practice which has been followed in most cases is the adoption of licensing principles which are similar to other development projects, such as of conventional power plants. However, the case of RET presents many complex issues and complications. The need for rapid entry of RET in the energy systems as well as specific features of these applications turn their licensing process into a complicated task.

This chapter examines a particularly interesting case: the RET licensing system of an EU Member State, Cyprus, which has several peculiarities: the country has small, insular energy system, and has no previous experiences in this area. In addition to these parameters, the state witnessed the destruction of 60 % of its energy production system due to an accident that occurred few months after the adoption of its national action plan for the promotion of energy production from renewable sources (Directive 2009/28/EC), the 4th largest accidental explosion worldwide (Wikipedia 2013). How compatible are the existing RET licensing procedures in this EU state and what principles should be adopted in order for them to be more effective? What makes RET licensing so much time-consuming and which alternatives could be put in place to attribute the Government the power of speed and reliability when it comes to energy sustainability?

2 Cyprus Energy System After Accession to the European Union

Cyprus, a small Mediterranean island European state, possesses an energy system with explicit characteristics. The main feature of the energy system of Cyprus is the fact that it is an insular energy system. This isolation, coupled with the fact that Cyprus is a state oriented to service provision with limited technological potential, led to the development of an energy system based solely on conventional energy sources. For many decades the power industry in Cyprus developed on the basis of available technology and know-how, and today it constitutes a key sector of the economy. Until 2004 the Electricity Authority of Cyprus (EAC) was responsible for the generation, transmission and distribution of electricity in Cyprus.

This situation, however, changed and the electricity market in Cyprus is now open. A Regulator's Office and a Transmission System Operator have been appointed and new participants are expected to join the electricity sector in the future. However, at the moment, EAC is still the sole conventional producer of electricity on the island (Electricity Authority of Cyprus 2012). The reason has to do with the fact that the development of infrastructures in a small system requires economies of scale, which the size of Cyprus does not allow. Also Cyprus does not yet possess its own energy resources, resulting to high fuel import cost. Indicative is the fact that Cyprus and Malta today have the highest energy production cost in the EU27 (Directorate General for Energy, EU 2011). All these reasons make the feasibility of new conventional power stations in Cyprus doubtful.

The accession of Cyprus in the European Union was a milestone for the energy system of the island. Cyprus is obliged to follow European directives in the field of energy, resulting in the mandatory upgrade of its energy system (Koroneos et al. 2005). The consequences of these changes were many, with the main being the following:

1. The liberalisation of the energy market
2. The establishment of national policies and action plans for the promotion of renewable energy
3. The adoption of energy saving strategy.

This chapter examines the ways in which the RET projects licensing procedures in Cyprus cause problems and how this may be accelerated. A presentation of the current energy status in Cyprus, in terms of the increase rate of RET projects is conducted. The story as well as the consequences of the major accident at Mari are discussed, and the policies that were adopted after the accident to speed up the entry of RET in the energy system of Cyprus are analysed. The existing licensing process of RES projects in Cyprus is also presented. Two alternative models for RET project licensing are introduced and discussed. These models are compared as to their effectiveness and practicality. We have indicatively chosen to demonstrate a model and the choice of a specific model is finally justified by means of a simplified numerical model, based on the queuing theory.

3 Cyprus Transition to a New Energy Regime

With regard to renewable energy sector, Cyprus since 2004 proceeded to adopt a range of measures to promote alternative sources in its energy system. The state has harmonised its legislation and adopted all European policies related to this field. In the power sector, the promotion of renewable energies was realised in 2006 with the licensing of the first wind farms. From 2010 onwards, four wind farms with a cumulative power of 140 MW, that cover 2 % of the island's needs, are licensed and operated. The biomass sector followed with the licensing and operation of 8 MW power plants, using biogas derived from livestock waste.

Photovoltaic (PV) presents another RET alternative which contributes significantly to the energy system of Cyprus. The energy department of the Republic of Cyprus moved in 2009 to the authorisation of medium- and large-scale PV power plants. So far, 12 MW PV systems are installed and operated. The outlook for the coming months includes the licensing of at least another 80 MW of PV installations by the end of 2013. The production from this installed capacity is expected to reach 130 GWh per year, which is almost the 3.5 % of the annual energy requirements in Cyprus.

The Republic of Cyprus, recognising the rapid development of the RET applications sector on the island, moved to the adoption of a very ambitious target for 2020. In the context of the action plan for the promotion of RET in the energy system of Cyprus under European Directive 2009/28, a plan that provides the production of 16 % of electricity needs from renewable sources by 2020 was placed. This plan was approved a few months after the adoption of the Directive on promotion of the renewable energy sources and specifically in mid-2010. The provisions of the action plan refer to the promotion of

- wind farms up to 300 MW,
- PV installations up to 192 MW,
- biomass plants up to 17 MW and
- solar thermal power plants at 75 MW.

The Cyprus national action plan provides precisely defined annual time steps for the application of the different installations. In the first 3 years of its implementation, the national action plan has been applied with great precision.

4 The Big Shock of Mari

Having achieved so much in the energy sector in such a short time, the Cypriot society experienced a major setback in July 2011. A large accident that occurred near the power station of the EAC in Mari incapacitated about 60 % of the installed capacity in the island. The accident was caused by the explosion of a cargo of munitions that had been seized a few years earlier, and were kept at a military facility, adjacent to the power station. The explosion severely damaged hundreds of nearby buildings including all of the buildings in the island's largest power station. According to a list of the largest artificial non-nuclear explosions, the explosion was the 5th largest accidental explosion and 8th largest in the world.

The accident at Mari created several major problems in the power generation system of Cyprus, namely:

- The remaining power stations were not able to meet the full load needed to meet the energy needs of the island. As a result, much of Cyprus was without power in the immediate aftermath of the incident and rolling blackouts were initiated in order to ration out the existing capacities.
- The fact that the explosion occurred in the summer further exacerbated the problem. Cyprus, as a summer dominant environment, has its peak loads in the summer, due to cooling demand. Therefore, the accident was at that time of year where energy needs were at the highest.
- The accident happened in the tourists' high season, resulting to significant consequences for the tourist section.
- Due to the cost of the damages, as well as the temporary change of the energy production fuel (diesel) the cost of electricity increased considerably. This was

another financial shock to the island's economy that coincided with ripples of the European financial crisis.

All the above factors contributed to the need for a quick response, in order to recover the energy deficit in Cyprus. Unsurprisingly, all the attention focused on the renewable energy sector. Having already the intention to cover 16 % of its energy needs with renewable energy sources, and due to the already great interest demonstrated by individuals in the field, the vast majority of the stakeholders supported the acceleration of the licensing procedures for RET projects (Cyprus Energy Regulatory Authority 2011). The interested parties came though across a small surprise: the state was not able to determine precisely the number of projects that were already licensed as well as the number of projects that were still in the process of getting all their licenses. The answer to this question was also not possible to be given, in view of the very tangled way of licensing RET projects in Cyprus. This fact led to considerable concerns, and initiated a discussion on licensing procedures of RET projects. It was more than obvious that a great necessity existed for a single stage licensing procedure.

5 Current RET Projects Licensing Procedure

The first RET projects in Cyprus were licensed in accordance to the procedures applicable for any construction project in the state. The competent authorities decided that for each RET project, following procedures would be secured (Kordatos et al. 2012):

1. Planning permission from Town Planning Authority
2. Building permit from Construction Authority
3. Final approval from Land Registration Authority

Beyond the above licenses, especially for RET projects, the following licenses would also be requested:

1. Construction and operation licenses from the Cyprus Energy Regulatory Authority
2. Connection permit with the central network from the transport network regulator
3. Contract with the EAC for disposal of the produced energy at the electricity production cost
4. Subsidy agreement with the state fund for supporting RET projects.

For several of the above licenses, the competent authorities should seek opinions from third services. For example, the issue of planning permission requires the collection of views from numerous other departments, such as the Department of Environment, the competent community, the military authorities, the Chamber of Engineers, the Energy Service, the Water Department, etc. Also it is fairly common, an authority to be asked to give its opinions on the same task in more than one occasions. The local community for example which will host a project will be asked to give its views on the project at least in four different occasions. All departments that obtain some development consent for RET projects in Cyprus are listed in detail in Kordatos et al. (2012). As a result, the average time to issue a comprehensive permit for a RET project is between 12 and 24 months. This greatly decelerates the development of RET projects in Cyprus, resulting to a series of problems and negative consequences for the investors in the renewable energy sector, the most important being the following:

1. Having no clear timetable for the implementation of their projects, the investors face significant difficulties to secure the funding for their projects. Also in this context, investors are forced to set aside funds for a significant time period, resulting in economic losses.
2. Due to the rapidly changing technology in the field of RET, the market developments in the technology of the equipment are forcing investors to redesign their works in progress, as there is a long interval between the beginning of the licensing process and the implementation of the project.
3. A climate of mistrust in the market to investments in the renewable energy technologies has been established, due to the unclear framework of their promotion.

It is evident that a reference point, an authority to coordinate the licensing of projects, is currently missing. When for example an interested party submits an application for a planning permission of a wind farm, this is not in the knowledge or approval of any other authority. The competent authority, e.g. the Energy Service will not have a record of the number of individuals and companies that have applied to Town Planning Authorities for licenses, and vice versa. This has apparently resulted in the creation of a small chaos in the permitting process, which was in practice revealed after the accident at Mari.

Based on the facts that followed the explosion at Mari, it was more than evident that the development of RET projects could not continue following the existing licensing procedure. The existing licensing procedure, while it has proved in the past to be very effective for building construction and other development projects, in practice it was completely ineffective for RET projects. The reason was the fact that this process was developed for another construction sector, the building construction sector. Therefore the principles of this process were tailored to a different region, without taking into account the peculiarities of the renewable energy sector. Though a relevant public discussion is still not drawn in Cyprus, the necessity for the establishment of a "one-stop-shop" licensing procedure was more than obvious. Based on which principles should a single stop shop be established and what powers and duties should a single licensing authority of RET energy projects have? Are the principles of a single stop service sufficient to boost the licensing procedure of renewable energy project in Cyprus or should another solution be discussed?

6 Single Stop Licensing Procedures

6.1 "One-Stop-Shop" Licensing Procedure

The core principles of a "one-stop-shop" service are:

- The issuance and control of renewable energy projects that seek to be licensed
- The acceleration and simplification of procedures, involving an reduction in expenses for the interested party, as well as for the state to examine the applications
- The support and consultation of stakeholders in developing such projects.

6.1.1 Issuance and Control

The example of Cyprus is particularly remarkable, regarding the potential integration speed of RET in the energy system of a country. Cyprus, before its accession to the European Union in 2004, had not a single kWh to show in its energy balance produced by RET. Contractual obligations towards the Community Acquis forced Cyprus to move rapidly and to adopt very ambitious targets. However, the high engagement of electricity produced by RET faces in general some challenges.

The accommodation of the RET produced energy requires the existence of a conventional back-up, especially in energy insular systems, mainly due to the fact that renewables are weather dependant. In such cases, the so-called cycling occurs (Bennett et al. 2011). *Cycling* is the restriction of the conventional production in favour of renewable generation. Cycling disturbs the steady and regular operation of power plants to the extent that the benefits of employing renewable generation are counteracted. Another unwanted possible outcome of cycling may be the plant damage that would lead to tremendous repair costs (Scott and Lee 2011). Based on the aforementioned reasons, the control of the rate of entrance of RET energy in the energy system of Cyprus deemed necessary, mainly to ensure the stability of the energy system of the island, as well as the avoidance of cycling effects.

6.1.2 Acceleration and Simplification of Procedures

The average time to issue all necessary permits for a RET project in Cyprus is currently lengthy. A "one-stop-shop" would lead to simplification of procedures and reduction of the necessary licensing duration. According to the data which will be analysed in the next section, the average time needed for permitting a renewable energy project could be less than 3 months, while the corresponding duration today is over 1 year. It is remarkable the fact that even in 2013, more than a year after the accident at Mari, and despite the intention of the state to expedite the process, complex procedures still result in time-consuming and ineffective procedures. The acceleration of processes is expected to result in their simplification. Cases where redundancies and duplications were observed could be eliminated. Also the simplification of the licensing procedures would provide transparency and better control of renewable energy projects.

6.1.3 Consulting and Support

A centralised service which records all cases and licensing requirements for green energy projects could also provide answers to any relevant questions. This should lead to an improvement in the quality of the licensing process and the timely response to all the questions of interest in green energy projects. Also such a service could guide stakeholders towards projects that are of greater national interest but for which there are is not enough interest. For example, in Cyprus it is already known that the objectives in the field of PVs for 2020 have already been exceeded based on the interest shown and the projects that are to be licensed. Instead, the 300 MW of wind projects that are targeted for 2020, are yet to be completed.

6.1.4 Cost Reduction

In a time of economic recession reducing the states operational costs is considered essential. Merging authorities which offer the same or similar services helps to rationalise the financial costs of the state. The function of a one-stop shop could have a positive impact on economic indicators of Cyprus by eliminating services that offer similar services and transfer staff to more productive and useful working positions.

6.2 Contra One Stop Shop Services

Despite the presented benefits of the "one-stop-shop" principles, there are also some considerations regarding their applicability for renewable energy projects. It is a fact that in most cases, the licensing of a renewable energy project requires in practice the involvement of more than one service. Let us take the example of a wind farm. Is there another authority, more appropriate than the Department of Environment of a state, to decide whether the operation of a wind farm can have serious negative impacts on the environment? A "one-stop-shop" service would have answered this question, either by establishing a second authority similar to the existing environmental assessment service for examining such questions or by applying the question to the existing governmental agency. Let's see for example the issue of the smooth integration of a solar thermal power plant to the transmission network. Who is responsible to decide on the technical adequacy of such facility to synchronise with the existing network and the necessary terms for its interconnection? Should, the "one-stop-shop" service create a department, staffed with technicians and engineers for this purpose? Or should the existing transmission system operator of the state be used to determine the technical compatibility and connectivity of the facility?

In answering the above questions, it is concluded that *while theoretically the "one-stop-shop" service simplifies licensing procedures, there are also many aspects of a RET project that cannot be effectively handled by a single stop service.* Both the technological demands in renewable energy projects, as well as specific particularities would have required the establishment of a new service with integrated functions. This in many cases would have result in replacing other similar existing services operating in all sectors, not only in the renewable energy sector. The second option would be the operation of a service centre that would manage and coordinate all agencies and all proceedings, and which would act as a reference point. Obviously between the two above solutions, the administrative centre seems to be the more viable and practical compared to the establishment of a new service.

6.3 Service Management Centre Versus "One-Stop-Shop"

Having referred in the previous section the basic principles that should underlie the operation of a service management centre, a debate over the powers of such a centre for RET projects is initiated.

1. The most basic requirement that a service management centre should fulfil is that it should be the only point at which an interested party for RET project licensing should contact from the first to the last day of the issuance process.
2. This centre would record in a transparent manner and should, at any time be able to inform the interested individuals and the state about the number of the submitted applications for each category, as well as the entrance of new renewable energy projects. It should also be an independent, non-profit body, where decision making should be publicly available for scrutiny.
3. The service management centre would also ensure the reduction of time, through the identification of overlapping procedures as well as the coordination of services. The centre would be able to guarantee a maximum time of examination of each application, which realistically would not exceed few months. Also through the operation of such a centre, it could be able to be ensured that no application would enter the evaluation process if not all necessary examination information is attached.
4. The service management centre would also be able to work as an information point, based on the same rationale described for the one-stop shops.

Obviously the service management centre should not obligatory be a state agency, but could be a private entity which would have the responsibility to fulfil the above mission, as the image of public bodies in the states of the European south is rather negative. The centre could also function as a research and development organisation, as well as a dissemination carrier for spreading out knowledge on the sector of RET. Successful operation of such a service management centre could be adopted by other Member States of the European Union, which at this stage face similar problems, namely multifaceted responsibilities and lengthy licensing procedures.

7 Queuing Theory Models for Establishing Single Stop Procedures

Having explained the necessity of establishing single stop procedures, the importance of upgrading, the existing procedure of permits issuance for RET projects will be examined by means of quantitative analysis tools. For this purpose, the queuing theory model will be applied. Queuing theory is a well-established mathematical study of waiting lines, aiming to manage queue lengths and waiting times.

7.1 Queuing Theory Principles

In queuing theory, service systems are usually classified in terms of their number of servers and the number of service stops that must be made. A single channel system, with one server, is typified by the drive-in service that has only one open service point. If, on the other hand, the service has several service points on duty and each applicant waits in one common line for the first available service point, we would have a multi-channel system at work. A single phase system is one in which the customer receives service from only one station and then exits the system. A multiphase system requires the customer to visit more stations before leaving the system (Render et al. 2011).

The licensing procedure for renewable energy projects in Cyprus is distinguished in the following two cases:

1. Current renewable energy projects licensing procedure, considered as a single channel, multiphase procedure.
2. Suggested one stop shop licensing procedure or service management centre, considered as a single phase, single channel system.

Other assumptions that are made for the modelling procedure are the following:
Applicants

1. Applicants are served on a first-in first-out basis.
2. There is no baulking or reneging.

Arrivals

1. Arrivals are independent of preceding arrivals.
2. The arrivals rate remains constant, and it is described by a Poisson probability distribution.
3. Arrivals come from an infinite large population.

Service times

1. Service times vary from one application to the next
2. Service times are independent of one another,
3. Service times' average rate is known.
4. Service times occur according to the negative exponential probability distribution.

Service rate

1. The average service rate is greater than the average arrival rate.

7.2 Evaluation of Licensing Duration Based on Queuing Theory

Based on the above conditions, a series of equations that define the queue's operating characteristics is developed. Using the Kendall notation (arrival distribution/service time distribution/number of service channels) (Render et al. 2011), following cases are examined:

- Current renewable energy projects licensing procedure: M/M/1.
- Service management centre, considered as a single phase, multi-channel system M/M/s.

M means that arrivals occur according to a Poisson distribution and have exponentially distributed service requirements (service time). Number 1 and letter *s* stands for the number of service points, *s* meaning multiple servers.

For the queuing equations we let:

λ mean time of arrivals per time period
μ mean number of applications served per time period
m number of channels open
L average number applications in the line
W the average time an application spends being approved.

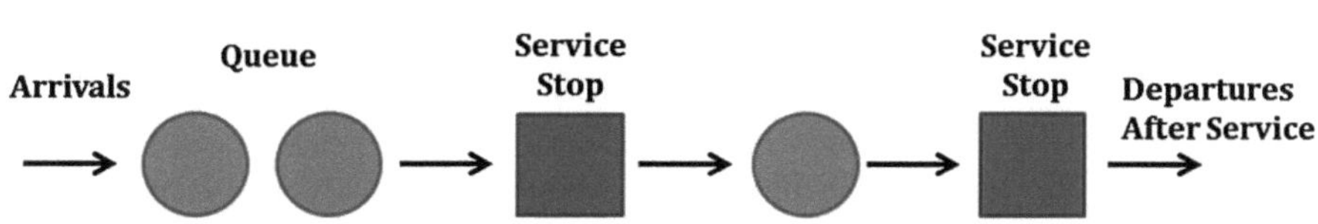

Fig. 1 Single channel, multiphase system

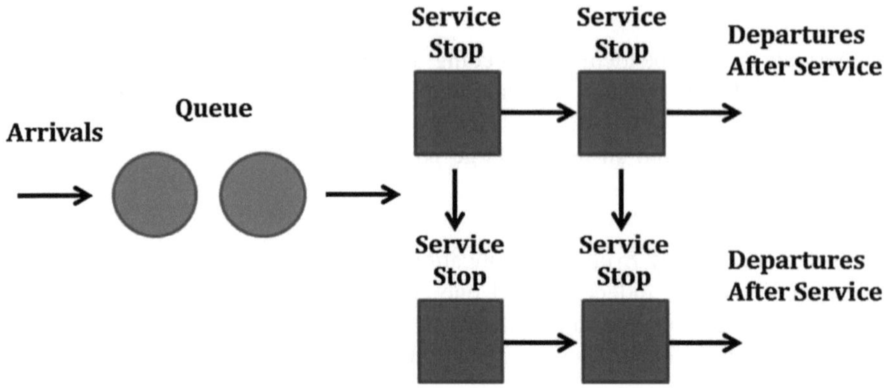

Fig. 2 Multi-channel, multiphase system

The average time an application spends in line and being served is

$$W = \frac{L}{\lambda} \tag{1}$$

$$L = \frac{\lambda\mu(\lambda/\mu)^m}{(m-1)!(m\mu-\lambda)^2} P_0 + \frac{\lambda}{\mu} \tag{2}$$

for M/M/s and

$$W = \frac{1}{\mu - \lambda} \tag{3}$$

for M/M/1 (Figs. 1 and 2).

The application of the above proposed model justifies the necessity of implementing the service management centres principle. The solution of the Eqs. 1–3 for a series of input data reveals that by adopting the (M/M/s) model, the average time an application spends being approved is significantly reduced. Also the proposed model (M/M/s) seems to be indifferent from mean time of arrivals per time period, whereas for the multifaceted model (M/M/1) it is important to have large difference between the number of applications and the time of arrival in order to reduce the time an application spends being approved.

8 Conclusions

As time runs up, and sustainable energy should be already in place, endless, bureaucratic procedures are both costly and time-consuming and do not lead anywhere. The regulation of the permits issuance rate for renewable energy projects is a crucial parameter for the penetration of RET and the upgrade of the energy mix. The business as usual practice of construction projects licensing, based on a multifaceted process, may have worked in the recent past, but it is apparent that it cannot meet current requirements. The case of Cyprus and the ability of the state to issue new licenses after the accident at Mari is a representative case study. A novel approach might be therefore necessary, that takes into account, in addition to best practices, tools and methods of quantitative analysis.

In this chapter different licensing schemes, the current licensing system, the "one-stop-shop" scheme and the service management centre, have been presented and discussed. The use of queuing theory as a useful tool that can be used to evaluate the feasibility of different alternatives was also evaluated and discussed, aiming to the overall restructuring of the licensing process. According to the analysis the most practicable licensing procedure is the concept of the service management centre. Simple application of queuing models demonstrated the significant potential for the elimination of the necessary time to issue a permit through the service management centre approach, without actually requiring the operation of additional licensing service points.

References

Bennett P, Lieskovsky J, McBee B (2011) Impacts of intermittent generation. In: Barnes F, Lavine J (eds) Large energy storage systems handbook. CRC Press, Boca Raton, Florida, USA, pp 17–50

Cyprus Energy Regulatory Authority (CERA) (2011) In: A strategic plan for the promotion of renewable energy sources in the Cyprus electricity generation system, Final Report

Directorate General for Energy, EU (2011) Key figures. Market observatory for energy. Directorate General for Energy webpage. http://ec.europa.eu/energy/observatory/countries/countries_en.htm. Accessed 30 Apr 2013

Electricity Authority Cyprus (2012) Annual report, 2011. ISSN 0258-4875. Electricity Authority of Cyprus webpage. http://www.eac.com.cy. Accessed 18 May 2013

European Parliament (2009) Directive 2009/28/EC on the promotion of the use of renewable energy sources

Kordatos C, Constantinou T, Charalampides A, Charalampous A (2012) A break down of the various procedures and involved organizations for the permitting, financing and installation of renewable energy projects in Cyprus. In: Poster, International Energy Programme Evaluation Conference, Italy, June 2012

Koroneos C, Fokaides P, Moussiopoulos N (2005) Cyprus energy system and the use of renewable energy sources. Journal Energy 30(10):1889–1901

Render B, Stair R, Hanna M (2011) Quantitative analysis for management, 11th edn. Pearson, Upper Saddle River, New Jersey, USA

Scott I, Lee SH (2011) Battery energy storage. In: Barnes F, Lavine J (eds) Large energy storage systems handbook. CRC Press, Boca Raton, Florida, USA, pp 153–180

Wikipedia (2013) List of the largest artificial non-nuclear explosions. Wikipedia. http://en.wikipedia.org/wiki/. Accessed 11 May 2013

The Perplexed Technical Governance of Wind Turbines in Greek Islands

George Caralis, Arthouros Zervos, Gao Zhiqiu and Kostas Rados

Abstract There are several types of restrictions applied on the implementation of wind energy in islands. Technical, economical and land planning aspects are the most important and are discussed in this chapter. Electricity produced by wind energy is competitive against conventional power production in sites with high wind potential, but wind penetration is restricted due to technical constraints imposed to autonomous electrical systems for reasons of safe operation. On the other hand, hybrid solutions which combine wind energy with pumped storage systems may be applied in few cases for further wind energy penetration, but this solution is considered as a rather expensive one and requires large scale civil works for the topology of the reservoirs and huge water quantities for initial fill. Although these solutions could be competitive against the current high electricity production cost and reduce the energy dependence, they lead to wind energy resource exploitation only for partial local supply. In several islands, there is abundant wind potential which could be only exploited and transported to the mainland through the development of large scale wind farms and underwater interconnections to decrease national energy dependence and contribute to the achievement of national goals for renewable energy supply. In all these cases, land planning issues associated with other land uses and protected areas set additional constraints to wind energy development.

In memory of professor Kostas Rados (1965-2013)

G. Caralis (✉) · A. Zervos
School of Mechanical Engineering, National Technical University of Athens (NTUA), 64514 Zografou Campus 15704, Athens, Greece
e-mail: gcaralis@mail.ntua.gr

G. Caralis · G. Zhiqiu
Institute of Atmospheric Physics, Chinese Academy of Sciences, Beijing 100029, China

K. Rados
Department of Environmental Engineering, Technological Educational Institute of West Macedonia, Kozani, Greece

E. Michalena and J. M. Hills (eds.), *Renewable Energy Governance*, Lecture Notes in Energy 23, DOI: 10.1007/978-1-4471-5595-9_11,

1 Introduction

The electrical systems in the majority of Greek non-interconnected islands are based almost entirely on oil fuel generated electricity. Most of those islands experience high rates of increase of energy consumption due to tourism development and high variation of demand between summer and winter, day and night, and within the day. Consequently, the islands' conventional power stations operate with low load factors and the systems are characterized by high electricity production cost.

Most of the Greek islands have abundant wind energy potential (CRES 2001) and therefore there is a high interest for investments in wind applications. Wind farms operating in such autonomous systems are subject to output power limitations, also known as wind power curtailment.[1]

The non-interconnected power system of the island of Crete was one of the first isolated systems with significant wind penetration which experienced the curtailment of surplus wind. By the end of 2012, the wind installed capacity in Crete surpassed 181 MW, with annual energy contribution around 18 % and annual wind power curtailment of the order of 15 %. Given the current infrastructure and the technical constraints, the prospects of wind power to decrease both the system's electricity production cost and the dependence on oil are limited (Caralis and Zervos 2007a).

Wind energy with Pumped Storage System (WPS) comprised of wind farms, two reservoirs for the recycling of water, hydro turbines, pumps and penstocks is considered as a means to increase the wind installed capacity, substitute expensive fuel oil and reduce the required conventional installed capacity. The later is possible because the variable output of wind power is managed and transformed into a guaranteed power supply.

During last decades, besides technical and economical difficulties, such as deep waters, lack of knowledge, bureaucracy and lack of financial resources, interconnection of Greek islands with the mainland is predicated on political will for several reasons: Firstly, due to the current high cost of local power stations in the islands, the high rates of increase of energy demand due to tourism and the difficulty[2] in sitting of new local conventional power stations. Secondly, the fact that exploitation of abundant wind and solar potential is considered as a priority towards the achievement of the national goals to increase the use of renewable energy sources. Within this nexus, several investors have proposed the development of large scale wind energy projects in islands, suggesting, in parallel, to undertake the construction of the underwater interconnection with the mainland.

---

[1] Wind power curtailment is the reduction of wind power production due to inability of the power system to absorb wind power, set by the system operator in order to ensure the safe operation of the island's power system.

[2] Due to touristic development and reactions from local population against the development of new conventional power stations close to their location.

Improvements in cabling technology, underwater connections and decrease of their cost, mainly thanks to the development of offshore wind farms in the North Europe, make those plans realistic.

Obviously, the development of large scale wind farms always face several reactions from local population, which mainly concern the sensitive environment of islands and the probable negative effects on tourism. In parallel, in Greece, there is a new legislation related to land planning of renewable energy technologies, which is expected to decrease the difficulties of land selection and to lessen bureaucratic bottlenecks. On the other hand, the new legislation comprises a quite strict framework with many parameters and restrictions.

Within this Chapter, we analyse issues related with wind penetration in islands, hybrid systems and interconnections. This chapter intends to discuss the most important technical, economical and spatial issues related with wind penetration and also provide simple tools and approaches towards wind penetration that can be achieved under the development of different infrastructures. The target is to understand important technical and non-technical parameters that should be taken into consideration in order to be able to analyse different cases and achieve the best solution towards the maximization of renewable energy sources. For this purpose, initially a historical and legislative background of wind energy development in Greek islands is presented. Then the discussion focus on the limitations of wind penetration applied to non-interconnected islands, on the combined use of wind energy with pumped storage and on the issues related with interconnections. Finally, land planning issues are discussed to understand how the Greek legislation is applied imposing an additional important constraint towards wind farms sitting, but also providing a spatial reference for the maximum allowed wind capacity to be established in a wider area.

2 Historical and Legislative Background in Greece

Although wind energy technologies represent one of the most mature solutions and an essential tool towards the reduction of emissions related to greenhouse effect and global warming, its use has constituted a controversy over the past two decades between local communities, local authorities, the state and investors.

The Greek history on wind energy has a lot to narrate: Historically, the first wind farm in Greek islands was installed in the island of Kythnos in 1982 and similar applications were developed by Public Power utility in other islands. With Law 2244/1994, new opportunities for private investors came up and during the late 1990s, the first private wind farms were installed in Crete, Evia and elsewhere, while the first initial doubts and objections were formulated against wind turbines, mainly due to visual impact and the effect on the sensitive physiognomy of Greek islands.

Rich wind resources on islands together with incentives set by legislation, enforced investors interest and wind energy development quickly reached technical limits of carrying capacity in autonomous islands. At that time, the wind installed capacity was initially defined by law following an old general rule, and then specifically per island by the Greek Regulatory Authority for Energy (RAE 2003). Regardless the way of definition, the maximum contribution of wind energy could not exceed 15–20 % of the annual energy needs of each island without upgrading the current infrastructure.

At the same time, already in the late 1990s, local newspapers and associations in islands blamed wind turbines for lack of rain, water scarcity, effect on sexual behavior of animals and weathering the original character of islands. However, in some other cases there was a positive attitude towards wind energy as a result of citizens' awareness campaigns, involvement of local organizations and also direct and indirect benefits of wind energy technology upon local employment and economic activity.

That same period, on a scientific and political level started the debate about technical solutions to overcome restrictions and further exploitation of wind energy. To overcome technical barriers, pumped storage was intensively discussed as the most convenient techno-economically acceptable solution for energy storage. Despite several applications for combined plants of wind energy with pumped storage, mainly in Crete, Lesbos and Rhodes, only one project is today under construction in Ikaria island. The law 3468/2006 set the principles for the operation and the pricing of these systems.

During this period, the lack of a spatial framework for the development of Renewable Energy Sources (RES) caused trouble for investors and local authorities during the environmental licensing procedure. This created the urgent need to formulate and implement a spatial framework for RES, which would protect the character of islands while allowing the development of wind energy to a desirable extent. One such effort that filled a large gap is the specific spatial framework for RES (ministerial decision 49828/2008), which set the land planning terms for the development of wind energy in islands.

In the middle of the first decade of the twenty-first century, the new trend in terms of new applications is the development of large-scale wind energy investments in islands with underwater interconnections with the mainland. In this framework, most of the Greek islands become fields of such applications. Most of the local population in Cyclades, Dodecanese, Crete and North-eastern Aegean islands display negative behavior against wind investments due to the planning of large scale projects without any information or debate between investors and local communities. Far from that, the law 3851/2010 gives additional incentives (more generous feed in tariffs) for these kinds of projects, in case the investor undertakes the development of underwater interconnection with the mainland.

This general discussion on the framework and the background of wind energy development in Greek islands is followed by a further discussion on technical aspects. Moreover, practical simplified tools, which could be easily applied in several cases of islands without requirement for analytical data, are presented.

3 Wind Energy Penetration in Non-Interconnected Islands

3.1 The Technical Issue of Wind Penetration

As discussed above, Greek islands are characterized by an excellent wind-energy potential, but face curtailments by the system operator in case of surplus wind-generated power during periods of low demand. The ability of the local autonomous power system to balance energy demand and supply, and also balance the wind power output variations, defines the allowed wind capacity to be installed and the wind power to be absorbed by the grid. The management of the autonomous power system is affected by technical constraints of the conventional generating units, namely the minimum loading levels of the thermal units (technical minimum) and a dynamic penetration limit, applied for stability purposes (Papathanassiou and Boulaxis 2006). Finally, the demand profile and the sharp changes occurred between day and night and between summer and winter, make power system less flexible to absorb wind power. Then, wind farms operating in non-interconnected islands are subject to output power limitations, which causes income loss to wind farms' investors and may affect the feasibility of wind farms' investments.

Before 2003, a limit on the allowed wind capacity was imposed by law, permitting wind installed capacity up to 30 % of the peak demand load of the autonomous power system. At same time, the system operator was defining minimum guaranteed hours for wind power absorption. Since April 2003, this rule was replaced by an analytical methodology for the definition of the maximum allowable wind capacity to be installed in autonomous systems, proposed by the Regulatory Authority for Energy (RAE). The maximum installed capacity is derived in order to ensure at least a 'real[3] capacity factor[4]' of 27.5 % after curtailment losses. This value assures the feasibility of wind farms' investments, given the excellent wind potential, the investment cost (EWEA 2004), and the fixed price for wind power (feed in tariff). The experience gained by the systematic analysis of wind energy penetration in several Greek islands, leads to parameterized diagrams which could be used to estimate the allowed wind energy capacity to be installed, the estimated amounts of wind energy curtailment and the rates of wind energy contribution.

3.2 Methodological Approach

The target of this approach is to determine technical limits of wind installed capacity which can be installed in an autonomous island, given the current

[3] (real capacity factor) = (capacity factor) × (percentage of annual wind energy absorbed).

[4] (capacity factor) = (energy produced)/[(Rated power) × (8760 h)].

electrical infrastructure without storage facilities. Results will be interesting in demonstrating the potential of wind technology to decrease the energy dependence, to increase the use of local sources, but also in understanding the relative technical limitations.

In the next section two diagrams, which are based on steady state simulations of islands' power systems, are presented. These simulations make use of annual hourly time-series of demand and wind, typical characteristics of conventional units (maximum ability, technical minimums, and order of commitment) which are provided by the power system operator.

The formulation of parameterized diagrams is based on the results of the thorough analysis on three representative Greek case-studies islands (Caralis and Zervos 2007a).

The technical constraints described in the methodology of RAE have been used to calculate the wind power that can be absorbed directly by the electrical system. The simulations are realized for different values of wind installed capacity. In order to take comparable results between different in size and characteristics islands, the wind installed capacity is introduced dimensionless as a percentage of the annual mean load demand.[5] This is a preferable way to levelise wind installed capacity in different size systems instead of the peak demand supply.

As mentioned above, the diagrams are based on the application on three representative Greek islands.[6] The simulation is based on the non-dynamic analysis of the autonomous electrical system using annual hourly time-series of demand and wind. The number of conventional units committed are defined in order to ensure the power supply (even if all the wind power is lost), an assumption that guarantees the safe operation of the electrical system. Then, the grid's ability to absorb wind power is calculated taking into consideration the conventional units' technical minimums and the maximum allowed instantaneous wind penetration δ. Finally, the wind power absorbed and curtailed is calculated.

3.3 Parameterized Diagrams

Applying the above approach for several wind installed capacities in three Greek islands, the following diagrams are derived. These diagrams are parameterized by the annual wind speed (6.3, 7.2, 8.1, 9.0 m/s) and the dynamic penetration limit δ (30, 40 and 50 %). They can be used to estimate the wind energy absorption, wind energy contribution and actual capacity factor for a given wind installed capacity (from 0 % up to 200 % of the annual mean load) in a specific island.

An estimate at the safe side can be obtained by using $\delta = 30$ %, which is considered as a typical value for this parameter. In case the island's power system

5 The mean annual load demand is the annual electricity demand divided by 8760 h.

6 Crete, Lesvos and Serifos.

is comprised of flexible small conventional power units and the system operator allows higher instantaneous wind penetration, a higher δ, up to 50 %, could be used. The decision to increase the instantaneous wind penetration δ, which is a dynamic penetration limit applied for stability purposes, is up to system's operator. Among other factors, it is based on: any previous negative experience related with wind energy and safety of the system, the probability of occurrence of sudden loss of wind energy production, the size of the island, the spatial dispersion of wind installations.

Generally speaking, Fig. 1a shows that for low wind installed capacity, the wind energy produced can be absorbed almost 100 %. With the increase of wind installed capacity, the absorption rate is dropped dramatically, showing that wind energy is not effectively used.

3.4 Numerical Example

Example: The island of Crete is the biggest Greek island, with population 600,000, annual electricity demand 3.3 TWh and peak power demand 585 MW. Estimate the annual wind energy curtailment, for wind installed capacity 165 MW.

Solution:

An estimation of wind potential is required. This information is always provided by existing meteorological stations, measurements or Aeolian maps. The annual mean load in the island is 3.2 × 106/8,760 = 377 MW. For levelized wind installed capacity 165/377 = 44 %, and assuming a conservative annual averaged wind speed 7.5 m/s, and δ = 40 % (large area), the following rates are estimated from Figs. 1a and b:

- Wind energy absorption: 93 %
- Actual capacity factor: 30.7 %.

Then wind energy contribution can be easily calculated as:
165 MW × 30.7 %/377 MW = 13.4 %

3.5 Economic Evaluation

The results of wind energy supply (Fig. 2a) and the effect of wind energy on the electrical system's electricity production cost of Crete (Fig. 2b), has been calculated for annual average wind velocity 8.1 m/s and are presented for three different values of oil price and three different values of simultaneous wind penetration δ.

Figure 2a shows that for low wind installed capacity, wind energy supply is increased almost proportionally to the wind installed capacity, but then due to wind curtailment, increase of wind installed capacity is not associated with a

(a)

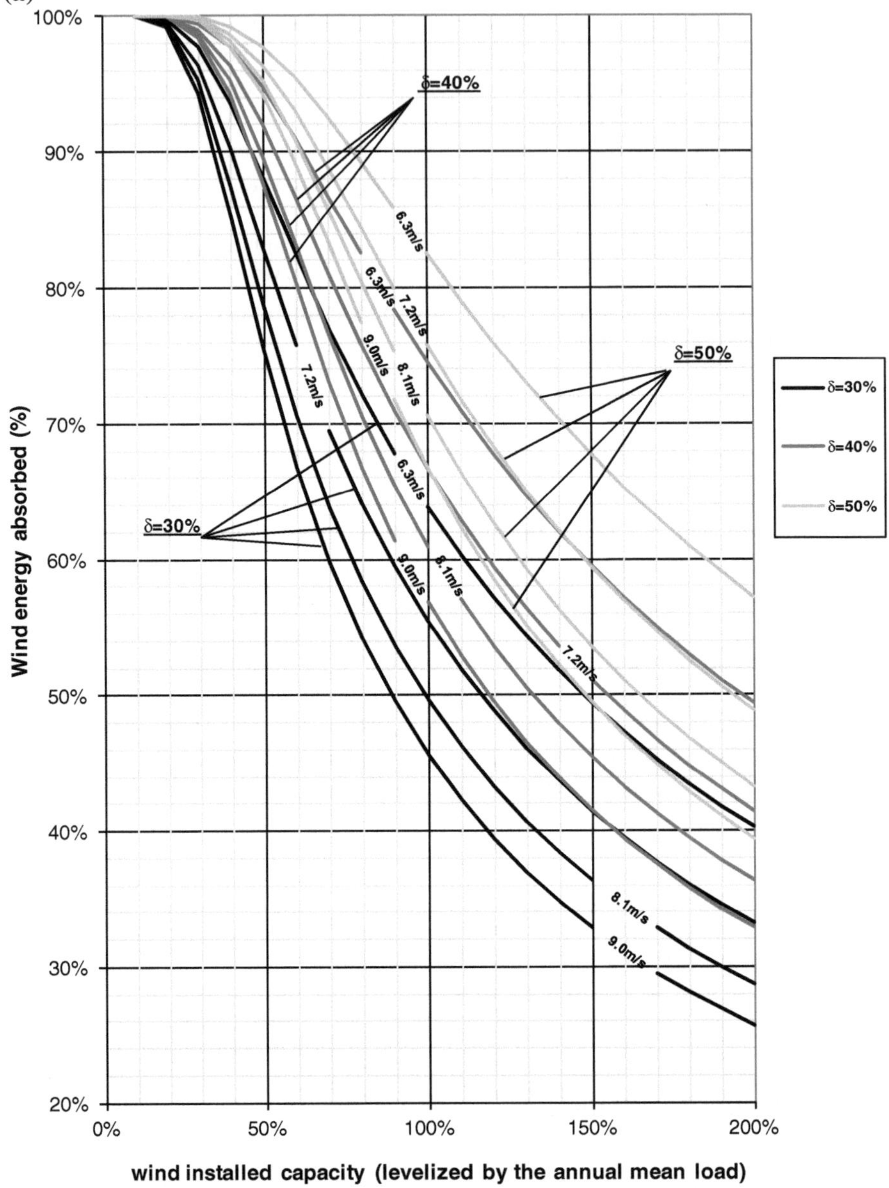

Fig. 1 **a** Parameterized diagrams of wind energy absorbed **b** Parameterized diagram of actual capacity factor

(b)

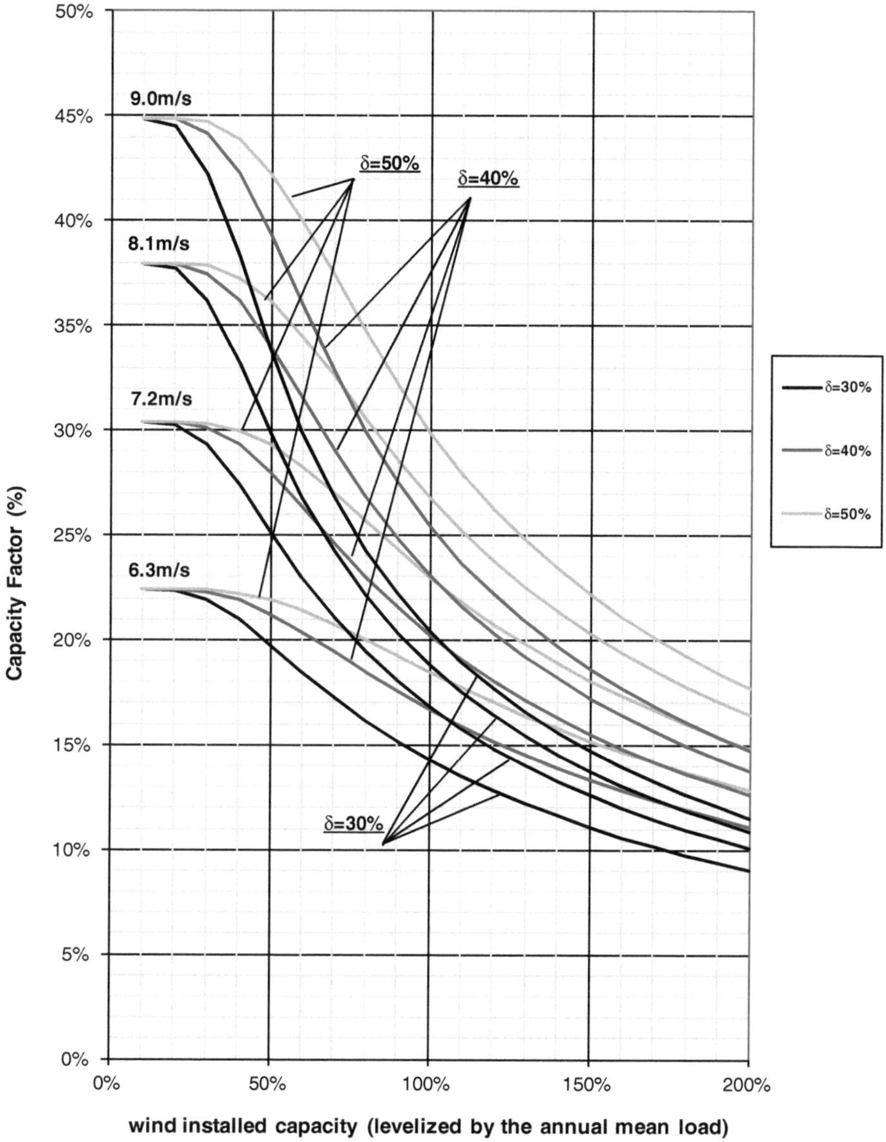

Fig. 1 continued

similar increase of wind energy contribution, showing that wind energy is not effectively used. This is an important conclusion of the analysis of wind penetration using the existing infrastructure, which proves that in non-interconnected systems only a small partial wind contribution is technically possible to be

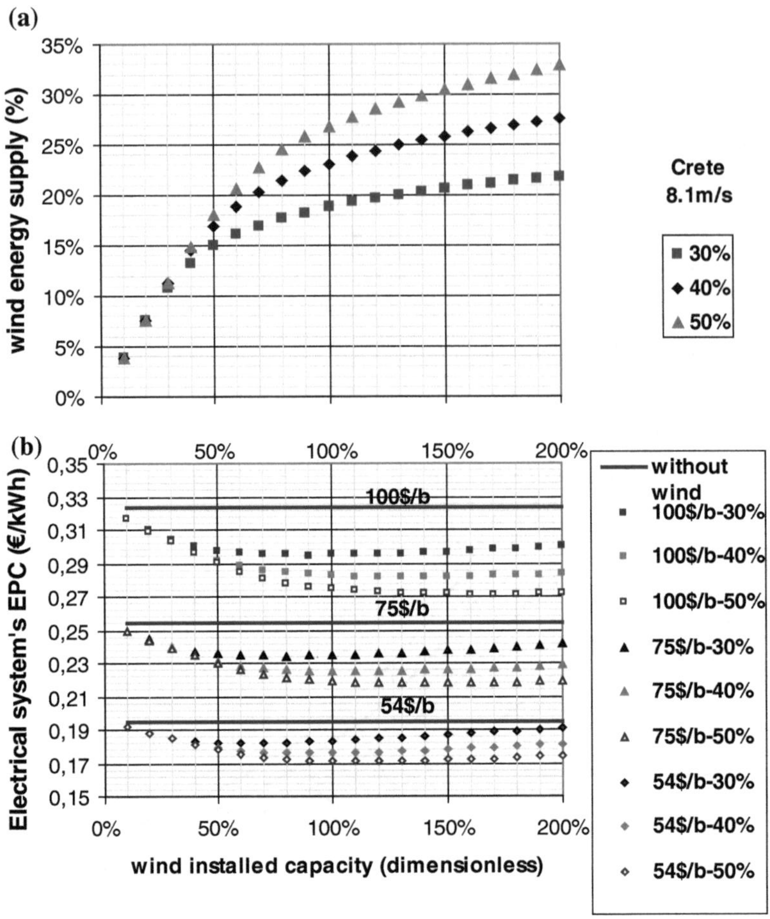

Fig. 2 **a** Wind energy supply parameterized by the allowed instantaneous wind penetration ($\delta = 30$, 40 and 50 %) **b** The effect of wind energy integration to the system's electricity production cost, parameterized by the oil fuel price (54, 75 and 100$/b) and the allowed instantaneous wind penetration (30, 40 and 50 %) (*Source* Caralis and Zervos 2007a, reproduced with permission of Wind Engineering Journal)

achieved, and further solutions such as pumped storage or interconnections should be evaluated towards large scale wind integration.

In Fig. 2b the effect of wind integration into the electrical system's electricity production cost, based on the wind energy investment cost, operation and maintenance cost and wind farms actual capacity factor, shows that the system's electricity production cost is reduced. In fact, there is an optimum wind installed capacity for which the economic benefit is maximized. With further increase in wind installed capacity, EPC is not possible to be further decreased, due to wind energy curtailment. Additionally, the optimum wind installed capacity is increased as soon as the price of oil is increased.

4 Wind Energy with Pumped Storage

4.1 Description

Given the current infrastructure and the technical constraints, the prospects of wind power to increase wind energy penetration are limited (Caralis and Zervos 2007a). Comprised by wind farms, two reservoirs for the recycling of water, hydro turbines, pumps and penstocks, a Wind and Pumped Storage (WPS) system is considered as the most effective solution to increase wind installed capacity, substitute expensive fuel oil and reduce the required conventional installed capacity in autonomous islands. Last years, WPS has been analysed by the scientific community for various autonomous islands (Zervos et al. 2000; Kaldellis et al. 2002; Katsaprakakis et al. 2005; Theodoropoulos et al. 2003; Boulaxis and Papadopoulos 2003; Caralis and Zervos 2006).

The experience gained from the systematic application and thorough analysis of specific case studies on three representative Greek islands (Caralis and Zervos 2007a) was used for the creation of parameterized diagrams towards the WPS initial design in non-interconnected islands (Caralis et al. 2010). The aim was to provide a tool which could be used to make an initial design and dimensioning of WPS for a specific Greek island.

All the case studies and the produced diagrams are based in the following design and operational policy of the WPS (Caralis and Zervos 2007b):

- connection of the Wind farms with the pumping station through the central grid,
- peak demand supply of the hydro turbine,
- consideration of the hydro-turbine as a spinning reserve to increase the direct wind power absorbed,
- double penstock and,
- complementary pumping using conventional power given the capacity of the committed conventional units.

The target of parameterization and generalization of results was an initial prospect of all previous studies (Caralis and Zervos 2007a, b) and always various attempts of dimensionless parameterization have been tested.

4.2 Dimensioning of Wind Farms, Reservoir, Hydro-Turbine

The required wind capacity is inversely proportional to wind potential (wind capacity factor) and to the efficiency of the WPS (the ratio of hydro turbine's energy production to the energy used for pumping), and proportional to the annual mean load and the load factor. The mean load is associated with the amount of demand and the load factor with the seasonal variation of demand. A power system with relatively low variations (high load factor) and one with large variations (low load factor) could have the same annual mean load. In the latter case, lower wind capacity is required. During the short peak demand period, the WPS could provide the guaranteed power thanks to the reservoir's stored energy, while the rest period of low demand, lower wind capacity is adequate. On the other hand, the required volume of the reservoir is inversely proportional to the available hydraulic head and proportional to the average efficiency of the pumping station.

In this connection, two indices related with the wind capacity index, δ_W, and with the reservoir's capacity index, δ_R, are defined:

$$\delta_W = \frac{P_{W,h,R} \cdot CF_{W,th} \cdot n_{PSU}}{\overline{P_L} \cdot LF} \tag{1}$$

$$\delta_R = \frac{V_{RESERVOIR} \cdot H}{3600 \cdot 102 \cdot P_{W,h,R} \cdot CF_{W,th} \cdot n_P} \tag{2}$$

where $P_{W,h,R}$ is the wind installed capacity in the WPS (MW), $CF_{W,th}$ is the wind capacity factor which introduces the wind potential in the island (%), n_{PSU} is the efficiency of the WPS (%), P_L is the annual mean load (MW), LF is the load factor of the island (%), $V_{RESERVOIR}$ is the capacity of the reservoir (m^3), H is the hydraulic head (m), n_P is the efficiency of the pumping station (%).

Figures 3 and 4 summarize collectively the results of a systematic application in several case studies in representative Greek islands. Apparently, the wind installed capacity and reservoir indices depend on hydro-turbine's peak demand supply α. As regards the reservoir, hydraulic head is also introduced as a significant parameter. Three curves have been adjusted to the available points:

$$\delta_R = \begin{cases} 298.17 \cdot a^{0.0341}, & for\ \mathrm{H} = 200\ \mathrm{m} \\ 217.74 \cdot a^{0.082}, & for\ \mathrm{H} = 300\ \mathrm{m} \\ 165.54 \cdot a^{0.0933}, & for\ \mathrm{H} = 400\ \mathrm{m} \end{cases} \tag{3}$$

$$\delta_W = 4.1889 \cdot a^{2.2133} \tag{4}$$

These diagrams and expressions can be used for the initial design and dimensioning of the WPS, given the basic data of the autonomous power system (annual peak demand, mean annual load of the system and the load factor).

To start with the dimensioning, one of the following parameters should be initially determined:

- the hydro-turbines peak demand supply,
- the reservoirs capacity,
- the wind installed capacity.

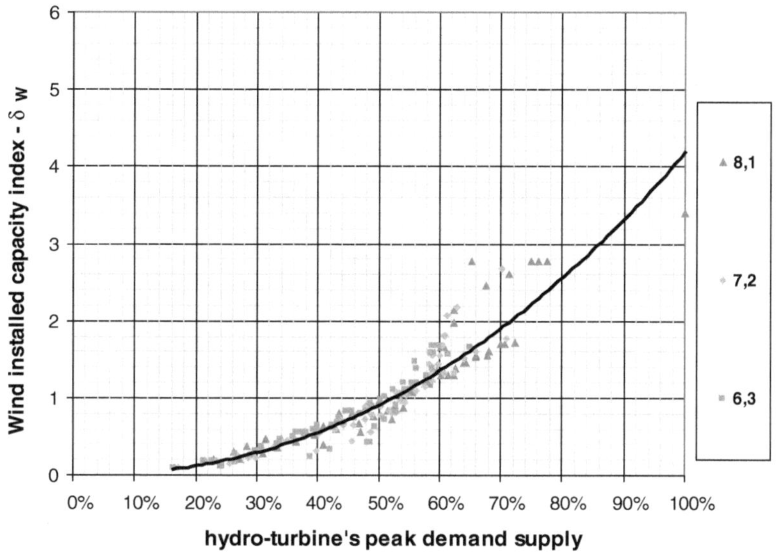

Fig. 3 Diagram for the definition of the WPS wind capacity (*Source* Caralis et al. 2010, reproduced with permission of Renewable and Sustainable Energy Reviews Journal)

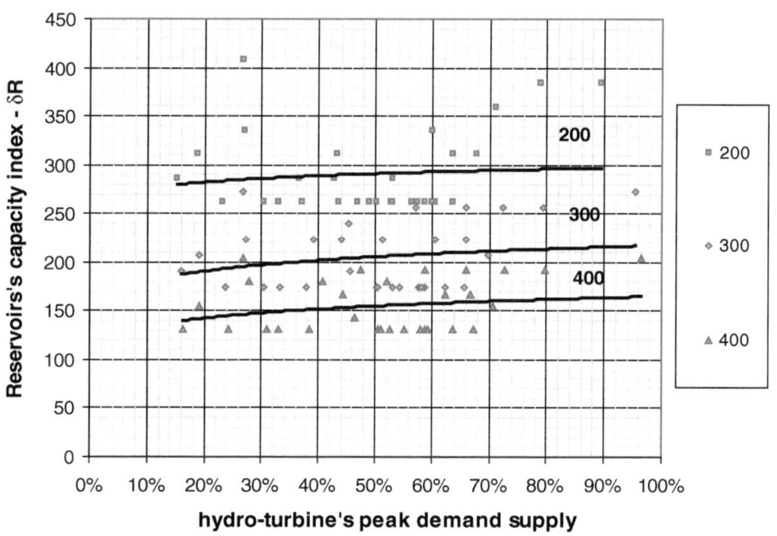

Fig. 4 Diagram for the definition of the WPS reservoir (*Source* Caralis et al. 2010, reproduced with permission of Renewable and Sustainable Energy Reviews Journal)

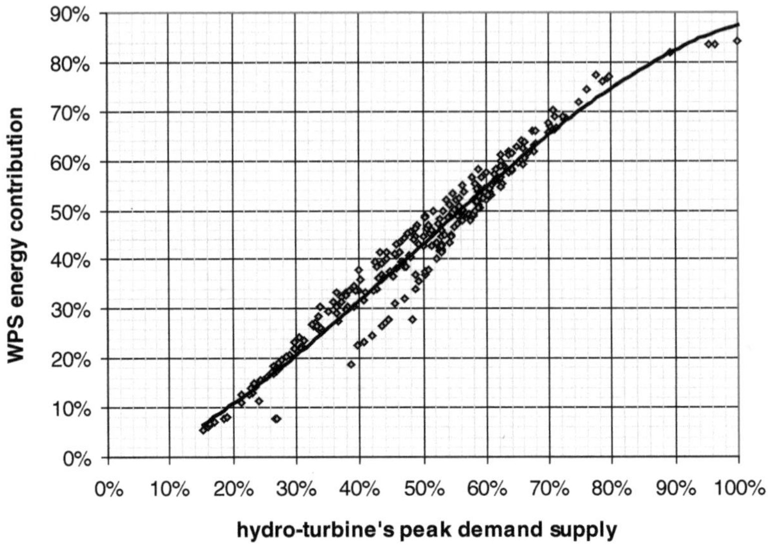

Fig. 5 WPS energy contribution (*Source* Caralis et al. 2010, reproduced with permission of Renewable and Sustainable Energy Reviews Journal)

Wind potential and hydraulic head are site-dependent features that strongly affect the attractiveness and profitability of the investment, but they do not affect the WPS energy contribution. Achieving a desirable WPS energy contribution or turbine's peak demand supply requires a specific wind energy amount combined with a specific storage capacity. The specific wind energy amount can be provided by lower wind installed capacity in a site with higher wind potential, or more wind installed capacity may be required in case of lower wind potential. Similarly, the energy storage capacity of the reservoir is determined by both the reservoir's water capacity and the available hydraulic head. In Fig. 5, although results are given for several case studies with various annual mean wind speed (6.3, 7.2 and 8.1 m/s), and various hydraulic heads (200, 300 and 400 m), the variation of the energy contribution with respect to the turbine's peak supply is not affected by either the annual mean wind speed or the hydraulic heads.[7]

[7] In Fig. 5, the points' dispersion is rather affected by the load factor of the power system and the correlation between wind and demand. Local summer north winds in Aegean Sea called "Meltemi", permits a higher hydro-turbine's peak demand supply for specific wind capacity and reservoir and specific energy contribution. During the short peak demand period, thanks to the correlation between wind and demand, WPS could provide more guaranteed power. During the whole year, even with lower wind potential, wind capacity is adequate for the medium loads. In other words, for specific wind capacity and reservoir, a bigger hydro turbine is justified and higher peak demand can be supplied.

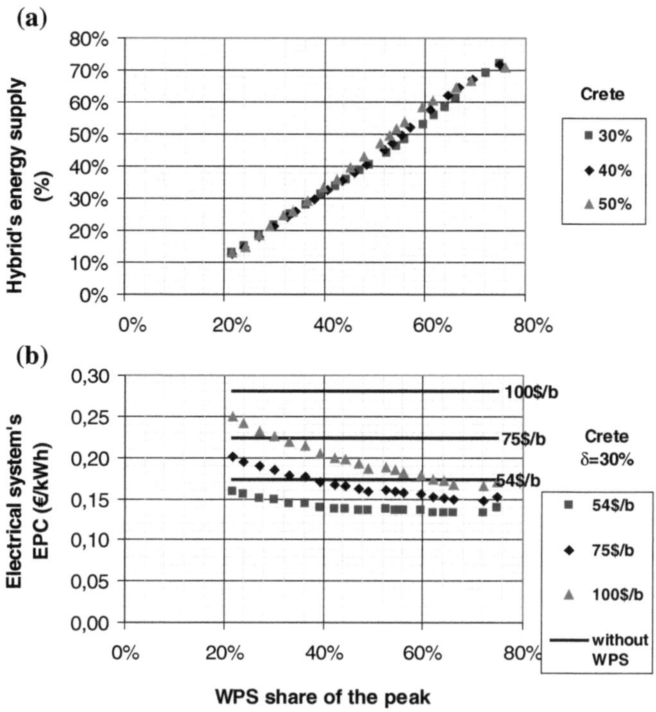

Fig. 6 **a** Hybrid's energy supply parameterized by the allowed instantaneous wind penetration ($\delta = 30$, 40 and 50 %) **b** The effect of combined use of wind energy with pumped storage to the system's electricity production cost, parameterized by the oil fuel price (54, 75 and 100$/b) for $\delta = 30$ % (Source: Caralis and Zervos 2007b, reproduced with permission of IET Renewable Power Generation Journal)

Next, the correlation between energy contribution ε_{WPS} and hydro-turbine's peak demand supply α is illustrated in Fig. 6 or by means of the following expression:

$$\varepsilon_{WPS} = -0.1868 \cdot \alpha^2 + 1.2702 \cdot \alpha - 0.1537 \quad (5)$$

4.3 Numerical Example

A numerical example for the island of Ikaria is given bellow. Initial information—data about this island is required:

Peak demand: 7.55 MW

Average load demand: $\overline{P_L} = 2.75$ MW

Load Factor: LF = 40 %.

Average wind velocity 8.1 m/s (source: Aeolian maps of CRES) and hydraulic head between the two reservoirs 400 m, then:

For hydro turbine's peak supply 60 %, the two indexes are calculated by formulas 3 and 4: $\delta_W = 1.3$ and $\delta_R = 160$.

The required wind installed capacity and the volume of the reservoir are estimated from Eqs. 1 and 2 (calculations for $C_{FW,\ th} = 38\ \%$, $n_{PSU} = 60\ \%$, $n_P = 80\ \%$):

$$P_{W,\,h,\,R} = 6.3\ \text{MW}.$$

$$V_{RESERVOIR} = 375073\ \text{m}^3.$$

The required hydro turbine is 4.53 MW (60 %×7.55), and energy supply of the whole system is 55 % (Fig. 5).

Similar approach is used in case that the wind capacity $P_{W,\ R}$ to be installed is defined, as it may happen sometimes due to land use restrictions. Likewise, if the reservoir's capacity V_R is defined, as it may happen due to site topography restrictions or due to the existing lower reservoir (Katsaprakakis et al. 2005), for the calculation of the rest components an iterative procedure is required, as soon as reservoir's capacity index is a function of wind installed capacity.

4.4 Economic Evaluation

Results from the economic evaluation of the combined use of wind energy with pumped storage systems in the island of Crete are presented for three values of oil price and three values of instantaneous wind penetration δ. Obviously, in comparison with the case of only wind (studied in paragraph 3.5, the contribution is not affected that much by the allowed instantaneous wind penetration. On the other hand, it seems that by the combined use of wind energy with pumped storage, a further wind penetration and high share of energy supply can be achieved, leading to a significant decrease of the energy dependence and also the dependence of system cost on the fuel cost becomes weaker. Obviously, these economic benefits are increased for higher oil price.

5 Large Scale Wind Integration and Underwater Interconnections

5.1 Current Status and Prospects

Interconnection of islands depends to a great extent on political will based on several factors:

- Exploitation of the islands wind and solar potential
- High cost of local power stations
- Improvements in the technology of cables and underwater connections
- High rates of increase of the energy demand
- Difficulty in the siting of new local conventional power stations
- Improvement of the quality of life and need for current of better quality
- Benefits for the consumers due to the market deregulation.

Generation of electricity in islands is generally made using expensive oil products. Hence, the average production cost in Greek islands is much higher than the consumers tariffs leading to an average annual surcharge of 600 M€, directly depended on the oil price. Consequently, their electrical connection with the large mainland power system using submarine cables is of considerable economic interest. In addition, interconnection gives the opportunity for better exploitation of the abundant wind and solar potential.

Currently, all big islands of the Ionian Sea in Western Greece are already connected to the mainland grid. Several interconnections also exist, mostly in the Aegean Sea for islands with relatively low peak demand situated close to the mainland.

Cyclades islands are of first priority for the forthcoming interconnections. This complex of islands has been partly interconnected to the mainland system and it is currently under study to be expanded to include most of the big islands of the region. It presents a special interest because of the large length of the cable lines involved and the considerable electric power demand of particular islands and the complex in total.

In 2010, the Hellenic Transmission System operator carried out a pre-feasibility study for island interconnections with the mainland. Minimization of the HV AC submarine cable length has been the main concern in designing the submarine cable interconnections for technical and economical reasons. In this study, the interconnection of almost all Greek islands was investigated for five main interconnections—island groups. This study considers the existing large scale wind farm applications:

- Cycladic islands (a part of this interconnection is in construction stage), with applications for wind capacity of 650 MW
- North Aegean islands, with an application of 700 MW
- Crete, with two applications of 2,200 MW in total
- Dodecanese, with expected wind power capacity of 700 MW
- A Stand-alone interconnection of Skyros with an application for wind power capacity of 333 MW.

Although investors, who apply for large scale wind integration in islands, wish to undertake underwater interconnection with the mainland with several positive impacts in local and national level, they face several objections by local populations.

6 Land Planning Issues

A ministerial decision for the physical planning and allocation of RES has been published in December 2008. It is the first effort to define land-planning priorities, rules and constraints for the development of RES and will be decisive for the next steps for RES development in the country.

A first important restriction is set for the maximum density of wind turbines. In mainland areas that are defined as Regions of Aeolian Priority, density should not exceed 8 % of the municipal area (i.e. one turbine per square km). In all the inhabited islands it cannot exceed 4 % at each municipality (i.e. one wind turbine per two square km).

Additional constraints are imposed related with other land uses:

- Distance 1,500 m from remarkable coasts
- 1,000 m from build-up areas
- 500 m from archeological sites
- 127.5 m from main roads (Fig. 7).

6.1 Example of Land Planning Application in Lesvos Island

The total area of Lesvos is 1632.8 km^2. The allowed maximum number of wind turbines due to the restriction of maximum density is $0.53 \times 1632.8 = 865$ for typical wind turbines with rotor diameter 85 m (rated power 2–3 MW), leading to a large capacity initially allowed to be installed.

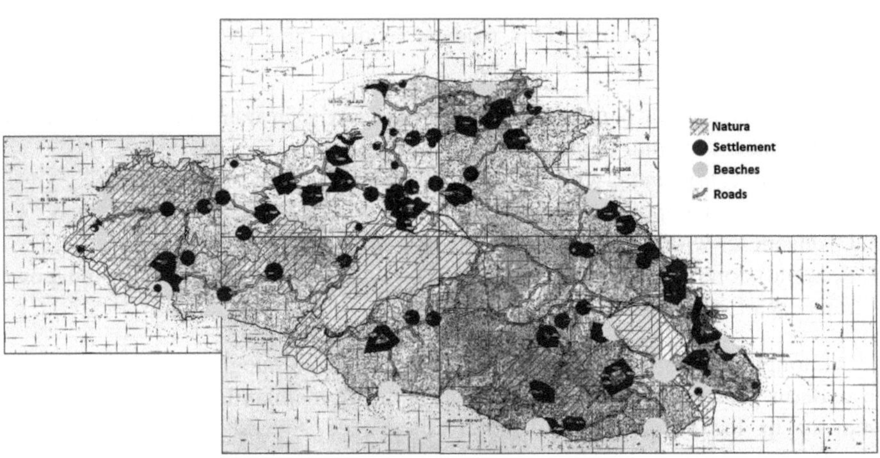

Fig. 7 Application of land planning restrictions in Lesvos

Additionally, macro-scale application which takes into consideration all restrictions, leads to results presented in Fig. 7. There is a big part of Lesvos excluded for wind applications due to urban areas, roads, beaches, protected areas (Natura 2000), forests etc. Also, there are land issues related with the wind potential (average wind speed and turbulence) and the wind technology, the slope of the land, forests that may further limit the areas to be exploited for wind energy development. Visual effects have to be considered as further restrictions that are usually applied on a case by case basis and in particular in sensitive areas of touristic interest.

7 Discussion-Conclusions

In this chapter special technical, economical and land planning issues related with wind energy development in islands were discussed. There are several constraints imposed by wind resource and wind technology, by the current electrical system infrastructure, the features and the cost of conventional power plants, and finally by land planning and land use. The issues discussed in this chapter are based on systematic analysis of specific case studies in Greek islands.

Some of these restrictions are more or less flexible and could be encountered. Some others require new infrastructure and they are expensive to overcome. Land planning restrictions are enacted, aiming to preserve the Greek natural environment from wind turbines' disturbances. In fact, wind energy is one of the most mature technologies and one of the few weapons that we have against climatic change and greenhouse effect.

This chapter provides an overview of all these different types of restrictions, aiming to provide some understanding and practical tools in the framework of local governance and decision makers. One of the main conclusions is that the capacity of non-interconnected islands to absorb significant amounts of wind power is determined from technical and economic constraints. In some cases, combined use of wind energy with pumped storage could increase wind energy penetration and supply. On the other hand, prospects of large scale penetration and interconnections face technical difficulties as well as restrictions imposed by land planning issues.

Interconnections with the mainland offer significant advantages to Greek islands to overcome technical barriers and achieve energy self-sufficiency and even export electricity generated from wind or sun to the mainland. However, realization of these ambitious plans constitutes a technical and financial challenge. They could only be realized by collaboration of public and private sector.

In any case, wind potential of the islands is considered as a national resource that should be exploited not only to meet local energy needs (this argument is now limited), but also to contribute towards the achievement of the national targets for CO_2 emissions reduction.

The development of interconnections between the islands and the mainland sounds like a vision similar to the one of the electrification of the whole country in 1960s and 1970s. This ambitious plan could be financed only through the development of wind farms and provided that political will is stable and local objections are overcome.

Acknowledgments This work is partly supported by postdoctoral fellowship for young international scientists provided by Chinese Academy of Sciences in 2012 to George Caralis.

References

Boulaxis N, Papadopoulos M (2003) Assessment of the contribution of hybrid systems in renewable energy penetration in islands. In: ISAP 2003, Lemnos Greece, 31 Sept, Paper ISAP03/141

Caralis G, Zervos A (2006) Prospects of wind and pumped storage system's integration in Greek islands. In: EWEC 2006, Athens

Caralis G, Zervos A (2007a) Analysis of wind power penetration in autonomous Greek islands. Wind Eng 31(6):487–502

Caralis G, Zervos A (2007b) Analysis of the combined use of wind and pumped storage systems in autonomous Greek islands. IET Renew Power Gener 1:49–60

Caralis G, Rados K, Zervos A (2010) On the market of wind with hydro-pumped storage systems in autonomous Greek islands. Renew Sustain Energy Rev 14(8):2221–2226

Center of Renewable Energy Sources (CRES) (2001) Development of an information system for the assessment of the technically and economically exploitable potential of Renewable Energy Sources. Oper Programme Energy Measure 3(4):3

EWEA (2004) Wind energy the facts: an analysis of wind energy in the EU-25. EWEA, Brussels

Kaldellis J, Kavvadias K, Vlachou D (2002) Electricity load management of APS using wind-hydro solution. In: Proceedings of MedPower, 2002, Athens, Greece

Katsaprakakis D, Betzios G, Christakis D (2005) Combined management of water sources and wind energy in Lesvos: is the target of 100% renewable electricity supply in Lesvos possible? In: 17th Conference on Pan-Hellenic Network of Ecological Organizations, Lesvos-Greece, 7–9 October 2005 (in Greek)

Papathanassiou St, Boulaxis N (2006) Power limitations and energy yield evaluation for wind farms operating in island systems. Renewable Energy 31(4):457–479

Regulatory Authority of Energy (RAE) (2003) Methodology for the assessment of wind penetration in non-interconnected islands. http://www.rae.gr/k2/ape-penetration.pdf, http://www.rae.gr/K2/deliberation-ape.html (in Greek)

Theodoropoulos P, Mantas Z, Zervos A, Betzios G (2003) Integrated power system of Serifos Island with high RES penetration using pump-storage. In: International Conference on RES for Island Tourism and Water, 2003, EREC, Crete-Greece

Zervos A, Caralis G, Gorgoulis M, Zografakis N (2000) Implementation plan for the large scale deployment of RES in Crete-Greece. In: Altener 2000 Conference, Toulouse-France

Environmental Impacts of Renewable Energy: Gone with the Wind?

Viktor Kouloumpis, Xiongwei Liu and Elspeth Lees

Abstract Wind energy is constantly gaining ground, especially in the UK, helping to tackle climate change and support energy security as the country wants to become less dependent to imported fossil fuels like coal and gas. Nevertheless, wind farm life cycle environmental impacts are not negligible and the construction and operation of wind turbine generators can cause several environmental impacts to the area where they have been sited. Therefore, it is reasonable to question, (a) whether there are "hidden" environmental impacts from the use of renewable energy technologies, and (b) whether supporting wind energy in order to displace fossil fuels just substitutes one environmental problem with another one (or more). This chapter uses UK as a case-study to describe thorough processes of environmental assessment and the environmental impacts related to wind energy. This lecture creates a pallet of potential issues that should be taken into account with regards to implementing environmental energy governance practices. The two methods used here, are (a) the Environmental Impact Assessment approach (used to identify on-site impacts that wind farms have on the environment) and (b) the Life Cycle Assessment (LCA) approach, which connects the energy, material, wastes and emissions with a wide range of environmental impact categories. The findings could change the way we think about wind energy and might make it easy to understand why there are still people who are opposed to the development of wind farms.

V. Kouloumpis (✉) · X. Liu · E. Lees
Faculty of Health and Science, University of Cumbria,
Fusehill Street, Carlisle CA1 2HH, UK
e-mail: Viktor.Kouloumpis@cumbria.ac.uk

X. Liu
e-mail: Xiongwei.Liu@cumbria.ac.uk

E. Lees
e-mail: Elspeth.Lees@cumbria.ac.uk

E. Michalena and J. M. Hills (eds.), *Renewable Energy Governance*,
Lecture Notes in Energy 23, DOI: 10.1007/978-1-4471-5595-9_12,

1 Introduction

The main drivers for the development of renewable energy sources have been the challenges of climate change and energy security. The reason is that fossil fuels that dominate our energy mix have two main disadvantages: their combustion emit greenhouse gases that contribute to global warming and their reserves are diminishing, threatening the security of supply. Therefore, renewable energy's advantage is that can be utilised inside a country's territory according to the government's will and contributing to tackling climate change, which is the main environmental problem of our times. Nevertheless, their development so far has highlighted that they can be restricted by their associated resulting environmental impacts. Wind power has been the renewable energy that is increasingly gaining ground, especially in the UK, where its contribution to the electricity production mix has gone up from 2.5 to 4 % approximately within the last 2 years, contributing to the reduction of CO_2 emissions by 6 million tonnes annually. Since both onshore and offshore farms have been used for adequate years to allow the technology to come to maturity, the question that has arisen is the following: Which are their adverse environmental impacts of using renewable energy technologies and how can we maximise the wind energy contribution without harming the environment?

The development of Wind power in the UK can be examined better by distinguishing onshore from offshore wind farms. The reasons for doing that, is the fact that although they may use the same wind turbine generators they require different supportive infrastructure as well as different design, construction operation and maintenance. Moreover, the development of onshore wind started earlier and that led to reaching an early technological and institutional maturity level providing more secure choices with regards to the specific technology that can be used. This has resulted in the creation of more complete policies and regulations, streamlined investment processes, better awareness and education of the relevant stakeholders, etc. On the other hand, that means that the market is not virgin anymore and it is more difficult to develop additional wind power schemes restricting the availability of land and capital. In other words, there are *little or no low hanging fruits for the development of onshore wind farms, unless a new breakthrough in their technologies allows a new "renaissance" especially if this coincides with the retrofitting of older wind farms.* From an environmental point of view, the different terrain of onshore and offshore wind farms together with the different accompanying infrastructure result in different type or magnitude of impacts.

These differences are mentioned here so as to set a technological and policy framework through which the environmental impact section should be looked at (EWEA 2009). In the next paragraphs, two approaches will be followed to study the case of the UK wind power environmental impacts in order to answer this question.

2 Some Interesting Methods to Assess Environmental Impacts

With regards to the method, we have used two methodological approaches to identify the environmental impacts of wind farms, throughout the whole life cycle of the projects. So first it is useful to distinguish the environmental impacts according to the different phases of their development. The on-site activities which include installation and operation and maintenance phases can be studied separately than the rest (construction, transportation, disposal and recycling) as their main characteristic is the interaction with the public and the environment they are located, rather than the material and energy use or their waste and emissions. Of course, during installation, operation and maintenance, materials and energy resources are required but their impacts are much lower and less noticeable or disturbing than the activities taking place on the particular site, such as the traffic of vehicles, the explosions for the excavations and the utilisation of heavy machinery. For example, the greenhouse gas emissions emitted during the operational phase for the production of electricity are less disturbing than the noise and the potential negative effects on the biodiversity (e.g. birds). Therefore, it is useful to study the environmental impacts of wind power in two different ways.

2.1 On Site Environmental Impact Assessment

The first way is to follow an approach similar to an Environmental Impact Assessment process which is mainly required for large projects, but also for even smaller projects if there is a strong probability that they will have significant effects on the environment and that depends on the nature, size and location of the specific project. According to the European Union, Environmental impact assessment is a procedure for the environmental implications of decisions to ensure that plans, programmes and projects likely to have significant effects on the environment are made subject to an environmental assessment, prior to their approval or authorisation (European Commission 2011).

Apart from assessing the environmental impacts that arise from the production of the wind turbines and the other wind farm components which is an approach more or less product oriented, there are also other important environmental impacts that come as a result of placing and operating the wind farms in a specific location. These impacts resemble the ones that a common building/industrial infrastructure would have to an area like noise, land use change, impacts on birds and visual aesthetics, while others are more specialised and relevant to the operation of the turbines like impacts from electromagnetic fields as well as impacts to the anthropogenic environment, like problems to radars and ship collisions.

As already mentioned before the offshore wind farms are more complex as the whole infrastructure has to be placed and interact with the sea elements and therefore need additional components and work like platforms, underwater cabling, interconnectors and substations, dredging, underwater construction and explosions. Even the operation and maintenance is more complex as the structures and the wind turbines have to be protected against corrosion and the inspection has to utilise boats or helicopters for the transportation of the personnel.

2.2 Life Cycle Assessment of Environmental Impacts

Therefore, the second way is to follow a Life Cycle Assessment (LCA) which is a structured and standardised method that takes into account all the phases of a product or service from the raw materials extraction, production, transportation, operation and maintenance to the end of life treatment (disassembly, disposal and/or recycling) and provides a systematic view for the flow of materials, energy, waste and emissions. This approach provides a transparent and objective way to map the inputs of material and energy and outputs of emissions and wastes with regards to the product or service under study. The LCA methodology according to the ISO 14040 standard includes four stages, as explained below (ISO 2006).

In the first stage, at the *Goal and Scope definition*, we define the aim and set the boundaries of the product/service that we want to study. For example, our goal can be to assess the environmental impacts that come from the electricity of 1 kWh generated from a wind farm, to these that come from a coal fired power station. Our boundaries could include the wind turbines, the substations and exclude the connection to the national grid. Later on, based on the study requirements and on more information about the breadth and depth as well as data availability, we may need to redefine these boundaries. For example, if we cannot find enough data for the substation, we should restrict our study to the wind turbine.

In the next stage, the *Life Cycle Inventory analysis*, we analyse the product to its components and map the necessary materials, energy and processes used during all the phases of the product/service. For example, we would break the wind turbine down to its components (tower, foundations, nacelle, blades etc.) and these into the materials and processes they need. So for the wind turbine's tower we would take into account the amount of steel required the galvanisation process it has to go through the fact that it will be transported from the factory which could be in Germany, Denmark (if we are talking about a Siemens or a Vestas wind generator) and so on. In that way we compile and quantify the inputs and outputs against the functional unit, which in simpler words shows what is the need that our product or service covers. This analysis ideally leads to basic flows of material/energy.

Then we proceed to the next stage, the *Life Cycle Impact Assessment*, where knowing the inputs and outputs we can match them with their associated environmental impacts and evaluate the overall impact of the product/service.

Let us consider a very simplified case where we know that the carbon dioxide emissions for the transportation of 1 kg for 1 km by boat would be 10 g and by truck 130 g. Then we could add the carbon dioxide emissions for the transport of the 200 ts wind turbine from a Chinese port 17,900 km away (by boat) and to the place where it will be sited 100 km away (by truck) and find out that the total carbon dioxide emissions from that would be 38.35 ts which will contribute to the global warming potential by 38.35 ts CO_2 eq. On that point, we should mention that it is not only the Global Warming Potential (GWP) that we are interested in, but also other environmental impact categories like Ozone Depletion potential or Human toxicity potential.

Finally, in the *Interpretation* stage, we check the results and understand what they mean with regards to the original aim of the study. We could see which part of our product contributes more to the environmental impacts and analyse different scenarios for the use of alternative materials and process which would make our product or service more environmental friendly.

The Houses of Parliament in the UK, through the Parliamentary Office of Science and Technology (POST), published two reports on Carbon footprint of electricity generation (POST 2006, 2011). The latest results show that the life cycle GWP of the electricity coming from coal ranges between 786 and 990 gCO_2 eq/kWh and from gas between 365 and 488 gCO_2 eq/kWh, while for onshore and offshore wind the ranges are 38–96 and 5.2–13 gCO_2 eq/kWh, respectively. This is not bad if we take into account that the variety of the low carbon energy sources could achieve almost zero emissions or do not usually exceed 150 gCO_2 eq/kWh. Of course, the results vary among the different literature sources and that depends on the exact characteristics of the power station or wind farm under study, but even these estimations highlight the dramatic decrease in the environmental impacts that an increase of wind power contribution to the UK electricity mix would have.

3 Turning Theory into Practice: Case Studies from UK

3.1 Environmental Impact Assessment Case Studies

Before a wind farm project proposal gets consent in the UK an Environmental Impact Assessment should take place and therefore many documents can be found for cases from small/medium wind farms to large ones. Through studying these documents someone can get information about the potential impacts of a wind farm project, but the easiest way to find the cases of rejected wind farm proposals can be found in newspapers, especially in the local ones as it is the local councils that mainly reject the projects after the intervention of the people who are/will be affected. Those people can be affected in a number of ways:

3.1.1 Noise Impacts

Usually, a wind farm during its operation produces low to moderate noise ranging from 35 to 45 dB, where 35 is the noise similar to that of a library and 45 to that of an open plan office and this can be either because of the mechanical noise that comes from the generators and the gear boxes or the aerodynamic noise from the blades. Noise, is affected by the turbine characteristics, the landscape, the layout of the farm, the topography and of course the wind's speed and direction. In depth research has been undertaken from the especially set up Working Group on Noise from Wind Turbines whose findings are depicted in the ETSU-R-97 report (Department of Trade and Industry 1996) and which are taken into account into the guidelines for national planning of renewable energy. Nevertheless, in 2011 a Lincolnshire couple sued locals, on whose land some of the turbines have been sited, as well as the companies which own and operate the 2 MW wind turbines as written in the BusinessGreen (Shankleman 2011).

Apart from the easily perceived noise that humans can hear wind turbines also produce infrasound which is low frequency noise at a range below 200 Hz, inaudible to humans and which allegedly could cause symptoms ranging from irritability and sleep disturbance to anxiety and memory disturbances (Berglund et al. 1999). The UK Health Protection Agency acknowledges the lack of evidence supporting wind turbine generated infrasound as a health risk and more research is necessary (Centre for Sustainable Energy 2011).

3.1.2 Land Use

Like every construction project, wind farms change the use of the land and this may have important environmental impacts. Equally important environmental burden can be placed on peatlands and natural reserve areas, which are included in the surroundings and in general natural habitats that should be conserved which are included in the Ramsar and Natura 2000 sites.

Peatlands in particular are very important and if the main driver towards developing wind farms is global warming mitigation then they should be a place to avoid. Plans to build one of Europe's largest onshore wind farms in the Outer Hebrides were formally rejected in April 2008 after Scottish ministers ruled the £500 m scheme would devastate a globally significant peatland. A Scottish MEP claimed that Scotland's 1.9 m hectares of peat and bog were part of the planet's "airconditioning system" and this coincides with the scientific review which states that peatlands could be the most important terrestrial carbon store which has the ability to store so much carbon that could offset the annual UK carbon dioxide emissions many times (Worral et al. 2010).

3.1.3 Impacts on Birds and Bats

The Royal Society for the Protection of Birds (RSPB) takes into account the impacts of wind farms on birds and classifies them in the following four types: collision, disturbance displacement, barrier effects and direct habitat loss (Langston and Pullan 2003). The two most important effects are collision and disturbance displacement, which may also include the barrier effect. These are quite intuitive as such vertical structures with moving parts could easily be resembled to giant scarecrows that keep the birds away reducing the availability of flight paths for resident species and disrupting the paths of migratory birds. Also intuitive is the fact that birds may collide on a height structure as it usually happens with big buildings. The direct habitat loss could also be very important though, in the cases that very rare species are affected and of course attention should be given to placing a lot of wind farms in a small area. In the case of the £1.5 billion project of a wind farm scheme near Norfolk which was scrapped over fears that it could kill 90 birds a year as written in the Guardian (Macalister 2012) the RSPB admitted it had opposed that wind farm but supported other schemes in the area.

Additionally, for the offshore wind farms the disturbance from ships and the adverse impacts to the shallow water ecological areas (which host the habitats for breeding, resting and migratory seabirds) has to be taken into account, too. Furthermore, offshore wind farms use higher wind turbines and cover greater surface and include higher abundance of sensitive large bird species, in their area. As observed in Blyth Harbour, bird collisions with rotor blades are rare events even despite the wind farm is located within a Site of Special Scientific Interest and Special Protection Area, under the Birds Directive.

Similarly to birds also bats are affected by wind farms, and several mortality cases have been reported, especially since their migratory paths are not well known like birds and more research would be needed but since bats are protected species, there are certain policies in place and specific guidance is provided by Natural England (2012). Based on such policies, as broadcasted from BBC the District Council rejected the scheme near North Devon because of the impact on bats (BBC 2008).

3.1.4 Impacts on the Anthropogenic Environment

Visual impacts subsequently affect both the landscape and sea scape as the introduction of a manmade tall structure in the environment is not aesthetically welcomed by some people especially when this is placed in areas famous for their beautiful surroundings like touristic areas, archaeological sites, or in general areas of outstanding natural beauty. As mentioned in the Courier (Reoch 2013) in the case of a wind farm, the Scottish Natural Heritage consider it to be inappropriate, and the Council's development quality manager, recommended refusal of the plan, based partly on the "siting, size of turbines, prominence and visual association" with existing and approved wind farms having a major "adverse" cumulative impact on the existing landscape and visual amenity. On the other hand, offshore wind farms may have taller structures but they are usually far from human vicinity and therefore they are not so disturbing.

Another adverse effect that may take place, is the phenomenon known as 'shadow flicker' which is observed easier during sunny days and has to do with the shadow that the blades are casting on a specific area when they hide the sun during their rotation. This can be annoying but according to research there is no sound evidence supporting that this causes seizures to photo epileptic people (Smedley et al. 2010).

The presence of high vertical metal structures like the wind turbine towers as well as their blades may cause false radar responses or mask the genuine aircraft returns and cause misidentification or miscalculation of the location of the aircrafts, posing potential threats to aviation safety as well as the national security (DTI 2002). The case of Hoyle offshore 60 MW wind farm though which has been the UK's first major offshore renewable power project and which covers an area of 10 square kilometres, started its operation in 2003 and has shown (Howard and Brown 2004) that it does not significantly compromise marine navigation or safety.

3.2 Life Cycle Assessment of Wind Farm Environmental Impacts

Although Environmental Impact Assessments (EIA) are mandatory for the development of wind farm projects of significant size, a LCA is not legally required for the development of any product or service in the UK. This fact, together with the time and money required for an LCA, has restricted the existence of these studies only within the scientific community and the manufacturers of wind turbines. Moreover, the external validation and the interpretation of the results can increase the complexity and discourage the use of this method.

Nevertheless, in Denmark where the offshore wind has been developed earlier than the UK, a couple of LCA case studies exist (Vestas 2006) and provide us with interesting findings. In addition, LCA studies and reports for the wind power can be found in various sources. There are European and nationally funded projects reports, like the ones from ECLIPSE, the NEEDS, the CASES and the SPRIng (Azapagic et al. 2011) or Environmental Product Declarations from wind turbine suppliers like Vestas and Siemens. Scientific journals like the Journal of Cleaner production, the International Journal of Life Cycle Assessment and the Journal of Industrial Ecology have published interesting articles relevant to LCA of wind power too. Last but not least, professional databases like the Ecoinvent and the ELCD provide very useful material (SCLCI 1998), which can also be used for modelling and which allows further incorporation of the specific characteristics of each wind farm.

According to those references, the phases that are usually covered during the whole life cycle of a wind farm are described as follows. Firstly, *the construction of the different components* of the wind turbines like the tower, the nacelle and hub, the blades, the foundations and the grid connection cables and substations. This takes into account the extraction of the raw materials and manufacturing of these components or building of the specific structures. It has to be noted that the wind turbines might be the same for both onshore and offshore farms but offshore wind farms require usually larger wind turbines and more sophisticated foundation and transmission infrastructure and that increases the total environmental impacts.

Then, comes the *installation phase* where the components of the wind turbine are assembled. In some cases this may include the on-site building works like the foundations and substations mentioned in the construction phase, especially important for the offshore wind where boats, drilling and dredging might be included.

The *transportation phase* includes a wide range of activities like the transport of the raw materials to the manufactures, the transport of the different components among the suppliers and to the assembly place, the transportation of the technicians required for the operation and maintenance and finally the transport of the different parts after the end of life for disposal or recycling. The operation and maintenance phase may include the technical inspection and fixing/replacing of parts, cleaning and the lubrication and that might include the use of trucks, cars, boats and sometimes even helicopters. Finally, the end of life or decommissioning phase includes the dismantling of the wind turbines, disposal for landfill or incineration and recycling of the different parts.

As described in the previous paragraphs the wind energy is "doing well" with regards to their life cycle impact assessment comparing to coal, gas and falls near the mean value of the rest renewables. Nevertheless, these results can be very generic and may be restricted in terms of the breadth and depth of the LCA and they may not express the environmental impacts in more categories other than the Global warming potential.

So, the following question arises: *If we exclude the Global Warming potential, is wind power better than fossil fuel generated electricity for the rest of the environmental impacts?*

Based on LCA modelling of the UK Electricity mix that was performed using GaBi software and Ecoinvent database, at the Life Cycle Assessment Research Hub of the University of Cumbria, wind power electricity generation seems to have less impact on the Acidification, Eutrophication and Ozone Layer and Photochemical Depletion, Freshwater and Marine Aquatic Potential *but it seems to have more impacts on and Terrestrial Ecotoxicity as well as Abiotic depletion potential.* This practically means that we have to pay attention and address specific areas that could create problems in the future and they have to do mainly with the materials for the following reasons: The toxicity issues are a result of the use of materials like zinc, tin, etc. that can be toxic and which are mainly used for the anti-corrosion treatment of the metal parts which is necessary for the increase of the wind turbine's life time. The abiotic depletion potential has to do with the fragile abundance of materials like steel which is the main material used (the tower itself may contain 200 ts of steel) as well as with the scarcity of special materials called rare earths (like neodymium and dysprosium) which are used in the magnets of the generators.

The next question that can be answered through the LCA study is which phases during the whole life cycle pollute more. It seems that the construction phase accounts for the majority (around 80 %) of the impacts. The transportation usually accounts for the least and the rest 20 % is divided between operation and maintenance and end of life phases and this depends on the disposal and recycling options used. Nevertheless, it has to be noted again that these results are site specific but a general suggestion can be drawn that it might be appropriate to focus on the improvement of the construction phase.

4 Discussion and Concluding Remarks

Following both approaches (EIA and LCA) for the evaluation of the environmental impacts of the wind farms, a lot of issues are highlighted that may put pressure on the actors involved. Hopefully, scientific progress and policy development might alter the existing technical and institutional framework by providing mitigation or adaptation opportunities that will facilitate the renewable energy governance.

A lesson that can be derived from the case studies of the rejected wind farm proposals as well as the LCA modelling, is that renewable energy may be the solution to climate change and fossil fuel reserves depletion but a *successful governance* should also include the aspects of a good design and intense development effort, so as to minimise the potential adverse effects that might come up.

Tackling the adverse environmental impacts of wind farms can be more complex and less uniformly since they are very much site specific. Nevertheless, the following generic suggestions could be adopted. With regards to the noise, this can be mitigated by insulating the nacelle that contains the mechanical parts and by setting the wind farms at a certain distance away from residential areas. For the land use issues, abiding the guidelines and policies and including the local stakeholders early in the wind farm project development can be really helpful. Aviation species issues can be more complex and the need of experts such as ornithologists should be sought as the mitigation measures vary by sites and species but as a rule of thumb important zones of conservation and sensitivity areas must be avoided.

Concerning the anthropogenic environment impacts, proper siting of wind farms could also help again interference with radars and communication signals. Thus, national and local authorities should include a various range of stakeholders like the Environment Agency, the National Grid, the Ministry of Defense, the Civil Aviation Authority and of course representatives of the local citizens and non-governmental associations during planning.

For both the natural and anthropogenic impact, proper design and location can help to avoid any potential problems, but if the "perfect place" cannot be found or issues are realised after the wind farms have been installed, there is a variety of solutions that could be applied but these should always be balanced against cost implications. To this point, early stakeholder involvement can be the key as the cost of retrofitting gets higher the closer the project comes to an end. So although the Government has all the infrastructure in place to protect local habitants of an area, it seems that it lacks insights and link with local communities, highlighting the gap between national and local planning and implementation. But EIA is not enough, because there are "hidden" environmental impacts from using wind power, which can only be demonstrated through new scientific methods, such as LCA. The majority of the environmental impacts identified from the LCA approach could probably be mitigated by using alternative materials so as to avoid the ones that are or will become scarce. For example, very strong, low cost, environmental friendly, permanent magnets could be manufactured by iron nitride $Fe_{16}Ni_2$, alleviating the need for rare earth inputs, such as neodymium (University of Minnesota 2010). For more common but high demand materials, like steel, apart from their substitution by other materials, a more efficient manufacturing design can be developed, minimising the material inputs requirements to the production system (Allwood and Cullen 2012).

Nevertheless, the wind farm development boom is happening **now** and the toxicity brings up the concern whether *we create one problem* (*water poisoning and ecosystem deterioration*) *in our effort to solve another one* (*climate change*). The scarcity of materials on the other hand is connected with economic problems, security of supply and potential increase of greenhouse gas emissions due to the increased transportation and raw material acquisition processes that have to be implemented to make up for the demand.

Therefore, the issues highlighted by studying the development of wind energy farms should be taken into account and involvement of all the relevant actors should be sought at first place. Furthermore, the planners and developers should always keep themselves updated for the solutions that science and engineering advances offer and use their imagination for developing the answers to the questions that will arise. For example, what will happen if climate change is dethroned from the top of the environmental challenges agenda?

References

Allwood J, Cullen J (2012) Sustainable materials: with both eyes open. UIT Cambridge, Cambridge

Azapagic A et al (2011) Assessing the sustainability of nuclear power in the UK. http://www.springsustainability.org/downloads/SPRIngReport.pdf. Accessed 22 March 2012

BBC (2008) Wind farm rejected over bat fears. BBC News, 29 February. http://news.bbc.co.uk/1/hi/england/devon/7270438.stm. Accessed 24 May 2013

Berglund B, Lindvall T, Schwela D (1999) Guidelines for community noise. Geneva: World Health Organization (outcome of WHO expert task force meeting: London). http://whqlibdoc.who.int/hq/1999/a68672.pdf. Accessed 28 Feb 2013

Department of Trade and Industry (DTI) (1996) The assessment and rating on noise from wind turbines. ETSU-R-97. Working Group on Noise from Wind Turbines. Final Report

Centre for Sustainable Energy (2011) Common concerns about wind power. http://www.cse.org.uk/downloads/file/common_concerns_about_wind_power.pdf. Accessed 22 Dec 2012

Department of Trade and Industry (DTI) (2002) Wind energy and aviation interest interim guidelines. http://www.apere.org/docnum/recherche/view_docnum.php?doc_filename=doc1288_aviation.fiche115.pdf&num_doc=1288. Accessed 22 Dec 2012

European Commission (EC) (2011) Directive 2011/92/EC of the European Parliament and of the council on the assessment of the effects of certain public and private projects on the environment. http://eur-lex.europa.eu/LexUriServ/LexUriServ.do?uri=OJ:L:2012:026:0001:0021:EN:PDF. Accessed 22 Dec 2012

European Wind Energy Association (EWEA) (2009) Wind energy—the facts: a guide to the technology, economics and future of wind power. http://www.wind-energy-the-facts.org/documents/download/Chapter5.pdf. Accessed 22 Dec 2012

Howard M, Brown C (2004) Results of the electromagnetic investigations and assessments of marine radar, communications and positioning systems undertaken at the North Hoyle wind farm. http://www.dft.gov.uk/mca/effects_of_offshore_wind_farms_on_marine_systems-2.pdf. Accessed 22 Dec 2012

International Organization for Standardization (ISO) (2006) ISO/DIS 14040 environmental management—life cycle assessment—principles and framework. Geneva

Langston R, Pullan J (2003) Wind farms and birds: an analysis of the effects of wind farms on birds, and guidance on environmental assessment criteria and site selection issues. http://www.birdlife.org/eu/pdfs/BirdLife_Bern_windfarms.pdf. Accessed 22 Dec 2012

Macalister T (2012) Wind farm scrapped over fears for birds. The Guardian, 6 July 2012. http://www.guardian.co.uk/environment/2012/jul/06/wind-farm-scrapped-fear-birds. Accessed 24 May 2013

Natural England (2012) Bats and onshore wind turbines: interim guidance. Technical Information Note TIN051 Sheffield http://publications.naturalengland.org.uk/publication/35010. Accessed 22 Dec 2012

Parliamentary Office of Science and Technology (POST) (2006) Houses of Parliament (UK), Carbon footprint of electricity generation, Number 268. http://www.parliament.uk/documents/post/postpn268.pdf. Accessed 22 Oct 2012

Parliamentary Office of Science and Technology (POST) (2011) Houses of Parliament (UK), Carbon footprint of electricity generation, Number 383. http://www.parliament.uk/documents/post/postpn_383-carbon-footprint-electricity-generation.pdf. Accessed 22 Oct 2012

Reoch P (2013) Mull Hill wind farm plan rejected, The Courier, May 23 http://www.thecourier.co.uk/news/local/perth-kinross/mull-hill-windfarm-plan-rejected-1.95903. Accessed 24 May 2013

Shankleman J (2011) Couple seek £2.5 m damages over noisy wind turbines. Business Green. http://www.businessgreen.com/bg/news/2086053/couple-seek-gbp25m-damages-noisy-wind-turbines Accessed 24 May 2013

Smedley A, Webb A, Wilkins A (2010) Potential of wind turbines to elicit seizures under various meteorological conditions. Epilepsia 51:1146–1151

Swiss Centre for Life Cycle Inventories (SCLCI)(1998) Ecoinvent database. http://www.ecoinvent.org/database. Accessed 22 Dec 2012

University of Minnesota (2010) Iron nitride permanent magnet, alternative to rare earth and neodymium magnets. http://www.license.umn.edu/Products/Iron-Nitride-Permanent-Magnet–Alternative-to-Rare-Earth-and-Neodymium-Magnets_20120016.aspx. Accessed 28 Feb 2013

Vestas (2006) Life cycle assessment of offshore and onshore sited wind power plants based on Vestas V90–3.0 MW turbines. http://www.vestas.com/Files/Filer/EN/Sustainability/LCA/LCAV90_juni_2006.pdf. Accessed 22 Dec 2012

Worrall F, Chapman P, Holden J, Evans C, Artz R, Smith P, Grayson R (2010) Peatlands and climate change. http://www.iucn-uk-peatlandprogramme.org/sites/all/files/Review%20Peatlands%20and%20Climate%20Change,%20June%202011%20Final.pdf. Accessed 22 Dec 2012

Champagne and Metal Flowers: Who is Invited to the Wind Generation Party in France?

Céline Burger and François Mancebo

Abstract Sustainability has gradually become fundamental to the regional planning and decision-making process of French local governments. Local authorities are supposed to incorporate sustainability into their mode of governance. They do it, formally, but, the very idea of sustainable development implies a breakdown of the systems of reference for public action, and of individual and collective practices. So combining sustainability with preexisting governance uses is all but evident, as it appears with wind power plant development. In the Champagne-Ardenne Region, local authorities are caught in the cross fire among developers, farmers, landowners, environmental activists, without any tool to help decision-making.

1 Introduction

When discussing renewable energy development everybody implicitly presupposes a functioning local governance and sustainable policies, a type of "government of compromise" (Taiclet 2007). François Hollande, during his recent October 29, 2012 speech specified that the energy transition issues will now placed under the responsibility of local authorities (mainly communes and inter-municipal organizations) within the new decentralization bill (Gaudin 2002). However, in Champagne-Ardennes, as in other French regions, policy coordination between local authorities is rather poor: hence there tends to be haphazard local governance actions (Pasquier et al. 2007). For example, in the process of land acquisition for

---

C. Burger (✉) · F. Mancebo
Laboratory EA 2076 HABITER, 57 rue Pierre Taittinger 51096 Reims Cedex, France
e-mail: celine.burger@univ-reims.fr

F. Mancebo
e-mail: francois.mancebo@univ-reims.fr

E. Michalena and J. M. Hills (eds.), *Renewable Energy Governance*,
Lecture Notes in Energy 23, DOI: 10.1007/978-1-4471-5595-9_13,

wind farm implantation programs, the decisions made by local authorities are directly and almost uniquely dependent on the immediate foreseeable financial gains, to the municipality hosting wind farms.

Champagne-Ardennes, the study site, was the first French region undertaking the production of wind power with an installed capacity of 838 MW in 2011, 21 % more than in 2010 and 130 % more than in 2009. The regional policy and political will proved decisive in this evolution (Le Gales 2006). But, on the other hand, Champagne-Ardenne region is also the only region in France in demographic decline, so the local authorities want to develop activities which attract people. Moreover, it faces the competition and attractiveness of very large urban areas of the Paris region -Ile-de-France- and of Alsace with Strasbourg. Thus, Champagne-Ardennes suffers from a poor self regional brand image, despite its rich history. Nowadays, the perception of the region by its inhabitants is not good at all, because the territory has gone through lots of economic crises which have not helped people to build a link with their living territories, much people have lost their job.

Is it possible that wind farms installations generate a positive or negative impact on this perception? Could wind farms contribute to a "repopulation" of this region? At least, could the process of RE implementation itself—with its conflicts and its pitfalls—tell us something about the social and political dynamics that entails this negative perception?

These are the questions this chapter addresses, and which includes an inventory of regional information and local actions concerning wind power plants, followed by the analysis of both public and private local actor's interviews as well as anti- or pro-local news articles. The interviewees that I took included citizens directly concerned by wind farms (neighbors, farmers, etc.), local authorities and wind farms developers. Considering that local decision-making about the positioning of wind farms impacts concerns mainly local identity and territorial construction, *this raises issues of local governance and the articulation of decision-making with search of sustainability*.

2 Promoting Renewable Energies Production in Champagne Ardenne: Mode of Land Acquisition and Participatory Governance

Wind farms construction requires massive investments, as any industrial branch, and involves many stakeholders. Wind power market is therefore severely constrained by laws and regulations. Though financial issues strongly influence local stakeholder's choice to install wind power units, the economic benefits are definitely *not shared among the household concerned by the wind projects* (reasons will follow). On the other hand, the situation generates important conflicts at a local scale; that local authorities usually try to handle with participatory governance procedures (Ost 2004).

2.1 *Which Strategies for Wind Farm Developers?*

In France, the installation cost of a wind turbine of 2.5 MW is around 3 millions €-—75 % of this sum concerns construction, 5 % the development cost (impact studies), and the remaining 20 % the earthworks, etc. This large cost makes it almost impossible for an individual, a household, or even a small village to undertake by themselves the installation of a wind turbine. Therefore, it is necessary that developers, with their investment capital, are involved in the wind farm development.

However, there are big differences among the developers: they don't share the same objectives and practices. Most of them are, at the same time, both developers and operators of wind farms, such as Ostwind or Enel Green Power. They handle the construction, the operation, and the maintenance of the equipment. This type of developer-operator is very active in Champagne-Ardenne. Those developers put big pressure on local authorities to foster new projects. However, it must be the local authorities who should control the decision making with process of participation and negotiation with all actors (Mancebo 2007b).

On average, a wind farm needs 15 years of exploitation to be profitable. As a result, some companies—Ostwind for example—keeps operating a number of their wind turbines in order to ensure stable financial resources in the long run, while the other companies sell to release cash so that they can reinvest. The new owner of a wind farm, usually outsources the operating part to a subcontractor, they are financial investors. The benefit to the new owner is in the differential between the operating costs and the income achieved from selling the produced electricity to the grid.

Often, the operator commissioned by the new owner is the original developer-operator who constructed the plant: it is a win–win deal with economies of scale for both partners. This system allows the developer/operator to charge operating fees to the investor, who—in the reality—might be a branch of the same company. These fees are a financial input, usually used to buy back the installations previously sold, when they become profitable. Besides, these developers/operators have strong ties and partnership with wind turbines manufacturers, like Repower, Enercom, or Vestas, and which are responsible for the installation of wind turbines on the selected sites.

The manufacturers associate the installation of wind turbines with maintenance contracts on their name, for the next 10 or 15 years. In case of underwriting this type of contracts, the maintenance is provided by the manufacturer during the first years of the operation. Therefore, some wind turbine manufacturers tend to become developers too as economic difficulties recently appeared in the wind turbine industry: therefore, they become developers/manufacturers aiming at a growth management strategy.

This happened with Gamesa group. This Spanish manufacturer has created a development branch. This branch of the company is in charge of the wind farm development, which wind farm will sold before begins operation. These wind farms are often sold by Gamesa through Wind Prospect, a wind farms broker for big companies like IKEA.[1] Such a configuration assures the manufacturers that their wind turbines will be installed on upcoming wind farms once the building permit is obtained. There is a double interest here: If finally the developer/manufacturer is mandated to operate the wind farm, he receives payment for the exploitation of the farm and farm was created with the specific group turbines.

In the latter category of developers, design and technical consultants engineering offices can be called "developers/engineering offices": they got involved only recently in wind power development. These developers only manage the wind farms during the construction phase and their expertise usually doesn't enable them to go beyond this phase. An engineering office is supposed to deliver studies and reports for the companies that commission them. Developers/engineering offices can though undertake studies by themselves when they identify a site that is suitable for a wind development. Once these studies are done and if they prove positive, they sell their report either with the building permit granted or ready to be submitted to manufacturers or local authorities.

The business is profitable: usually, the price for a report with the building permit granted is about 250,000 €/MW if the wind farm is well located, and 100,000 €/MW if the location is less attractive. The benefits depend on many parameters, such cost of studies, cost of turbines, interest tax, and of course the production of farms which depend of physical aspect (numbers of wind hours, wind capacity, etc.). Nowadays, the reports without the building permit are naturally considerably less interesting because the risk for the buyer is not to be granted the permit afterwards.

There are several steps that precede the erection of a wind farm: the site choice, the feasibility study, the land acquisition, the constitution of the impact study, and the submission for the building permit. Purchasing (or renting) the land is the most delicate undertaking. It is commonly called by the developers in France, *sécurisation du foncier* (securing land use). This terminology makes sense: this is when the first confrontations between the different stakeholders take place. The conflicts that initiate at this phase of the project tend to intensify later, as public stakeholders come into the picture as supervisors (Torre et al. 2006). Short-term interests and personal or financial bias should not impact the decision-making process, neither should allow political shifts after election… but they usually do (Brunel 2010).

In the field of heritage development, many studies show that initiatives are frequently blocked when funders and public authorities do not coordinate their decision-making: they have trouble in imposing decisions to all the stakeholders (Cettolo et al. 2002): Paradoxically, when installing wind turbines, particular interests (often totally disconnected of the real issues of wind farms, such as to step up production of renewable energy to conform with European agreements, of renewable energy production), have a disproportionate impact on decision-making

---

[1] IKEA owns wind farms to produce approximately the electricity equivalent consumed by IKEA group.

(Mormont 2001). Indeed, wind farms usually settle in rural locations, therefore the communes that are concerned by wind projects often amount to less than 1,000 inhabitants. In such small populations, intense political rivalries, land use, and neighborhood conflicts between individuals or microscale social groups and families and/or vendettas, frequently affect decision-making process (Hirschman 1986).

Elected officials are commonly directly concerned by the projects, either as beneficiaries (when they rent their land to developers, for example), or negatively (when they are, for example, neighbors of wind turbines and have to support the, supposed, noise and view nuisance, or the impairment of their properties value). NIMBYs conflicts, fuelled by ancient grievances, often annihilate the efforts to implement wind generators, and the introduction of sustainable policy (Mancebo 2011).

Within all this development of "sustainable projects", as soon as a site is identified as potentially adequate for hosting a wind farm, the relations between wind developers routinely enter a phase of conflicts. Routinely, their relations are rather fraternal, as wind developers are all members of the *Syndicat des Energies Renouvelables* (SER—Renewable Energies Board),[2] and are united particularly when they deal with the initiating departments. Relations turn confrontational on the spot, when developers rush to gain permits for the same site. To arbitrate potential conflicts among actors, a *charte de l'éolien* (chart for the wind industry) has been established. It consists of a code of good conduct that has been ratified by all members of the SER. Which means this is not a contractual document, and, as a consequence, some developers decide not to respect this ethical code, even if they are members of the SER.

According to French wind industry's professionals,[3] it is the foreign companies that do not follow the unwritten code that says: "contact the mayors before the landowners." When they identify a potential plot to host a turbine, they meet directly the landowners, without necessarily contacting the mayor beforehand. For example, this practice not offends in Germany, where the practices are different. But, in France, French developers consider that this introduces a bias against them. Thus, more and more French companies adopt the same behavior.

2.2 *Respecting the Protocol for Land Acquisition*

The protocol for land acquisition allows the basis of the negotiation phase. It encompasses the successive steps of wind farm projects, defines everyone's obligations, and proposes a set of contract models (lease, extract of the cadastral

[2] SER is the spokesman of the wind business in France toward public authorities and media. Its creation is due to the French government who focus on wind power industry to reach the objective of the law *Grenelle II de l'environnement*: to produce 23 % of its energy out of renewable sources.

[3] Comment from fieldwork among project developer of wind in second quarter of 2012.

map, lease of construction). It can also be considered as a reference document for good practices to install a wind farm.

As soon as a site is identified by developers, developers get in contact with the mayor of the municipalities concerned by the project. Identifying a site is the preliminary step and the feasibility studies are often conducted in-house by the development companies themselves. The principal criteria to define a site as suitable are: (a) available space and (b) environmental sensitivity. Getting in touch with the mayors is often made by phone. Developers present themselves, ask if the local community is interested in a wind project and whether they have already been contacted by another developer. A few years ago, if the mayor was already in touch with another developer, it was an established (though unwritten rule) that he should stop soliciting. Now, things are quite different: potential sites are becoming fewer and rarer, which induces strong competition among companies and developers outbid each other.

Nowadays, the cost for the land renting increases as there is more competition among the many developers: for example, it is now up to 5,000 €/MW in Champagne-Ardenne instead of 2,000 €/MW 2 years ago.

Once the mayor is convinced that the wind farm would be beneficial for the community, he usually acts in two different ways: either, he convenes the city council and provokes a debate without presenting his own choice, he then follows the council decision; or the developer he has chosen is directly invited to present his project during a normal city council session.

Some foreign companies which are not familiar with this way of business doing, short-circuit the mayors and get directly in touch with the landowners of the potential sites. These landowners are very easy to find, since the *cadastre*[4] of every French commune is on the Internet. It lists each lot and parcel with owners' names and addresses.

In the *Mont des quatres Faux* project in Champagne-Ardenne, the developer Winvision proceeded this way, contacting the landowners before the local authority. During the public inquiry for a wind farm installment, Denis Rousseaux—the mayor of the commune of *La Neuville en Tourne à Fay*—exposed in an interview that the developers began contacting the landowners and started trying to coax them. As a local elected politician, he felt at odds with such actions. Indeed, local elected politicians are in a precarious position: being against a wind farm project while landowners have given their approval, which might mean they lose votes and having trouble with the local population. But accepting the farm might create conflict with other local groups in the future who will be disturbed by the visual or noise impact of the farm.

For example, the *Mont des quatres Faux* project involves 47 turbines and expands more than 200 m height. The turbines are offshore models which are installed onshore! This means this is a huge construction work planned to last for

[4] A cadastre is a public record legal survey of the value, extent and ownership of land for taxation or administrative purposes.

the future 3 years. Winvision always launch this type of project use the offshore turbines, like the Company did in Estinnes (Belgium).

At the same time the Company Winvision contacts the landowners and developers to ensure land security of the parcel. As we have seen previously, a potential area is previously defined for the project. The first identification phase is not very precise (to avoid great expenses). This potential area is therefore much larger than the area where the turbines will be actually positioned. Developers' work consists, at this point, in securing the land by having a lease agreement signed by all the concerned owners and renters.

The term "land security" makes perfect sense, when considering that developers voluntarily extend the area beyond their real needs, in order to ensure that no competitor will come near, thus strengthening their territorial arrangement. Among, the landowners who sign the promise to lease, some never will see any turbine on their fields. So that, when comes the time to announce who will have (or not) wind turbines in their property, the time for conflicts also comes along. Animosities arise between those who have turbines and those who have not, resulting in opposition movements against the installation.

Beyond all the justifications, it is usually the frustration that fosters the opposition movements (Hirshman 1970). The choice of the plots that will be hosting wind turbines is determined before submitting the application for a building permit. This choice can change thirty times a year, depending on the environmental impact studies. The environmental impact studies take 2 years, on average, sometimes more—up to 6 years—which means that promises to lease maintained for many years (for the land security), may become null all of a sudden, ensuing in big financial loss for the unlucky leasers.

At this stage, the policies of developers to avoid frustration are different and depend on the company. Some companies do not consider the promises of lease as engagements, thus they do not propose any compensations to the landowners. Others propose right the way agreements, passage agreements, overhang agreements, or installation of delivery substations (one for five turbines) to the landowners that do not have turbines on their fields: so they will receive a rent that will reduce their loss.

Other companies, finally, prefer to develop a general system of equalization payment to the landowners called ZDE (Zone de Développement Eolien—Wind power Development Area), where all the landowners inside the securing area ZDE receive a rent in direct proportion to the size of their lot, whether they have turbines or not. ZDE was formalized and generalized by the law *Grenelle II de l'environnement*. The municipalities usually establish ZDE where they want to allow the installation of wind farms. In order to respect these ZDE wind power development areas and under the article 10 of the law n°2000-108 about the modernization and development of the public service of the supply of electricity (modified by the law n°2005-781 defining French energy policy), only the wind turbines built in ZDE areas may benefit from the obligation imposed to EDF (*Electricité de France*—French national power operator) to buy the electricity they produce.

In fact, the tariffs were very high and regulated to be fixed at the national level, which meant a comfortable and constant source of revenues for the wind power operators. Since July 13th 2007, only the electricity produced within ZDE are included and the tariffs have been re-estimated regularly to be closer to the energy market realities: for wind power onshore the price is 0.082 €/kWh linked to the inflation rate during the ten first years of production. It then varies between 0.028 and 0.082 €/kWh during the next 5 years depending on the location of the production site.

A wind farm owner or operator who installs his turbines outside a ZDE cannot ask for the green energy-purchasing obligation by EDF under the advantageous conditions fixed by the CSPE (*Contribution au Service Public de l'Electricité*—Contribution to the electricity public services). Connecting a wind turbine to a power network, outside a ZDE, and without a responsible entity for electricity purchasing, has become illegal. This law was repealed the 12th March 2013.

Once the competitive procedure is launched by the municipality, the choice of the developer is mainly based on the renting price he is ready to pay for the land and the financial investment he is ready to make for the thriving of the municipality. Therefore, the contract award decision depends heavily on pure financial issues as will be discussed in the next section. These issues usually result in another type of conflict, following those evoked before (choice of the turbines implantation sites, and promises of lease) but of a completely different type.

3 Interesting Financial Issues

Green energy development is encouraged by financial incentives like the feed-in tariff (see above). In France, the wind power market is doing well. Professionals in the wind power sector—especially developers—bring with them strong financial arguments when they want to convince municipalities, tenants, or landowners to accept wind farms on their land.

3.1 The Profit Motive for the Municipalities (There is Good Money Here, but Whose Pocket is it Going to Go into?)

The wind farms are usually located in rural areas. In order to be allowed to be installed, a wind farm must be located at least 500 m away from houses. Within the farm, there must be at least a distance 800 m between each turbine. Considering that a wind farm must have at least 5 turbines, the minimum area covered by a wind farm is about 2.132 km^2. Furthermore, for the production to be optimized the wind must circulate regularly in the area. Because of these restrictions, it is almost impossible to install wind farms in urban environments. In Champagne-Ardenne a lot of rural communes can provide optimal sites for wind farms.

These, are small municipalities with around 300 inhabitants. Since the building and installation of a wind turbine cost is around three million euros, they cannot develop farms. Thus, when a mayor is prone to creating a wind farm, he absolutely needs developers' assistance and their financial strength. Developers make a financial commitment to the communes where they install wind farms. Apart from the rent received after the wind farm installation and the royalties received to accept the wind farm, this financial commitment may take two other forms:

- Specific endowment funds to municipalities supporting a project development. Though this is a legal provision, this type of payment raises a problem: If the project doesn't become a reality, the company (developer or operator) does not have to hold this financial commitment.
- The sponsorship through the contribution to an association promoting wind energy (for example 5,000 euros during 5 years). The wind energy's developers use more and more this possibility. Within the farm project called ≪*Mont des 4 faux*≫, the developer Winvision proceeded according to this very exact way.

Due to the financial support provided to the commune, developers hope to influence the local decision-making, fostering a positive bias toward their projects. The installation of a wind farm in a commune usually generates debates and opposition among the population: it may easily become a major political issue. And it is a long way from the blueprint to its realization: once local authorities give their approval for the implementation of a wind farm project, 26 public services have to be consulted before the building permit is delivered. It does not always work out, which usually ends in a backlash against the mayor in the next local elections.

Furthermore, in the Region Champagne-Ardenne, mayors are usually landowners and farmers, and so are the members of the municipal council. They are directly interested in the installation of wind turbines on their properties. They mostly judge the pros and cons of the project, which sparks all kind of reactions and concerns between opponents to the project and the municipal council during the public inquiry.

Statements in the local press usually exacerbate the conflicts. For example, in April 1, 2011 *L'Union* -the Champagne-Ardenne favorite local daily newspaper—wrote on its first page: "*Non aux fleurs en feraille?*" (*No to the metallic flowers*). Then, on November 18 2011, an article supported the claims of the «*Association de sauvegarde des paysages de la vallée de la Coole*» (Valley of Coole landscape preservation association) against wind farms. On November 22nd 2011, the newspaper published the answer of GAMESA—the developer having undertaken the project in the Vallée de la Coole—presenting the goals of the project: which implies a production that optimizes the potential of all the turbines.

Recently, there has been an additional reason that makes the wind farms projects rather difficult to realize: the reform of the tax system, stipulating a TPU—*Taxe Professionnelle Unique* (unique business tax) within each intercommunality.[5]

[5] It is a group of commune. You can enter in intercommunality with other commune which implies the sharing of power to many competences.

This means that each and every commune counting less than 5,000 inhabitants has to be part of an intercommunality by 2013, and that the wind farm tax will be shared between the different municipalities of the intercommunality according to their population or surface. The consequence is that every commune concerned wants the wind turbines to be installed on different sites and different communes of the intercommunality; the idea being to spread the nuisances around since profits are shared.

Concluding, the wind turbines' placement cannot be made according to the sole desire of the mayors or the administrative limits of the communes. And wind turbines location is a real bottleneck for the installation of wind farms, since so many conflicting parameters which have to be considered both on the technical and on the human level. The local authorities love to present their arguments (use by person who is in favor of wind energy) from the point of view of the economic interests. That is why, on April 10th 2010, the mayor of the commune of Coole announced in *L'Union* that he will use wind farm's economic spinoffs (rent, subsidies, etc.) to bury all electricity lines in the village. *A project that, according to him, "costs a lot"*.

The development of wind power is a real financial support for rural municipalities counting a few hundred inhabitants that cannot cope with some vital works with the small budget they have. The financial argument is also crucial to obtain a positive decision from the Prefect, the final, and only real decision maker.

In the commune of *Epine aux Bois*, the construction permit for Wolkswind Company's wind farm was delivered by the Prefect, who chose not to consider the unfavorable opinion expressed by the concerned local authorities and the public enquiry commissioner. And this is why the local authorities usually prefer to conciliate with developers in order to maximize the advantages they can obtain from the developers, rather than to adopt a passive attitude or to refuse the project, which can end in being forced to accept it finally but with less benefits; or, worse, to endure the nuisance of a wind farm installed in a neighbor commune without having any advantages.

The local authority position is ambiguous; they are rather "facilitators" than real decision-makers. Still, they are directly impacted by the economic dimension of the project, positively or negatively.

3.2 No Redistribution, Big Frustration

The landowners and the operators are the most important beneficiaries out of the installation of wind farms, whether they are involved, or not in the project. In Champagne-Ardenne the first wind farm was built in the department of *La Marne* in 2002 as a private initiative of a local farmer, Mr Huet. In fact, wind potential in Champagne-Ardenne is rather poor. So, in the beginning, nobody believed in this project. Developers were not interested. But Mr. Huet insisted and finally succeeded in building a wind farm.

Because of this first success, wind farms started developing swiftly in Champagne-Ardenne. Today, more than 95 % of the wind farms in Champagne-Ardenne are created by developers, while the farmers who were the pioneers finally sold their turbines to private providers; like Mr Huet, who sold his wind farm to *Eole Generation* in 2007. This company extended the surface of the farm and changed its name into *Mont de l'Arbre*. Mr Prince, another farmer—now a shareholder in the wind farm *Chapelle Vallon*,—created the second wind farm of the region, in the department of *Aube*. He explains that his sudden interest in wind power was due to the evolution of the European CAP (Common Agricultural Policy).

In fact, French farmers, after the EU enlargement, thought that the CAP would be threatened and that the subsidies will diminish greatly. Thus, compensatory indemnities on their classical framing activity could turn inadequate to ensure the agro-profitability of their establishments. Green power development, like bio-fuels and wind power, appeared as a good way to diversify the activities of some farmers. Besides, the regional council of Champagne-Ardenne, provides a big support to the Chamber of Agriculture, which promotes wind power industry as a way of diversifying farmers activities without losing agricultural land (unlike with what happens with ground photovoltaic energy for example). In such a context, wind farms are emerging as an important issue for Champagne-Ardenne's economy.

At the same time there are more and more norms to respect when installing a wind farm (ZDE, ranking ICPE, minimum size, etc.). It results from this burgeoning regulatory framework that the financial cost to create a wind farm increases steadily. Thus, wind power branch is now an industrial sector, where only big companies can play.

The rent to the landowners and to the operators is around 3,000 €/MW, but when the farmers bring their contract to the *Chambre d'agriculture*, they voluntarily hide the rent. In theory, the rent is distributed for 1/3 to the operators, and 2/3 to the landowners. In fact, as observed with *Enel Green Power Company*, the distribution is equal between operators and landowners, sometimes even more for the operator. The lease is usually signed only after important negotiations between operators, developers, and landowners. During the construction work, there is a special compensation, since transporters of wind turbines cannot only drive in the edge of the construction area but have to pass through it. Thus, the operator receives an agricultural compensation for the damages caused: compensation for the rights of way, compensation for things such a ruts in the fields from the large transporting lorries.

Once the wind turbines are installed, a lot of money flows between the operators and the farmers. Oppositions against wind farms are often due to the fact that the wind power sector is a hyper lucrative sector but the benefits only go to the owners and the developers, with nothing left for the inhabitants. In order to avoid problems, some developers pay all the owners and operators of the ZDE. But this has apparently proved not enough to calm down the tensions. In the Lorraine Region the pooling the wind farm's rent is almost systematic. In Champagne-Ardenne this way of proceeding does not work, most farmers have a capitalist attitude more individualistic for land management and less practiced sharing action of benefits.

Most landowner farmers in Champagne-Ardenne have big agricultural holdings in open field.

Paradoxically, in Champagne-Ardenne the opponents of wind farm are still few in number in comparison to regions like Picardie where the opposition is systematic. The "*Association de Sauvegarde des paysages de la vallée de la Coole*" is one of the two associations to be opposing to wind farms installation in Champagne-Ardenne, this association can be considered as structural stakeholders group for the social relation (Chia et al. 2008). The members of this Association consider that financial gains are the only reason for wind power installation. For the people who oppose wind farm installation, wind energy exploitation is responsible for the deregularization of the land.

The price for renting a land with a wind turbines is 330 times higher what the exploitation of a same agricultural land can produce with traditional activity (land cultivation) according to the leader of the movement, Mrs. Robert: ≪for a very small parcel of a land, it is too much≫, ≪an agricultural land is made for feeding people, the size of a French department disappears every 10 years because of city's extension, many hectares of agricultural lands disappear due to the installation of wind turbines≫.[6]

Mrs. Robert also argues that the mayor of Cernon—the commune where she lives, in the Marne department (in French "département")—supports the wind power project only for financial reasons: ≪during his commune's meetings, he supported this project because it brings money; it's what he always says in these meetings≫; ≪nowadays mayors are keen for turbines because they want to invest in other projects and they need money≫.

Still, the opponents argue that ≪small rural towns will become ghost towns, without anyone but with new schools and city hall≫: Naturally, the money generated by wind farms could induce the creation of new services in rural areas, therefore attracting people, but the members of these associations don't agree with this idea: according to them, wind turbines fuel neighbors conflicts, and the opponents feel misunderstood and rejected by the other inhabitants. They also think that developers ≪win money with wind turbines who make embarrassment for people≫, and are worried about the value of their land: ≪what if we will not be able to sell our houses because of that (the wind turbines)≫; ≪you have worked all your life and someone puts that in front of your house≫.

4 Discussing the Whole Situation

A territoriale governance considered territory as an area of collective project (Lagagnier 2002) governed by rules allowing concerted public action (Le Gales 1995). However, this type of governance has limits, as the wind energy project, the

[6] Interview with Mrs. Robert, the 17th July 2012 at Cernon.

multitude of stakeholders, and the difference of their interest, compromised power regulator of local authorities and the decision-making is affected. To solve the problem we can use the participatory governance (Mancebo 2007a). But, regional distinctiveness and the individuality compromised the utilization of participatory governance, as in Champagne-Ardenne where the individualities are strong. These observations, asked the question of what type of governance can we used, to reconcile different challenges and posed the question about values and principle compatible with sustainabilities (Buclet 2011).

In France, and through the green energy-purchasing obligation of EDF, the government subsidizes de facto wind power industry. This turns wind industry into a very profitable business. Even as the purchase tariff constantly decreases, there is always the same huge number of wind power companies and developers who try to enter the market. This purely financial profitability is an issue very frequently criticized by wind power opponents. What is the real link among frustration, envy, and opposition? If the wind power industry was not such a profitable business, would it ever have so many opponents? On the contrary, if someday the wind game becomes much less profitable; would the developers disengage from the wind power industry? Finally, what is the real impact of financial issue on the wind farms installation?

When you look at the French wind farm installation procedure in Champagne-Ardenne, the divergence between the actors' point of view is obvious. The development of wind power is supposed to respect the general European agreement, to produce 23 % of green energy. But, besides the European aspirations and financial aspects, sustainability is never evoked by local authorities, when they are interviewed about the reasons why they want to install a wind farm on their commune and how they operated their decision-making process. The European policy in renewable energy shows the difficulty to change the scale when applied policy, notably with the example of the National Renewable Energy Action Plans (NREAP) (Michalena and Hills 2012).

Moreover, during the local consultation, at the level of the communes, strong individuality usually monopolizes the debate. However, the real power of the communes in deciding over a wind farm is not so important. The real decision level in France is the *département*, to be more specific the Préfet de départment who is finally the one that delivers (or not) the authorization for building.

The context is very different in other European countries. In Germany, for example, there is a big pressure toward green power anchored in the public opinion and citizens. In Denmark, the production of wind power energy per capita is much bigger than in France, by 20 %. And Germans target no fossil energy production by 2050. Of course, to reach those objectives, those countries follow simpler rules to install wind farms.

References

Buclet N (2011) Le territoire entre liberté et durabilité. PUF, Paris
Brunel S (2010) Le développement durable. PUF, Paris
Cettolo H, Lefebre A, Mancebo F (2002) Gouvernance territoriale et pratique culturelle de mobilité: deux exemples de réseaux de tourisme culturel. Département Etudes et Prospectives de la DATAR, Toulouse
Chia E, Torre A, Rey-Valette H (2008/4) Conclusion: vers une «technologie» de la gouvernance territoriale! Plaidoyer pour un programme de recherche sur les instruments et dispositifs de la gouvernance des territoires, vol 209. Norois
Gaudin J-P (2002) Pourquoi la gouvernance? Presses de Sciences Po. coll. La bibliothèque du citoyen, Paris
Hirschman AO (1986) Exit and voice. An expanding sphere of influence. Rivalsviews of market society and otheressays. Harmondsworth, Penguin
Hirschman AO (1970) Exit, voice and loyalty: Responses to decline in firms, organizations and states, Cambridge, M.A., Harvard University Press
Laganier R, Villalba B, ZuindEau B (2002) Le développement durable face au territoire: éléments pour une recherche interdisciplinaire. Développement Durable et Territoire. Dossier n° 1: Approches territoriales du développement durable
Le Galès P (2006) Gouvernement et gouvernance des territoires. Problèmes politiques et sociaux. n° 922 mars 2006, La Documentation Française Ed., Paris
Le Galès P (1995) Du gouvernement des villes à la gouvernance urbaine. Revue française de Science Politique 45(1):27–95
Mancebo F (2011) Ville durable et changement climatique, environnement urbain/ urbanenvironment, vol 5. VRM, Montréal
Mancebo F (2007a) Le développement durable. Armand Colin, Paris
Mancebo F (2007b) Le développement durable en question(s). Cybergéo, Epistémologie, Histoire de Géographie, Didactique article 404
Michelena E, Hills J (2012) Renewable energy issues and implementation of European energy policy: the missing generation? Energy policy 45:201–216
Mormont M (2001) Gouverner l'environnement, dans DebuysT (F) et al., Savoirs et jeux d'acteurs pour des développements durables, Louvain-la-Neuve, Bruylant-Academia, pp 221–236
Ost F (2004) De la pyramide au réseau: un nouveau paradigme pour la science du Droit? dans SupioT (A), Tisser le lien social (dir.), Paris, Maison de la Science de l'Homme, pp 175–196
Pasquier R Simoulin V WeisbEin J (2007) La gouvernance territoriale. Pratiques, discours et théories. Droit et Société, vol. 44, Paris
Taiclet AF (2007) Le développement économique territorial au regard des hypothèses de la gouvernance territoriale, dans Pasquier R, Simoulin V, Weisbein J (dir.), La gouvernance territoriale. Pratiques, discours et théories, Paris, LGDJ, coll. Droit et Société 44:109–127
Torre A and al (2006) Conflits et tensions autour des usages de l'espace dans les territoires ruraux et périurbains. Le cas de six zones géographiques françaises, Revue d'Economie Régionale et Urbaine n°3:415–453

Part IV
Renewable Energy Governance: Food for Thought

Renewable Energy Governance Challenges Within a "Puzzled" Institutional Map

Maria Kottari and Panagiotis Roumeliotis

Abstract Environmental sustainability, a highly politicized issue, constitutes one component of global energy security. Awareness of mutual interdependence between global societies and their resources is supposed to facilitate the multilateral cooperation approach towards energy and climate issues. The transition to extensive renewable energy (RE) use, globally, is a key instrument for the passage towards a more sustainable energy system. The global energy governance's institutional architecture has been transformed in order to respond effectively to this goal. In particular governance has being enhanced with the formation of a series of institutions committed to promote the appropriate expertise for a regulatory and legislative framework for further expansion of RE technologies. This Chapter aims at the study of certain institutions in the establishment of RE governance principles. The Chapter indicates that RE international governance is time consuming and still lacks the necessary coherence, which is capable of compromising the public and private interests for the good implementation of RE at a global scale. The variety of the existing national and international institutions dealing with RE offers a fragmented rather a global approach of RE governance. This is further demonstrated by the insufficient RE integration at an EU level, despite the adoption of a very ambitious environmental regulation package. The analysis of the RE policy immaturity in the EU will be shaped, in this Chapter, around the analysis of the Directive 2009/28/EC and its weaknesses regarding the RE local implementation.

M. Kottari (✉) · P. Roumeliotis
Panteion University of Social and Political Sciences, Athens, Greece
e-mail: maria.kottari@gmail.com

E. Michalena and J. M. Hills (eds.), *Renewable Energy Governance*,
Lecture Notes in Energy 23, DOI: 10.1007/978-1-4471-5595-9_14,

1 Introduction

Given the multifaceted nature of the energy related issues, requiring an interdisciplinary approach when assessed and studied, the issue of global energy governance arose as a multidimensional subject of study. The prominence of global energy governance as a new study area for energy policy scholars and policy makers, but also as a national and international policy priority issue, is based on two affirmations. Firstly, the environmental sustainability and the climate change questions are being developed as politically charged ones along with the traditional geopolitical concerns of fossil fuels. They cover subjects such as the security of energy supply and demand, the functioning of energy markets and price volatility, to mention only a few from a continuously updated list. Secondly, national governments and the private energy stakeholders understand the existing interdependence between the energy systems (an increase in the energy demand in Asia impacts on the energy supply systems of the European Union).

Achieving energy security, widely defined as the ability of an energy system to encompass the aspects of affordability, reliability, security of supply and environmental sustainability (Sovacool 2011), is no longer estimated as a *zero-sum game, in which one country's energy security is another's lack* (Goldthau and Witte 2010). Achieving energy security is rather a game where the cooperation and activities coordination could result in mutual beneficial gains and advantages. Such an approach requires the existence of institutions that *place the rules for this game to be played among governmental and non-governmental actors* (North 1990).

The basic notion of governance is formulated on the principle that governments and governmental institutions are not the sole actors of the international system able to impose formal rules or informal principles. There exist also numerous other initiatives, stemming from the civil society and private sector as well, able to enhance the regulatory framework of a certain international policy domain or procedure, or to override certain processes as well. Energy is a prime example of a sector where public and private interests are in continuous interaction with ongoing controversies and cross-border impacts, like the climate change. Achieving energy security does not only include decisions made just from one national government but depends also on the overall smooth and balanced functioning of the energy market involving primarily the private sector and the energy companies.

International institutions have a great power as they could facilitate the functioning of energy markets, lowering the transaction costs, confining their shortcomings and setting certain rules and standards. An additional important purpose of the international institutions is limiting the negative consequences of the energy related activities. An example is the following: In' 70s the oil shocks and the imbalance caused in the energy supply-demand chain led to the establishment of the International Energy Agency (IEA) which adopted a specific mechanism in order to manage the supply risk management: the obligatory creation of emergency oil reserves among its member states.

The extensive environmental pollution and the phenomenon of climate change make the pursuit of environmental sustainability a key energy trend of the twenty-first century's global energy governance agenda. The global governance of environmental degradation is addressed in a versatile way; the existing international institutions incorporated the sustainable development in their policy making procedures. Further, new institutions have arisen devoted exclusively in the promotion of renewable energy (RE) to combat the negative externalities of the current energy system which is highly dependent on fossil fuels.

In this chapter, we will discuss the role that the IEA (an inter-governmental organization) and the G8 Group (a summit process) are playing in the renewable energy game. Furthermore, we will discuss some additional governmental mechanisms that have been especially created for the promotion of renewable energy and more precisely the International Renewable Energy Agency (IRENA), the Renewable Energy Policy Network for the twenty-first century (REN 21) and the Renewable Energy and Energy Efficiency Partnership (REEEP).

Throughout this discussion, we intend to draw attention to the institutional inconsistence of RE global governance which is mainly due to (a) the differentiation on organizations' constitution purpose and goals and (b) the absence of a broader global deal on environmental issues and climate change combat.

In this Chapter we will, additionally, study the paradigm of RE implementation in the European Union's (EU) energy policy, although the adoption of an ambitious binding regime for the RE implementation and the maturity of RE policy within EU which cannot be described as fully successful. At the end of the Chapter, we will offer some brief conclusions on RE global governance failures and positive perspectives and a brief assessment of the EU's RE policies and future prospects.

2 The Renewable Energy Global Institutional Governance

The principal international RE institutions and partnerships are summarized at the Table 1.

At the following sections, we present the analysis of each one of those institutions and their role into international RE Governance.

2.1 The International Energy Agency's RE Agenda

Since its establishment, in 1974, the IEA is widely considered to be the most credible and high estimated energy think tank and energy policy authority. It was initiated by the US government, following the oil shocks of 1973–1974, with initial and main objective to watch the oil market functioning and coordinate the oil reserves amongst its member-states, IEA remains the most influential energy

Table 1 Summary of RE governance institutions

Institution	Date of creation	Headquarters	Functions brief description
IEA	1974	Paris, France	Inter-governmental organization with 28 member-states with main task the oil market watch and the creation of an emergency oil-stock mechanism. The most influential energy institution worldwide
G8	1975	–	Summit process aiming at the policy coordination of the participant states. RE issues are addressed through climate change commitments
International Renewable Energy Agency (IRENA)	2009	Abu Dhabi, United Arabic Emirates	Intergovernmental organization with 118 member -states devoted to the RE deployment globally through cooperation and synergies amongst the various RE stakeholders in a regional and international level
Renewable Energy Policy Network for the twenty- first century (REN21)	2005	Paris, France	Policy multi-stakeholder network providing an information exchange platform in order to facilitate the rapid global transition to RE
Renewable Energy and Energy Efficiency Partnership (REEEP)	2002	Vienna, Austria	International NGO gathering governments, NGOs, banks and industry actors focusing on RE projects deployment in emerging economies

institution worldwide. Over all these years, IEA has not changed its membership status as it remains an inter-governmental organisation limited to OECD countries. IEA's regular publications, like the annually released World Energy Outlook, present the most accurate data for the rest governmental and private actors of the world energy markets thus, influencing their energy policy orientations and decisions.

The gradual increase of the RE share in the global energy mix, to tackle the climate change, constitute a major challenge for IEA's future operations and policy orientations. IEA disposes the adequate expertise and political leverage in order to play a pivotal role in the global energy governance institutional scheme, taking into consideration the arisen environmental challenges. The current global energy trends require from IEA to set a broader target: to include the environmental sustainability in its priorities together with the oil market smooth functioning.

Besides acknowledging the coming era of RE market, IEA has been preoccupied with RE within the context of global energy governance notably after the establishment of the IRENA in 2009 which counts 118 members already instead of IEA's 28. IEA's and IRENA's roles are supposed to be complementary, but in reality their cooperative scheme is only limited to a joint approach of RE data and statistics and in expertise sharing in the fields of technology and operation. These initiatives, though, do not really open an institutionalised dialogue about the future of RE further deployment amongst the member-states of the two organisations, nor does it support the active engagement of the RE private actors.

Still, IEA's "prestige" in the shaping of energy governance agenda should not be negligible. The G8 Summit of 2005 (Gleneagles 2005 Summit), having dealt thoroughly with energy and climate change, has asked IEA's collaboration on energy efficiency and RE policies recommendation. IEA has submitted a report entitled Toward a Clean, Clever and Competitive Energy Future (IEA 2007). The very late development of IEA's involvement in RE global governance is the cooperation with the United Nations for the recently launched "Sustainable Energy for All" initiative, which calls for a global target of doubling the share of RE by 2030. IEA collaborates with the United Nations on this target's baseline definition and progress monitoring (IEA, World Energy Outlook 2012).

2.2 G-8 and the Renewable Energy Task Force

The G8 is a summit process initiated by the governments of France, Germany, Italy, Japan, UK and USA in 1975. Canada joined the summit in 1976 and Russia became full-member in 1997 giving at G8 its current form. The summit's participant governments aimed initially at the coordination of their microeconomic adjustments following the collapse of the Bretton Woods monetary system and the oil price volatility caused by the shocks of' 73–74 (Barry 2005). The G8 lacks the traditional institutional form of a typical inter-governmental institution (absence of charter, secretariat, headquarters, membership criteria etc.). However, since the summit's initial meeting, the member states are gathering on annual basis to deal with a variety of major political and security affairs (Florini and Sovacool 2009). The G8 exerts a considerable influence over the traditional global governance institutions, thus *placing G8 at the centre of global governance* (Barry 2005). The other organisations' decisions and actions usually follow discussions made during the summit's meeting. Still, even if the G8 Summit enjoys, admittedly, influential power, we can observe that the summit's decisions do not always lead to political commitments.

For example, the city of Okinawa, in Japan, hosted the G8 meeting in 2000. In the aftermath of Kyoto Protocol, in 1997, the summit asked for the creation of a Renewable Energy Task Force assigned with the task of concrete recommendations preparation for the promotion of RE use in developing countries. Held in 2001 in Genoa, the next summit repeated the importance of RE for supply diversification and sustainable development. However, the US administration, under George Bush presidency, didn't support the further implementation of the Task Force's recommendations.

The year 2005 present a turning point in summit's energy preoccupations. Tony Blair's summit's presidency, in Gleneagles, UK, put strongly on the agenda the issues of climate change. More precisely, this summit introduced the Cleneagles Plan of Action which contained 63 commitments on climate change, clean energy and sustainable development. The summit asked, also, IEA's assistance on producing alternative energy scenarios and IEA prepared an alternative energy scenario report which was published in 2008. The Cleaneagles Plan of Action was the platform for G8 members' involvement in the UN Climate Change Conference held in Montreal in December 2005, the largest intergovernmental climate change conference after the adoption of Kyoto Protocol. The G8 member states signed various agreements on emission reductions; nonetheless, emissions continue to grow in G8 states the subsequent years. The next G8 Summit didn't pay the same attention on climate change issues as the 2005 summit.

In fact, the 2006 summit of St. Petersburg paid special attention at global energy security introducing the Joint Oil Data Initiative (JODI) for the improvement of collection and report of market data oil and other energy sources. It was only in the 2007 summit, held in Heiligendam, Germany, which commented positively on EU, Canada and Japans decision on emission reductions by 2050.

The city of Hokkaido–Toyako, in Japan, hosted the G8 summit of 2008. During this summit the member states expressed their commitment to developing a partnership with the emerging countries on energy efficiency, implementing IEA's recommendations, though so far no real developments have been recorded. Albeit, the 2008 summit agreed on the adoption of UNFCCC goal of achieving at least 50 % reduction of global emissions by 2050 and welcomed the establishment of the International Partnership for Energy Efficiency Cooperation (IPEEC).

The inclusion of environmental sustainability and RE issues in the agenda of the G8 summit, was a decision that brought together the most powerful world economies featuring a high diplomatic influence on global affairs, which represents a significant development in the RE arena. Nowadays, the issue of sustainable development and the RE contribution has been upgraded into a major playing field of the international political scene requiring multilateral cooperation. Nevertheless, since the conclusions and the declarations are not followed by concrete actions towards the RE expansion worldwide, G8 summit's contribution on RE governance cannot be classified as a productive one.

2.3 International Renewable Energy Agency (IRENA)

The first official proposal for the creation of an international agency exclusively covering the issues of RE can be found in 1981 at the United Nations Conference on New and Renewable Sources of Energy, Nairobi. During 2000, RE started gaining the global energy interest; the World Summit for Sustainable Development in 2004, and the G8 Gleneagles Summit in 2005 addressed RE issues while the 2004 Bonn International Renewable Energy Conference concluded on the need of the establishment of IRENA. IRENA finally came to life in 2009, after a founding conference taken place in Bonn and counts already 118 member-states. IRENA defines itself, as stated in its official website, as a potential central hub gathering and complementing the already existing services and activities in RE sector. The initiative for IRENA's establishment gained strong support by the governments of Germany, Denmark and Spain. IRENA is, also, the first international inter-governmental organisation that USA participates in over the last 15 years. Hitherto, China, Brazil, countries with great technical prospective in RE, still abstain from IRENA's body. China's absence is the most eminent given the country's pioneer position in RE technologies worldwide. Since China has an already a successful RE industry, one could argue that IRENA's assistance could be highly needed as it could set standards, thus, impose limitations to China's RE strategy. However, it is argued that is not the budgetary lack from China' behalf that leads to such an absence, but the lack of trained staff which hampers the management of such a membership (Van de Graaf 2012a, b).

However, China has, very recently, expressed its willing to join the organization. This development has a great importance as China is leading the global RE industry; it is the world's largest manufacturer of wind turbines and sofirst inter-governmental sectorallar panels (Bradsher 2010). China's turn towards a green economy is double sided. It aims at the reduction of external, unstable, energy source dependence and the augmentation of its final energy consumption from RE sources to 15 % by 2020 compared to the current 9 %. China's development of RE industry is part of a wider development plan. The RE industry is adding job posts rapidly while offering to the country the opportunity to become a world technology leader very quickly, thus boosting its external trade. In 2012, many Chinese RE industries had significant activity in Africa and South America. The World Trade Organization (WTO) has been preoccupied by China's complaints about anti-dumping tariffs imposed by US on Chinese solar panels (Tong and Go 2012). USA and Barack' Obama presidency called for redoubling the incentives in favor of the development of the country's RE industry but, the US RE industry haven't reached yet the Chinese levels[1] (Bradsher 2010).

IRENA comes with important shortcomings; its budget is modest, its functions are not clearly defined and the adoption of UN governance procedures, especially the consensus rules about decision making, may generate obstacles in IRENA's vision to lead the way towards a more RE oriented global energy market. The new IRENA's head, Adman Amin, pursues the mitigation of organization's deficiencies. He is setting as a priority the creation of an advisory, information exchange and capacity—building support framework for the governments to enhance their decision- making and technical capacity in RE sector.

Still, IRENA's creation has led to a reformation of the global energy institutional structure. Firstly, IEA responded jealously to IRENA's creation by setting-up a separate Division with specialized RE staff. Furthermore, IEA included RE energy production perspectives in its future energy scenarios. Finally, in 2011, IEA published a study book entitled "Deploying Renewables: Best and future Policy Practice" (IEA 2011).

We have to state, though, that IEA, while trying to preserve its dominant position in the international energy arena, has tried to prevent IRENA's establishment which created an oxymoron as the main initiative for the creation of IRENA has been taken by three IEA's members, Germany, Spain and Denmark. *Germany played a leading part in establishing an organisation which would be independent from IEA and exclusively devoted to RE governance issues, mainly accusing IEA as a closed group promoting fossil fuels and nuclear power* (Van de Graaf 2012b). IRENA is the first intergovernmental sectoral institution created in order to provide exclusive diplomatic and technical support for RE deployment globally. Still, the establishment of IRENA contributes to the further fragmentation of global energy governance architecture. IRENA should ensure that the appropriate structures will be established in order for RE to play a pivotal role in the shaping of the future global energy system.

2.4 Renewable Energy Policy Network for the Twenty-First Century

REN21 is an international and multi-stakeholder network, based in Paris, aiming at the promotion of RE through the connection of a wide range of public and private energy actors. These actors are governments and international organizations but also industry associations, academia and civil society. There are no restrictions regarding the membership status and any legal entity or natural person aligned

---

[1] REN 21, Renewables 2012 Global Status Report: http://www.map.ren21.net/GSR/GSR2012.pdf

with REN21 purposes can join the network. The EU, many EU member-states and emerging economies, like China and India participate in the REN21 Steering Committee as well as important institutions among them the IEA, the Asian Development Bank and IRENA.

The primary focus of REN21 is the distribution of RE policy—relevant information to governments, legislators and private energy actors and assisting them in the establishment of RE regulations and standards. REN21 seeks to work as an interconnection platform for multi-stakeholder actors from RE field for the further deployment of RE worldwide. REN21 publishes global and future status reports about RE market, industry and RE policy updates. REN21 has, also, launched the REN21 Renewables Interactive map, an online research tool offering an updated picture of RE developments global status. Finally, REN21 facilitates the holding of International Renewable Energy Conferences (IRECs), hosted by alternate governments every 2 years, gathering governments, private sector and NGOs engaged in high-level dialogue on RE field.

IRENA and REN21 sealed their cooperation through a Memorandum of Understanding (MoU) on education and training, including REN21 in IRENA's Renewable Energy Learning Partnership (IRELP). The REN21's global network is expected to complement IRELP's database and increase the public awareness of IRELP's platform.

The information exchange platform implementation and the network's popularity among national governments, academia and private energy stakeholders, render REN21 a significant element of RE global governance coordination. Albeit, the information coordination does not imply coordinated activities in RE field respectively. REN21 has a rather subsidiary role; it lacks a decision making platform, while its office operation located at the United Nations Environment Programme (UNEP), in France, raises questions regarding the REN21 Network's operational independence.

2.5 Renewable Energy and Energy Efficiency Partnership (REEP)

The United Kingdom and other partners launched the REEEP in 2002 at the Johannesburg World Summit on Sustainable Development (WSSD). REEEP is widely considered as a UK initiative. Gathering 358 member organizations including NGO's, governments, industry associations and banks, REEEP became an international non-governmental organization in 2004 moving its headquarters to Vienna, Austria. The mission of REEEP focuses on the reduction of greenhouse gas emissions and the acceleration of global market for sustainable energy primarily in developing and emerging countries markets (Parthon et al. 2010). REEEP is trying to identify which RE source and energy efficiency technology is the most appropriate for a specific region, advising its partners accordingly.

REEEP agency is differentiated from the other global energy agencies in three ways; REEEP has a different, more flexible, structure compared to the international governmental organizations, the majority of REEEP partners are non-governmental organizations (NGO) heads and it includes members from the RE industry and the financial institutions. The positive aspect of REEEP is the engagement of RE public and private actors, which, however has downsides: Regional and sectoral partners are willing to promote RE technologies at their benefit but they abstain from REEEP annual funding program, a fact that obliges the organization to be only be capable for short-term planning projects (Florini and Sovacool 2009). Besides, the mission of REEP is applicable to a smaller scale compared to IEA or IRENA (Florini and Sovacool 2009). Thus REEP's activities focus on a more regional level.

Finally, REEEP follows an interesting funding process trying to fund exclusively the projects that are based in scale up business models and could be implemented in different countries, energy markets and regulatory frameworks. REEEP receives its funding from the engaged governments, basically from the UK and Norway, but also from banks, NGO's and business units. For 2013, REEEP announced the funding of 28 new projects in selected developing countries in Africa and Asia including, for the first time, Peru, under Swiss funding, in order to increase Peruvian population's access to energy supply.

IRENA and REEEP formalized their working relationship with a Memorandum of Understanding signed between the two bodies in 2011. Their co-operational framework will be based in the information and expertise exchange for a common synergy towards the enhancement of public awareness in favor of RE adoption.

3 RE Governance in EU

Achieving sustainable development and assuring the path for a more clean energy future is a core goal ascribed to the EU energy policy. The EU renewable energy strategy is a key instrument towards a sustainable energy mix and the reduction of fossil fuels dependence and it is placed amongst the most ambitious ones in a global scale.

Beyond the internal measures in favor of RE expansion, EU is a pioneer RE global actor, participating actively in the transatlantic dialogue on climate change mitigation and sustainable development (Vogler and Stephan 2007; Carpenter 2012). Hereby, some important highlights:

As a proponent of the Kyoto Protocol and the United Nations Framework Convention on Climate Change (UNFCCC), the EU made its first commitment on emissions through a reduction to 8 % below the 1990 levels, for the period 2008–2012. This commitment made in 1997, aimed to deliver the target mainly through the increased use of RE. The European Commission initiated a public–private partnership named Global Energy Efficiency and Renewable Energy Fund (GEEREF), in 2004, in the context of the Johannesburg Renewable Energy

Coalition (JREC). Sponsored by the EU, Germany and Norway and advised by European Investment Bank Group, GEEREF currently has an €108 million budget at its disposal to support private equity funds that focus on renewable energy and energy efficiency projects and/or technologies in emerging economies in priority.

All the sectors of the European RE technology and industry have been united under the umbrella organization European Renewable Energy Council (EREC) in 2000, representing an industry with annual turnover €70 billion corresponding to 550,000 job positions. EREC acts as an information exchange and consultancy platform for RE European technology Besides, EREC produces research documents on RE issues and participates in a series of international RE projects and events. Despite the EU's international initiatives and commitments on RE further development and the ever growing European RE industry, the EU RE policy has not been fully implemented. The intended goals have not been achieved neither from the overall perspective nor regarding the policy implementation in EU Member States (Michalena and Hills 2012). There still exist great discrepancies between the EU Member States (Reiche and Bechberger 2004).

The significant support of the EU, and its Member States, for the RE expansion is reflected in the Directive 2009/28/EC (hereafter the Directive) launched by the European Commission in 2009. This Directive, introduced binding legislative measures for the implementation of 20-20-20 set of goals, defined by the European Commission's Communication entitled "An Energy Policy for Europe" in 2007. The Directive includes *inter alia* the increase of the share of renewable energy, in the total gross final consumption (Segers 2008) of the EU member—states, to be 20 % by 2020. The "EU package on climate and energy", which was agreed by the European Parliament and the European Council in December 2008, passed into law in June 2009.

According to the Eurostat data the share of the RES in 2010 in EU reached 12.5 %—an increase of 4.4 % from 2004 levels. Given, also, the fact that until 2008 RE policy was driven by a *loose legislative framework,*[2] the available data provide a steady upward trajectory towards the overall achievement of the 2020 goals.

However, RE implementation progress varies greatly across the EU member-states. Germany, Sweden and Austria are ranked in the first places of both RE targets implementation and energy production from RE across the different RE technologies.

Second, as said above, the Directive introduces, for the first time, mandatory national targets. Each member state should enhance the effectiveness of its RE policy, so as to bridge the gap between the Member-States having a well-developed RE industry and those who do not, and in order to create a stable investment environment for the development of all RE technologies. The holistic, target-based logic of this Directive seems, though, to set an important barrier towards the successful local implementation of RE policy and technologies (Michalena and Hills 2012).

---

[2] EC Communication (2011) 31 entitled Renewable Energy: Progressing towards the 2020 target.

Indeed, the issues related to RE are characterized by a great complexity stemming from the variety of potential RE sources and technologies. This complexity results from each member state's different capabilities and special conditions and factors favoring, or not, the development and the implementation of certain types of RE technologies. These conditions can be of economic- business nature: limited budget, volatile economic environment for investors, unstable financing sources. There do exist, also, important non-economic factors like the geographic and natural conditions, the administrative and legal barriers, and limited public awareness on RE benefits and issues related to implementation. All the above can present barriers for the facilitation or RE implementation at a local level.

Under the provisions of the Directive 2009/28/EC, member—states are required to draw up National RE Action Plans (NREAC) including the provisions of each member-state (Kitzing et al. 2012) to facilitate the RE development in order to achieve the 2020 target. According to the EU Commission Communication of 2011, entitled "Renewable Energy: Progressing towards the 2020", a review of NREACs shows a faster pace of RE growth for all member states in the years up to 2020. This fact could signify that the new approach starts bearing fruits.

Indeed, the Directive's article 6 ascertains that the decentralization of RE technology has many benefits: the utilization of local energy sources, local jobs creation and the increased local security of energy supply due to shorter transport distances and the reduction of energy transmission losses. However, the flexibility given to each member—state to implement the Directive as it fits to the national context, which often opposes the local benefits (Ostrom 1990; Ostrom et al. 1999), entails the risk of national failures to meet the 2020 targets at both national and EU level. The EU Communication 2012 (271) calls for a greater consistency in national approaches in order to avoid the internal market's fragmentation (European Commission, Communication 2012). The Communication pays special attention on the need of cooperation mechanisms to facilitate the trade of RE products amongst the member—states and the openness of RE market.

4 Discussion and Conclusions

The potential positive contribution of RE sources and technologies to climate change mitigation and global energy efficiency raises no disagreements within the international energy community. The future prospects of RE provide an additional element to the discussion of global governance structure and status. Still, the RE status within this discussion is open to heated debates due to the fact that the RE governance itself involves many gaps and inefficiencies. *Furthermore, although time and money is spent from the political heads responsible for the decision-making in the international institutions, the related discussions have not leaded yet to any major political commitment.*

The two major problems that the RE institutional framework has to deal with, are policy and collective action coordination amongst the different players of RE field and market (Bazilian et al. 2010). We can observe that among the RE international institutions, prevails an inconsistent degree of participation which is predominantly caused by diverging interests, goals and incentives among the different institutions but also the engaged states.

From a thorough study of the institutions analyzed in the Chapter regarding states engagement, we can notify that Germany and the UK participate in all while Russia, an important international energy player, abstain from all except for G8. Germany's pioneer role in RE politics can be explained by two facts; the country's great RE technologies production and its political background regarding Green parties creation since 1970s. As for Russia, the EU-Russia Energy Dialogue is the only initiative that Russia participates. The EU-Russia Energy Dialogue aims at increasing security, predictability and transparency of the energy relations of the two partners. However, the dialogue focuses primarily in gas supplies. Until now, we do not observe any remarkable achievement. The recently published Roadmap[3] on EU-Russia energy cooperation until 2050, refers to the underdeveloped cooperation between EU and Russia in renewables. The EU 2020 targets and Russia's immense RE potential could make a closer cooperation on the RE field mutually beneficial in terms of exchange of RE sources, technology transfer, financial mechanisms and RES legislation and standards approximation. The emerging economies (Brazil, Russia, India, China, South Africa[4]), whose growing energy demands stand for a current energy challenge, indicate a very low involvement in RE institutions. Striking, is the non-engagement of China in all the discussed institutions. Finally, even if USA participates in the majority of them, a leadership in RE global agreements is not recorded. The energy policy of the Obama administration[5] (Barack Obama website 2013) includes also a plan to combat climate change (Bang 2010) through the doubling of clean energy in the USA by 2020. Thereby, RE developers can beneficiate from the government's grants instead of tax credits.

Overall, consistent and uniform goals are needed as well as specific targets which should be discussed at a high level and ensure the enforcement of international agreements in favor of RE development on behalf of its member states. The establishment of IRENA could be a positive step towards this direction. However, this institution is still in its infancy; it is in a transition period after having faced serious budget and administration issues.

We have also observed that the trade competition between leading countries (such as China and USA) gives a further boost for international RE uniform governance and its institutionalization. At the same time, it could result to nationalistic trade competition (embargo imposition, import taxes, export taxes from EU to Chinese PV) that would not facilitate an overall governance institution.

---

[3] http://ec.europa.eu/energy/international/russia/doc/2013_03_eu_russia_roadmap_2050_signed.pdf.

[4] Also known as BRICS.

[5] Website of the US President Barack Obama: http://ofa.barackobama.com/climate/.

Indeed, the imposition of specific trade rules on RE industry exchanges and the development of technology spreading platforms is another issue the RE institutions should take into serious consideration. The creation of a stable international RE market would facilitate RE's penetration in global energy market on equal terms with fossil fuels.

The existing international RE institutions should further foster the participation and the engagement of the various RE actors, both public and private, at different levels. The active engagement of the private energy stakeholders in the race of a more sustainable energy future should be a major preoccupation of RE governance institutions. We know that many large fossil fuel enterprises started deploying their own RE programs (ex. BP webpage (2013),[6] Chevron webpage (2013)[7]). RE governance, though, is primarily a political decision requiring political commitments (Mallon 2006) which are durable over time. Political commitments can be developed within the institutional frame of RE organizations via the setting of general rules of RE governance game.

The paradigm of the RE integration in the EU energy policy is very illustrative to this issue. Political commitment is a core factor for the implementation of RE relative engagements. The three main EU energy policy objectives, security of supply, energy market competition and environmental sustainability, are equally considered (Dupont and Oberthur 2012). There is a clear political commitment to develop climate friendly policies and thus to develop the RE sector, but the EU energy policy does not prioritize the climate policy issues. The European Commission, fitting with the institutionalism perspective of EU integration, is the proponent of climate and RE policy integration. A new Directorate- General for Climate Action (DG CLIMA) has been established in 2010 in order to facilitate the 20-20-20 targets implementation and to lead the negotiations on climate change, in the international level. The European Commission assists the EU climate policies implementation with the various related documents; Communication, Directives etc. However, the translation of these EU-level documents to action on the ground is challenging and evidence has be collated that shows that *the process of subsidiary is the missing link between the EU 20-20-20 policy and the local RE implementation* (Michalena and Hills 2012).

References

Bang G (2010) Energy security and climate change concerns: triggers for energy policy change in the United States? Energy Policy 38:1645–1653

Barry T (2005) G8/G7 and global governance. Foreign Policy Focus. http://www.fpif.org/reports/g8_and_global_governance Accessed 21 Jan 2013

---

[6] BP Alternative Energy: http://www.bp.com/en/global/corporate/sustainability/the-energy-future/alternative-energy.html.

[7] Chevron Renewable Energy: http://www.chevronenergy.com/renewable_energy/.

Bazilian M, Outhred H, Miller A, Kimble M (2010) Opinion: an energy approach to climate change. Energy Sustain Dev 14:253–255

Bradsher K (2010) China leading global race to make clean energy. New York Times, Jan 30. http://www.nytimes.com/2010/01/31/business/energy-environment/31renew.html?pagewanted=all&_r=0. Accessed 2 March 2013

BP webpage (2013) Alternative energy: http://www.bp.com/en/global/corporate/sustainability/the-energy-future/alternative-energy.html. Accessed 6 July 2013

Carpenter A (2012) The role of EU as global player in environmental governance. J Contemp Eur Res 8(2):167–172. http://www.jcer.net/index.php/jcer/article/view/533/334. Accessed 1 Feb 2013

Chevron webpage (2013) Renewable energy. http://www.chevronenergy.com/renewable_energy/. Accessed 6 July 2013

Dupont C, Oberthur S (2012) Insufficient climate policy integration in EU energy policy: the importance of the long –term perspective. J Contemp Eur Res 8(2):228–247. http://www.jcer.net/index.php/jcer/article/view/474/338. Accessed 1 Feb 2013

European Commission DG CLIMA. http://ec.europa.eu/dgs/clima/mission/index_en.htm

European Commission DG energy, renewable energy- targets by 2020: http://ec.europa.eu/energy/renewables/targets_en.htm

European Commission EU-Russia Dialogue: http://ec.europa.eu/energy/international/russia/dialogue/dialogue_en.htm

European Commission (2013) Roadmap.EU—Russia energy cooperation until 2050: http://ec.europa.eu/energy/international/russia/doc/2013_03_eu_russia_roadmap_2050_signed.pdf. Accessed 7 July 2013

European Commission Directive 2009/28/EC on the promotion of the use of energy from renewable sources and amending and subsequently repealing Directives 2001/77/EC and 2003/30/EC: http://eur-lex.europa.eu/LexUriServ/LexUriServ.do?uri=Oj:L:2009:140:0016:0062:en:PDF

European Commission, Communication (2011) (31) Renewable energy: progressing towards the 2020 target: http://eur-lex.europa.eu/LexUriServ/LexUriServ.do?uri=COM:2011:0031:FIN:EN:PDF

European Commission, Communication (2012) (271) Renewable Energy: a major player in the European energy market: http://eur-lex.europa.eu/LexUriServ/LexUriServ.do?uri=COM:2012:0271:FIN:EN:PDF

Eurostat (2010) Share of renewable energy in gross final energy consumption :http://epp.eurostat.ec.europa.eu/tgm/table.do?tab=table&init=1&plugin=1&language=en&pcode=t2020_31

Florini A, Sovacool (2009) Who governs energy? The challenges facing global energy governance. Energy Policy 37:5239–5248

G8 Information Centre (2013) G8 Conclusions on Energy. Cleneagles (2005).http://www.g8.utoronto.ca/conclusions/index.html. Accessed 20 January 2013

Goldthau A, Witte J (2010) Global energy governance. The new rules of the game. Brookings Institution Press, Washington

International Energy Agency (2007) Toward a clean, clever and competitive energy future. http://www.iea.org/publications/freepublications/publication/GermanyG8.pdf. Accessed 14 Jan 2013

International Energy Agency (2011) Deploying renewables: best and future policy practice http://www.iea.org/publications/freepublications/publication/Deploying_Renewables2011.pdf. Accessed 14 Jan 2013

International Energy Agency (2012) World energy outlook. http://www.worldenergyoutlook.org/publications/weo-2012/#d.en.26099. Accessed 14 Jan 2013

Kitzing L, Mitchell C, Morthorst P (2012) Renewable energy policies in Europe: converging or diverging? Energy Policy 51:192–201

Mallon K (2006) Renewable energy policy and politics: a handbook for decision making. Earthscan Publications, Oxford

Michalena E, Hills J (2012) Renewable energy issues and implementation of European energy policy: the missing generation? Energy Policy 45:201–216

North D (1990) Institutions, Institutional Change and Economic Performance. Cambridge University Press, Cambridge

Obama B (2013) U.S. President webpage. http://www.barackobama.com/?source=action-bar

Ostrom E et al (1999) Revisiting the commons: local lessons, global challenges. Science 284(5412):278–282. doi:10.1126/science.284.5412.278

Ostrom E (1990) Governing the commons: the evolution of institutions for collective action. Cambridge University Press, Cambridge

Parthan B, Osterkorn M, Kennedy M, Hoskyns St.J, Bazilian M, Monga P (2010) Lessons for low-carbon energy transition: experience from renewable energy and energy efficiency partnership (REEEP). Energy Sustain Dev 14:83–93

Reiche D, Bechberger M (2004) Policy differences in the promotion of renewable energies in the EU member states. Energy Policy 32:843–849

REN 21 (2013) Renewables 2012 global status report. http://www.map.ren21.net/GSR/GSR2012.pdf.. Accessed 2 April 2013

Segers R (2008) Three options to calculate the percentage renewable energy: an example for an EU policy debate. Energy Policy 36:3243–3248

Sovacool B (2011) An international comparison of four polycentric approaches to climate and energy governance. Energy Policy 39:3832–3844

Tong I, Go P (2012) Renewable energy review: China, renewable energy world.com. http://www.renewableenergyworld.com/rea/news/article/2012/12/renewable-energy-review-china. Accessed 2 March 2013

Van de Graaf T (2012) How is IRENA reshaping the global energy architecture. Eur Energy Rev. http://thijsvandegraaf.be/wp-content/uploads/2012/04/20120329-EER-How-IRENA-is-reshaping-the-global-energy-architecture.pdf. Accessed 16 Feb 2013

Van de Graaf T (2012) IRENA should be a platform where anybody who has good ideas is welcomed. Eur Energy Rev. http://thijsvandegraaf.be/wp-content/uploads/2012/04/20120524-EER-IRENA-should-be-a-platform-where-anybody-who-has-good-ideas-is-welcomed.pdf. Accessed 16 Feb 2013

Vogler J, Stephan H (2007) The EU in global environmental governance: leadership in the making? Int Environ Agreements 7(4):389–413

Geopolitics, Climate Change and Energy Governance: A Grey Area in the Black Sea Region

Jeremy Maxwell Hills and Evanthie Michalena

Abstract The Black Sea region is diverse, not entirely autonomous and geopolitically tense; however it produces, or transports across its area, significant fossil fuels to the EU market and further afield. The twelve countries of the area which form the Black Sea Economic Cooperation (BSEC) have recently adopted Declarations which target climate change, green economy and renewable energy. The EU has long been operating in the region due to its strategic importance, primarily in the areas of energy, environment and governance; although direct programme success has been limited. The chapter focusses on the question: can a route be devised which uses the strengths of each organisation and effectively aligns both BSEC and EU to a regionalised approach to renewable energy (RE) development? The chapter considers the historical developments in the region by these regional players as well as the benefits and constraints of a regional renewable energy approach. The McKinsey 7S model is used to compare BSEC and EU in their RE approaches which demonstrates significant differences of intention, strategy, operation and capacity to implement. Based on these institutional differences, a third approach is developed in which these two organisations could work together based on their institutional strengths for RE regionalisation.

J. M. Hills (✉)
CTL Consult Ltd, 7 Mitford Road, Morpeth NE61 1RG, UK
e-mail: J.Hills@CTL-Consult.com

E. Michalena
ENeC Laboratory, UMR 8185, Paris-Sorbonne (Paris IV) and CNRS INSHS, Paris, France
e-mail: evamich@gmail.com

E. Michalena and J. M. Hills (eds.), *Renewable Energy Governance*,
Lecture Notes in Energy 23, DOI: 10.1007/978-1-4471-5595-9_15,

1 Introduction: Black Sea Regional Status

The Black Sea (BS) region is a very dynamic geopolitical landscape. The BS region countries have competing interests and are not always aligned with respect to strategies for energy governance and energy supply. Energy is the most important economic sector for many of the countries. However, to get fossil fuel to market requires cooperation and linkages to ensure efficient transportation. So there are needs for energy interdependence in the BS region (Karbuz 2010) although regional and international geopolitical developments impact directly the shaping of regional energy alliances.

Harmonious cooperation in the region is more of an aspiration than a reality as there are challenges of conflicting geopolitical interests in the area and clashes between various actors of numerous regional and sub-regional structures. For example, the region is a top priority for the North Atlantic Treaty Organization—NATO—which has already been involved in armed conflict in the region. The energy business necessitates inter-country co-operation, however, geopolitics can acts as a destabilizing influence. The BS region is still a region-in-the making and regional cooperation remains a nascent process (Pavliuk 2004). Within these regional frictions and tensions sits the Black Sea Economic Organization (BSEC).

BSEC is a multilateral diplomatic, political and economic initiative which commenced in 1992 and presently encompasses 12 countries: Albania, Armenia, Azerbaijan, Bulgaria, Georgia, Greece, Moldova, Romania, Russia, Serbia, Turkey and Ukraine. The goals of this regional-focussed initiative are to foster interaction and harmony among the Member States, as well as to ensure peace, stability and prosperity encouraging friendly and good-neighbourly relations in the Black Sea region.[1] Inevitably, one of the interests of BSEC is energy as it is a major structuring element of the regional economy (Dimadama and Timotheou 2010). As the majority of BSEC countries are eager in offering energy to the EU and the EU is thirsty to receive it, energy supply attracts high strategic interest in the area.

Apart from energy production, the Black Sea region is a vital transport route mainly for fossil fuels to meet the growing energy demands of Europe. The Black Sea region is marked by oil and natural gas transport routes and increasingly criss-crossed by pipelines. The regional collaboration, in maritime facilities and navigation, in agreeing transport routes across terrestrial borders and in the planning and fiscal mechanisms that are in place, display a proven but challenging regional architecture. The reasons for which EU is keen to provide energy lessons to neighbouring countries of the Black Sea region, is because those countries are sitting on the crossroads of energy demand (EU) and supply (e.g. Caspian Sea).

With fossil fuel exploitation and supply shaping and determining many of the internal and external relations, where does the RE agenda fit? What are the benefits, or constraints, of regionalising RE? Are the two regional organisations, BSEC and EU, overlapping or in competition? This can help determine the answer

---

[1] http://www.bsec-organization.org/Pages/homepage.aspx

to the ultimate question; can a route be devised which uses the strengths of each organisation and effectively aligns both BSEC and EU to a regionalised approach to RE development?

2 EU Efforts Towards Energy Regionalisation

The direct geographical footprint of the EU on the Black Sea region is small (just Romania and Bulgaria) and even the wider BSEC region only adds Greece. However, it is not the geographical location of the EU that drives interests but the relatively close vicinity of the main EU markets. Still, this means that the EU cannot deliver its agenda like it can within the EU Member States (through mandatory Directives and legislation), but through negotiation, incentive and agreement with non-EU sovereign states.

The EU has already proceeded with various policy initiatives to address regional cross-border security threats, including energy supplies. Within the energy sector, initiatives take place such as "*The EU Energy Policy: Engaging with Partners beyond our Borders*", which reflects the EU's external energy relations thinking and which suggests a new cooperation framework among the EU Member States and their energy partners. The *Baku Initiative* launched in 2004 is a dialogue aiming at enhancing the cooperation in the area of energy between EU, Black Sea and the Caspian Sea countries. Safety and security of energy production, transportation and supply and pursuing of sustainable development (including energy efficiency and RE) are part of the agenda.

The *Energy Community Treaty* (ECT), incorporating most BSEC countries, provides a legal framework and a dispute settlement mechanism for an integrated energy market between the EU and the contracted parties. The ECT which entered into force in 2006 provides the legal framework for the creation of an integrated energy market. Many countries of the BS have signed this Treaty. Through the accession of Bulgaria and Romania to the EU in 2007, the EU acquired a de facto geographic access to the BS region. This way, the EU gained increased geographical legitimacy to proceed with an agenda that will ensure regional security of cross-border nature.

The Black Sea Synergy (launched in Kiev, in 2008) placed the cooperation between EU and BS countries in the energy sector as top priority. However, even the EU accepts that results have "*so far been rather limited and no clear and comprehensive picture exists of the current implementation results of the BS Synergy, exposing the EU to criticism that it lacks a strategic vision for the region and that it is applying a fragmented approach to implementation*" (European Parliament resolution—EU Strategy for the Black Sea (2010/2087/INI)).

Other recent developments in the EU include the European Parliament resolution of 20 January 2011 on an EU Strategy for the Black Sea (2010/2087/INI). In this resolution it is noted that the Black Sea region forms a "*strategic bridge*" to the EU as the BS region has "*strategic importance for EU energy security and the diversification of the EU's energy supply*".

These initiatives suggest that the EU is trying to become an active agent in the region as well as perceiving the areas as important to secure energy supply. However, two most influential countries in the region, Turkey and Russian Federation (The Promotion of Knowledge about Poland 2011) are not signatories. The European Community Treaty reveals the competing visions around the Black Sea area: Russia and Turkey prefer policies that would serve their primary interests in the Black Sea area, namely, economic gains and greater political influence based on bilateral agreements within BSEC countries.

Maybe aware of this, the EU is also trying to access the region, not targeting energy directly, but operating through a range of initiatives and funding thematics on a range of "soft" targets. For example, the enviroGRIDS project (Building Capacity for a Black Sea Catchment Observation and Assessment Sytem supporting Sustainable Development) aimed to assess the water resources in the Black Sea catchment in the past, the present and the future, according to different development scenarios. The emphasis of this project was on impacts on ecosystems, biodiversity, agriculture, health and energy sectors. The project ran from April 2009 to March 2013 and had a total budget of €8.1 m of which €6.2 m is contributed by the EU.

The EU climate change, energy and environmental legislation provides a positive framework for accelerating climate change mitigation and adaptation in the region through directives on issues such as promotion of renewable energy, establishment of the EU ETS, reducing GHG emission from transport fuels, end-use energy efficiency, environmental impact assessment, environmental liability, chemical waste management, integrated pollution prevention and control or integrated watershed management. However, only Bulgaria, Greece and Romania are EU Member States and therefore tied to the EU environmental acquis communautaire and the EU climate, environmental and energy policy.

Of course, this might change as Turkey and Serbia are candidate countries and Albania a potential candidate country for EU accession. Maybe of more direct relevance at present is that Albania, Serbia, Moldova and Ukraine are part of the Energy Community, which extends the EU internal energy market to South East Europe and beyond. This aligns these participating countries with relevant acquis including key EU instruments in the area of electricity, gas, environment and renewable energy. Although the EU geographic footprint may enlarge in due course, the EU is already creating a forward "buffer zone" in which it acts as a lead player in a number of economic and environmental aspects of the BS region.

3 The Expanding Role of BSEC in Energy Regionalisation

BSEC is used to operating within the geopolitical friction in order to ensure regional and harmonious development. Production and transport in the BS region is reasonably well organized in terms of fossil fuel energy systems to provide a stable outflow to markets; BSEC may be able to take some or more of the credit for this. With the emergence of RE, and previous experience in fossil fuels, BSEC seems to be following displaying steps to regionalize RE in the same similar way as fossil fuels. For example, funds have been allocated in 2011 "*for renewable energy [...] to strengthen cooperation among its twelve Member States and to address the key energy challenges by decreasing the costs for greenhouse gas abatement, enhancing productivity and increasing competitiveness through economic benefits deriving from cost savings*".[2]

RE seems to be one of the tools in the BSEC approach to the green economy. The green economy is expected to experience a boost in the region, especially promoted through the BSEC regional initiatives and recently adopted declarations. The BSEC Green Energy Initiative has set up a Green Energy Task Force within the BSEC Working Group on Energy. The BSEC Working Group on Small and Medium Enterprises (SME) adopted an Action Plan for the BSEC focusing on green entrepreneurship and sustainability. The Black Sea Trade and Development Bank (BSTDB) is discussing the establishment of a multilateral Environmental Fund focusing on financing renewable energy and energy efficiency projects. The BSEC Business Council adopted the "Green Agenda" and is creating a "Green Business Network in the BSEC Region" focusing on renewable energy sources.

A step towards embracing the topic of climate change was also taken by the BSEC Council of Ministers of Foreign Affairs who adopted a Joint Declaration on Combating Climate Change in the wider Black Sea (26 November 2010). This political initiative started to promote new regional policy and development concepts under the motto "The Black Sea turns Green". The declaration recognizes the importance of developing and implementing regional approaches appropriate for combating the negative impacts of climate change as well as for reversing continuing trends of biodiversity loss and ecosystem degradation. It acknowledges that environmental technologies and practices aiming at protecting the land and marine environment, biodiversity and natural resources can secure the future welfare for the region. At the same time the declaration emphasizes that the response to climate change is an opportunity for promoting green economy, creating new prospects for enhanced regional cooperation and economic development in the region, by also focusing on a gradual turn towards new technologies, including cleaner energy and higher energy efficiency projects. Further developments in the BSEC Environmental Protection Working Group on climate change adaptation have also been supported by UNDP-BSTIP (Black Sea Trade and Investment Programme).

Towards the aforementioned goals, BSEC has developed a plethora of supportive processes for RE, green economy and for a mitigation/adaptation climate change response. However, to what extent BSEC appreciates the possibilities for widespread and systemic failures which blight RE implementation rather than

[2] http://www.insme.org/tenders-calls/441

assuming that regionalisation of RE policy can be fruitful based on the old "fossil fuel" model, is yet to be determined.

4 Why RE is Important for the Black Sea

There are a number of clear reasons for enhancing regional RE development which are concomitant with both EU and BSEC goals in the BS; these are related to reducing CO_2 emissions, low carbon and green economy development as a basis for future regional prosperity and environmental issues related to extraction and transport.

Whilst RE is related to the energy sector, which is dominated in the BS by fossil fuels, RE has a number of less obvious differential aspects which could potentially bring a number of benefits at both the national and regional level if effectively deployed. The advantages of RE for the region include:

(a) *Flexible deployment.* They offer the easy cancellation or modification of projects depending on a country's short and long term energy and economic strategic plans (unlike oil pipeline Burgas-Alexandropulis [partners: Russia, Greece, Bulgaria], where Bulgaria has cancelled its participation at the last moment and which created problems to Greece).
(b) *Grid inter-linkages.* RE can potentially strengthen countries' energy infrastructure and grids which allows the possibility of a more integrated grid which may transcend international boundaries or access directly into EU grid system.
(c) *Improved energy autonomy and security.* With electricity generation being dominated by fossil fuels produced outwit national boundaries, there is the possibility for enhanced national-level RE generation. This will help with national energy security, especially with the dynamic geopolitical backdrop of the region.
(d) *Diversification in the energy sector.* As oil and gas constitute the energy fundamentals of the BS Region. Diversification of energy supply, through RE deployment, could help buffer supply-related issues as well as fluxes in oil and gas market price.
(e) *Private sector entrepreneurship.* RE deployment and operation is mainly managed by private investors, in many cases these are not the giant corporate companies which dominate the fossil fuel extraction/energy generation operations. The development of entrepreneurship, private sector financing and clear and robust contractual arrangements, can help with the institutionalization and development of entrepreneurship within the wider green economy.

Effective deployment of RE thus has the potential to strengthen the infrastructure- and economic-modus operandi of the region as a whole and in individual nations as well. The benefits could accrue at the regional (within BS or BS to EU) level, at the national level and at the more local private sector level. Increasing involvement of the private sector in RE, and the processes that are allied to private sector development (e.g. investment, contracting and covenants, technological based-consulting) may help the region move towards a more mature, less-state dominated economy and enhance institutional thickness (Michalena and Angeon 2009).

5 Constraints for RE Regionalisation in the Black Sea

Geopolitically, BS countries diverge in their particular interests. For example, Russian oil cannot access the Mediterranean and EU markets if it is not transported through other BS countries territories. On the other hand, EU needs to diversify its nodes of energy supply. Ukraine needs to strengthen its bilateral ties with Russia, after their recent conflict over natural gas in 2009. Conflicts, like the one between Russia and Georgia (in 2008) do not help stability in the region. Bi-lateral trade creates geopolitical allies, for example, annual bilateral trade between Russia and Turkey is approximately 40 m USD and Russia is the biggest gas supplier to Turkey, covering 70 % of its consumption (Kottari et al. 2013). Turkey is the second most influential BS country due to its significant energy transit potential as neighbour to gas producing-countries like Azerbaijan, Turkmenistan and Iran.

The most influential Black Sea countries, Turkey and Russian Federation, seem to be following a more geopolitically motivated logic of energy security. Their fossil fuels strategy would influence their RE strategy accordingly, thus promoting a more interest-based set of policies through bilateral agreements instead of cooperation frameworks.

The development of RE is a key need for the region, but the way in which regional RE policy is approached, either centrally led from BSEC, promoted by the EU, or emergent from the nations, is a key consideration at this juncture. This is because global experience shows that RE is proving to be notoriously difficult to be implemented effectively when it comes to delivery. RE challenges the semi-monopolistic interests of fossil fuel corporate giants (e.g. PPC in Greece) and in most cases investor's demand guaranteed payments for renewable energy production to pay for investments which tend to lead to burdening consumers. RE also demands technological inputs, and although the region is generally literate and well educated it is not endowed with extensive technological capacity. There is also the need for grid infrastructure development to permit local generation nodes of RE into the grid system.

All above parameters necessitate RE regional approaches to be well formulated in order to be effective and efficient especially in the geopolitical context of the Black Sea.

Furthermore, although RE potential is high in almost all countries of the BSEC region, there are significant differences in the appropriateness of different technologies, for example, due to geographical features (e.g. large mountain ranges for hydropower, such as the Carpathians) and climate. Optimum exploitation trajectories across available RE technologies will thus be differentiated by country-level potential. These between-country differences can also be extracted from the National Communications on Climate Change prepared under the umbrella of the UNFCCC which most BSEC countries are required to develop. These reports illustrate the high variability in implemented RE, RE strategies and RE potential across the BS region.

6 EU and BSEC: Differing Modes of Governance

With apparent benefits of RE development in the BS region but a range of constraints associated with the regional governance and geopolitics, then the way which the main regionally active institutions, BSEC and EU, develop the RE agenda as driver for RE deployment is of paramount importance. The above analysis has identified differences in the agenda and approach between these organisations, however, these differences and commonalities are considered in more detail to try to uncover how progressive and effective action may emerge.

The institutional structures and modes of governance are considered in more detail using the McKinsey 7S model (Waterman et al. 1980). This model considers various aspects of one or more institutions in order to determine institutional effectiveness or strategic delivery through alignment of seven internal institutional characteristics. These seven characteristics are separated into three "hard" and four "soft" aspects.

The three "hard" characteristics are related to the upfront and stated goals and operations, and are defined in this example as:

Strategy: the plan devised to maintain and build regional development and competitive advantage.

Structure: the way the institution is structured and inter-linked in reporting processes.

Systems: the daily activities and procedures that staff members engage into get the job done.

The four "soft" characteristics are less tangible but related to people, culture and capacity:

Style: the style of leadership adopted.

Staff: the employees and their general capabilities.

Skills: the actual skills and competencies of the employees working for the company.

Shared Values: the core values of the company that are evidenced in the corporate culture and the general work ethic.

Based on documentation and authors experiences of working alongside both the EU and BSEC, the McKinsey 7S profile displays some notable differences between BSEC and the EU (Table 1).

The McKinsey 7S analysis helps to differentiate between these two regional players in regard to the energy sector. BSEC is closely aligned in all aspects in

Table 1 A comparison of the Black Sea Economic Cooperation (BSEC) and the European Union (EU) institutional and operational governance in the Black sea region through a lens of the McKinsey 7S model

	BSEC	EU
Strategy	Regional-focused to foster interaction, harmony, ensure peace, stability and prosperity encouraging friendly and good-neighbourly relations	EU Strategy for the Black Sea (2010/2087/INI) based onthree 3 pillars: security and good governance; energy, transport and the environment; economic, social and human development. Promote full compliance to EU Directives in EU Member States
Structure	Permanent secretariat (Turkey) with 18 Working Groups (WG) covering range of sectors including "Energy". Have externally funded Related Bodies and consultative Affiliated Centres. Commission of the EU has "observer status"	Multi-node operation through various EU initiatives and structures. Perceives European Grouping for Territorial Cooperation (EGTC), Cohesion and Neighbourhood Policy as strengthening role of EU. Promotes solid alliance to EU through closer integration to EU policies. Promotes and funds joint research and networks across BS area
Systems	WG are the main centers of sectoral activity with Action Plans and Statements/Declarations produced. These agreed through the Council of Foreign Affairs formed by Member States	Focus on (a) Promotion by EU of strategic linkages e.g. EU-Russia strategic partnership, (b) Funding or initiatives and projects in region, including energy and environmental governance
Style	Negotiating and championing developments through regional unified approach. Generally, strategic in intent	Creating multilateral and bilateral partnership on mutual issues. Using financial resources to promote collaboration and deliver initiatives; perceives specific budget lines and specific disbursements as a route to regionalisation
Staff	Secretariat formed of professional support and WGs have designated Executive Office. WG membership drawn from all Member States; attendance of diplomatic officials (Foreign Affairs) or sector specialists (national line Ministries)	Dispersed staff across range of EU structures working on a diplomatic basis or at the desk-officer level of administration and delivery of initiatives and projects. Resources and structures for embedding technical contract staff
Skills	Skills primarily in negotiation and promotion of unified and harmonious strategic intent. Technical specialist available for inputs through line Ministries etc	Good level competence within EU staff on diplomatic and administrative level but more importantly wider resources for commissioning expert/consultant inputs for technical appraisal as required
Shared values	Common belief in benefits or strengthening autonomy of the quite diverse region and agendas. Acceptance of complexity and difficulties of regional development	Acceptance of high-value of region and "strategic bridge" and integral part of the EU's broader foreign and security policy vision

terms of providing a uniform voice and the highest common denominator on issues to provide consensus and harmonies across the entire region. The Working Group (WG) sectoral structure, with diplomatic or line Ministry membership from all countries, provides a clear in country- to regional-level operational mandate. The WG outputs are further mandated by the higher-level organ of the Council of Foreign Affairs. The relatively strategic-level working of BSEC partly reflects the diverse and sometimes fractious nature of the region, within which commonalities and aligned strategies sit as principles and aspirations rather than operational guidelines. However, the upshot of this is that BSEC, with limited delivery capacity within its Secretariat in terms of Declaration and strategies, relies on subsidiarity and uptake in Member States. This ability to drive forward demonstrative change on the ground directly by BSEC may be its weakness, but on the other hand creating a stable, autonomous and harmonious region as a platform for delivery and deployment is an outcome in its own right.

The collectivism embodied by the EU is unsubstantial when in comparison to BSEC. Ultimately, the EU represents the interests of its Member States; only three of which (presently) fall within the BSEC countries. The BS region is an outlying neighborhood or "buffer" zone on EU's border extremities. However, the vast energy reserves in the BS region and transport routes through that region have necessitated a more deliberative policy dialogue and programme and project planning. The EU seems to be attempting to implant EU-approaches on non-EU areas as a way to gain leverage in the region, or attempting bi-lateral initiatives (e.g. EU-Russia strategic partnership) when wider collectivism would not be apparent. The approach is cushioned through availability of financial and technical resources, programme and project funding and administrative support functions which define much of the EU structure and function in the region as displayed in the McKinsey 7S analysis. Despite all efforts, this institution-driven approach of the EU has not yet been entirely effective.

The deliberative and agenda-related EU approach is in stark contrast to the attempted mutual co-working of BSEC. Within BSEC, countries have a balanced input into the workings of the organisation. With energy policy being a significant formulator of the EU approach, more substantive concerns are likely to orientate around countries which have significant fossil production or are "gate-keepers" of marine or pipe-line transport routes; suggesting a non-egalitarian approach across the BS countries. The BSEC structure and function is formulated around collectivism and regional autonomy, whereas the EU is agenda-orientated.

7 A Joint Regional Governance for National RE Deployment

This chapter has so far considered the advantageous and disadvantages of regional RE initiatives in the BS region and then carried out a comparative analysis of the two main regional players: BSEC and the EU. With this as backdrop this section considers what might be the most effective mode of governance to utilise the strength of the regional players and to promote RE deployment across the region.

BSEC has been creating an umbrella of Declarations on climate change response, green economy and RE deployment which are common to all countries in the BS area. The approach taken by BSEC seems to be one of an integrated trajectory, including low-carbon, energy efficiency, and climate adaptation as well as RE. With the geopolitical issues and diverse membership these agreements have the regional mandate. However, BSEC Secretariat and the various subsidiary organs, while they have the mandate, do not directly hold the power and resources to move forward with deployment. The Declarations reflect aspiration and intent, but are weak on delivery.

On the other hand, the EU has resources, both technical and financial, to move forward in developing demonstrative RE deployments, strengthening private sector involvement and maturing of business process and developing/enhancing governmental financial mechanisms for sustainable and long-term support of RE developers; especially in EU Member States in the BS region. The bilateral/multilateral pact/agreements, the way of working of some of the programmes, could offer a model for a focus on different technology nodes which reflect RE potential across the region. For example, various countries with high solar potential can operate as one node, other nodes can be in, for example, hydropower and wind. These nodes can provide demonstrative value in terms of actual technology but can also enhance the private and governmental working relations and management of private and financial transactions. Additionally, a number of functional RE technology nodes, each with operational multi-lateral connections, could help strengthen and stabilize the geopolitical fabric of the region.

If the EU can work within the BSEC umbrella, on areas which are of mutual interest to the EU, then with suitable strategic oversight an organized but fragmented involvement of the EU could be of benefit to delivery of BSEC Declarations. This fragmentation, rather than being a negative aspect, can help to optimize value across the diversities in terms of RE deployment. With clear signage of the overall approach and the role of the EU, then BSEC and their Member States can "fill the gaps" with national-level funding or alternative external funding sources (e.g. World Bank).

This approach in which EU are aligned to the BSEC intent requires formal oversight on the part of BSEC to ensure regional value and compliance to the terms of the Declarations. This is maybe a change away from the normal national-BSEC interaction which predominates, however, it is a role that BSEC could potential effectively undertake; moving it from a passive regional actor to more of an active regional governor with a collective mandate. In simple terms, negotiated agreements between BSEC and EU for future RE developments could be tested against the relevant Declarations. Once compatibility is ensured, BSEC then operate with an oversight function over programme planning and implementation to confirm that this thematic compatibility is maintained through EU project-level and programme-level delivery.

The advantage of this oversight approach is that the link between BSEC and national governments is close and governments themselves need to set up the licensing, consent, planning permissions, financial mechanisms and private sector agreements which are appropriate to their particular country. Whilst, RE deployment has to be a national-issue, regionalisation through sharing of expertise in relevant RE technologies and targeted inward support to stated Declarations could be of benefit in moving forward with RE deployment which meets EU, BSEC and national goals.

8 Conclusions

This chapter has tried to answer the question: can a route be devised which uses the strengths of each organisation and effectively aligns both BSEC and EU to an effective regionalised approach to RE development? Normatively, the two institutions are not a priori aligned with BSEC following a regional development approach and the EU a more targeted agenda. The deliberative agenda of the EU seems to have the potential to be molded to the collectivism and mandate of BSEC to support the significant national-level preparations necessary for instigation of a route to RE deployment. This approach works towards the strengths of both organisations, and also inculcates close-national linkages necessary for creation of a regulatory and financial platform for RE uptake by the private sector.

However, to what extent the EU would acquiesce to BSEC oversight, and become a delivery agency for aspects of BSEC climate change, green economy and RE within BSEC Declarations is not clear. Whilst country-driven and regional-driven approaches like BSEC are desirable to promote national priorities, egalitarian and equitable development, to what extent EU's hunger for fossil fuels obscures this vision remains to be seen. Whilst the BSEC approach is for the region, the EU approach is for the EU in the BS region; the difference between these positions is a grey area for the Black Sea region.

References

Dimadama Z, Timotheou A (2010) Greening the black sea: overcoming in efficiency and fragmentation through environmental governance. In: International Centre for Black Sea Studies (ICBSS), Policy Brief no. 21

Karbuz S (2010) Losing the energy battle: how and why the US and EU need to engage the black sea region, J Energy Secur 1(1):1–13. Available via http://www.ensec.org/index.php?option=com_content&view=article&id=255:losing-the-battle-why-and-how-the-united-states-and-europe-need-to-engage-the-black-sea-region&catid=108:energysecuritycontent&Itemid=365

Kottari M, Popovici V, Wisniewski J (2013) Energy politics, pipelines and the Black Sea Basin: on the route to diversification of EU energy sources. Strategy paper, vol 4, Young Researchers Series

Michalena E, Angeon V (2009) Local challenges in the promotion of renewable energy sources: the case of crete. Energy Policy 37:2018–2026
Pavliuk O (2004) The Black Sea Region—cooperation and security building. In: Pavliuk O, Klympush-Tsintsadze I (eds) East West Institute Editions, Stockholm
Waterman R, Peters T, Phillips J (1980) Structure is not organization. Bus Horiz 23:14–26

Times of Recession: Three Different Renewable Energy Stories from the Mediterranean Region

Sofia-Natalia Boemi and Agis M. Papadopoulos

Abstract The utilization of renewable energy sources is considered one of the crucial strategies for sustainability and economic growth. At the same time it can be a powerful tool for achieving economic, ecological, and social benefits on a community and local level. The goal of the European energy policy, namely to increase renewable energy by targeting a 40 % share in electricity production by 2020 and also to consider a more efficient energy use has to be considered as providing an ambitious aim, at least on a national and regional level, leaving to each member state the responsibility or the way in which the aim will be achieved. Although one may argue that target is certainly ambitious, especially in the light of the economic recession troubling the Eurozone, still, one cannot fail to notice that significant progress has been made in the utilization of renewables over the past 15 years, albeit not uniformly throughout Europe. It is in a sense puzzling that the Mediterranean countries, although featuring a rich RE potential, have lost ground compared to northern European countries. This chapter aims to discuss issues encountered during RE implementation with regard to governmental policies and financial and environmental aspects of renewable energy processes which seem to slow down the RE penetration and acceptance.

1 Introduction

Nowadays, the euro zone is in recession for six consecutive quarters, the longest recession since the area data began to be recorded in 1995. This is part of a greater recession which began in the United States. Fortunately, the United States has been

S.-N. Boemi (✉) · A. M. Papadopoulos
Laboratory of Heat Transfer and Environmental Engineering, Department of Mechanical Engineering, Aristotle University of Thessaloniki, Box 483 54124 Thessaloniki, Greece
e-mail: nboemi@gmail.com

A. M. Papadopoulos
e-mail: agis@eng.auth.gr

E. Michalena and J. M. Hills (eds.), *Renewable Energy Governance*, Lecture Notes in Energy 23, DOI: 10.1007/978-1-4471-5595-9_16,

doing significantly better than every single euro zone nation over that same period—the US economy is now more than 3 % bigger (after adjusting for inflation) than it was in 2008, a period that includes the recession.

In Europe, the crisis spread rapidly and affected much of the region with several countries already in recession as of February 2009, and most others suffering marked economic setbacks. The global recession was first seen in Europe, as Denmark was the first country to fall in recession. On average, the regional GDP fell 0.2 % in the first part of 2013 compared to the previous quarter and 1 % in the 2012. France's GDP fell 0.2 % in the first quarter of 2013, compared to the previous quarter, with a decrease for two consecutive quarters and therefore in technical recession. In the previous year, the GDP of France fell 0.4 %, down 0.3 % higher than that recorded at the end of 2012 compared with the end of 2011. Meanwhile, Germany was able to prevent entry into recession as it grew 0.1 % in the first quarter, but it is high below the forecasts of analysts who had expected growth of 0.3 % EIA US, 2013.

In the euro zone, 9 in 17 countries are in recession: Spain, France, Italy, Finland, Netherlands, Portugal, Cyprus, Greece, and Slovenia (country whose GDP is falling, although they have not been published for the first quarter data). Italy, the third largest economy in the euro zone, recorded the seventh consecutive quarter of decline, the longest since data started being recorded in 1970. Finally, the Portuguese economy fell for the ninth consecutive quarter and the decline is accentuated. In the first quarter of 2013, it decreased 0.3 % compared to the last quarter of 2012 and 3.9 % over the first quarter of 2012. These data are even worse than the European Commission predicted (The Washington Post 2013).

In the summer of 2013 it has become clear that the impact of the recession is particularly deep and lasting in Southern Europe, with Greece, Spain, and Portugal suffering severe drops of their GDPs that are treated, in a strict, doctrinal way in a pro cyclic pattern, by means of harsh reductions of public spending, hence enforcing this downward spiral even more. Italy is on the brink of entering a similar vicious circle, although its trade surplus is providing some hope for eventually avoiding it (Eurostat 2013). The problems of the French economy are not insignificant either, but the sheer size and importance of the country makes it rather improbable that a similar approach will be adopted.

The recession obviously affected the RES market which is now considered to be an "expensive" one, especially in those countries where state funding for investments is reduced to the absolutely minimum and where electrical consumption is reduced due to the recession. This applies particularly to Greece, Spain, and Portugal, which is why emphasis is been placed on those three countries. Although it may be true that the direct costs of renewable electricity generation are usually higher than the ones of conventional power generation, that statement fails to consider the macroeconomic and social benefits resulting from electricity from renewable energy sources (RES-E), including externalities like environmental protection and restoration costs, health costs, depletion costs of conventional resources, local growth, and development potential (DEPOIR 2007). It is, for example, an established fact, that RES-E pollutes much less than conventional (fossil-fuel-based) electricity and, therefore, negative environmental externalities are avoided (IEA 2009). Thus, RES-E contributes to the achievement of the climate change mitigation goals of the Kyoto Protocol (KP), while emissions of local pollutants are reduced as well. Apart from environmental benefits, RES-E provides several socioeconomic advantages, including diversification of energy supply, enhanced regional and rural development opportunities, creation of a domestic industry and employment opportunities (Rio and Gual 2004; Miera et al. 2008).

Setting up mandatory national targets for integrating energy from renewable sources into the gross final consumption of energy is an adopted policy tool to foster the further penetration of renewable energy into systems. Considering the EU targets, RES-E accounted for 62 % (17 GW) of the new electricity generation capacity installed in 2009 and, in addition to the existing capacities, it lead to producing 20 % of Europe's electricity consumption, something that is near the goals for 2020 but does not appeal to specific goals for each country (Michalena and Hills 2012; Giatrakos et al. 2009). If current rates of RE expansion are maintained, by 2020 up to 1,400 TWh of RE electricity will be generated. Other EU countries, like Denmark announced for 2020 that more than 60 % of electricity production will have renewable origin, while the target of the German Government (BMU 2011) is a 35 % share of RES-E in total gross electricity consumption and the Spanish government expects RES-E to contribute 42 % of total electricity demand (Moreno and Martínez-Val 2011). Referring to Greece, the first national renewable energy allocation plan foresaw an initial aim of 18 %, as determined by the aforementioned Directive 2009/28/EC. This was increased to 20 % of the total energy production, corresponding to 40 % of the national electricity production in 2010 (Law 3851/2010). But, this share of renewable energy sources does not include the non-interconnected islands which consist about 9.6 % of the total energy production (Lagie 2013).

Drawing from relevant experiences in electricity systems around the world, this chapter attempts a review of the existing policy supports for RES in Spain, Greece, Portugal focusing on the impact of the policies' mechanisms on the overall energy market structure and their performance in the short-and long-term. Given the rising importance of RES-E in systems everywhere, these impacts can no longer be overlooked. Those are the three more interesting cases in the EU Mediterranean area. France and Italy, despite being Mediterranean countries, are not analyzed in the specific chapter.

2 Current Situation

Governmental support is perquisite for RES due to the fact that (i) environmental costs are not adequately internalized for conventional electricity generation technologies; (ii) the market is often focusing on the capital intensiveness of the investments, which makes them less appealing to the private investor, especially in times of increasing capital cost; (iii) the small size of typical RES plants, at least compared to conventional ones, makes the market a clustered one, and (iv) the stochastic nature of their production of major RES, like wind power makes their penetration a headache to be dealt mostly from Government Boemi et al. 2010. Moreover, as most European countries are using guaranteed purchase contracts and Feed-in-Tariffs (FITs) to promote RES, the support for renewable electricity is finally paid by electricity consumers, and this is over the standard retail price. It is therefore not unexpected, that in countries experiencing a significant increase of RES-E generation, such as Spain, that there is a concern that this support places a significant and increasing financial burden on the final consumer. Consumer costs are the result of adding the RES-E support prices to the wholesale electricity ones. If, as with FITs, support is paid per kWh of RES-E generation to the electricity producers, then an increase in RES-E generation would translate into a greater amount of total payment and, thus, a greater financial burden for the consumer.

Failed attempts to use systems of voluntary purchases of green electricity by consumers and direct investment subsidies which led to significant market distortions, demand-side strategic deployment policies have emerged as the preferred instrument in most countries. FITs[1] become one of the most popular policies for supporting RES-E generation. FIT obliges electricity distributors (or the incumbent suppliers in a geographic area) to purchase electricity from any new RES-E plant in their service area and be paid with a minimum guaranteed tariff per kilowatt-hour that is fixed over a long period of time.

Another popular regulation measure for encouraging RES-E is the Renewable Portfolio Standards (RPS) which is also referred as renewable electricity standards (RES). These are policies designed to increase generation of electricity from renewable resources. These policies require, or encourage, electricity producers within a given jurisdiction to supply a certain minimum share of their electricity from designated renewable resources.

2.1 Greece: The Country with the Big Potential and the Wrong Decisions

The historical development of RES utilization in Greece can be divided into three main phases: The first period extended from 1985 until 1999, when RES fell within the responsibility of the state, the state owned Public Power Corporation

[1] *A feed-in-tariff* (FIT, standard offer contract, advanced renewable tariff, renewable energy payments) is a policy mechanism designed to accelerate investment in renewable energy technologies. It achieves this by offering long-term contracts to renewable energy producers, typically based on the cost of generation of each technology. Technologies such as wind power, for instance, are awarded a lower per-kWh price, while technologies such as solar PV and tidal power are offered a higher price, reflecting higher costs.

(PPC), and local authorities, with only minor initiatives by private entrepreneurs. During this period the results were rather meager (Kaldellis et al. 2012).

The substantial development of RES begun in the second phase, which started with the adoption of the European directive 1996/92/EC, about the liberalization of the electricity markets. The third phase started in 2010, when the Hellenic policy framework was reassessed and adapted to the development of the common European energy policy, referring to the ''20-20-20'' integrated strategic objective. In this line of approach, a series of supplementary laws and ministerial decisions were introduced, to create the framework for an efficient implementation of the aim. Hellenic legislation incorporated the aforementioned directives in a series of successive laws Papadopoulos, 2008.

The Law 3851/2010 on accelerating the development of Renewable Energy Sources in order to deal with Climate Change and other regulations (Governmental Gazette 85/4-6-2010) integrates Directive 2009/28/EC into the national legislation and sets the following national targets until 2020: 40 % RES penetration in gross electricity production, 20 % RES share in gross energy consumption for heating/cooling, and 10 % RES share in the transport sector. It is an extension of Law 3468/2006 and it also includes issues regarding the simplification of licensing procedures for RES-E units and establishment of a RES agency under the Ministry of Environment, Energy and Climate Change (MEECC) for advising RES-E investors. Additionally to this law the Ministerial Decision on Physical planning and allocation of RES has been issued (Governmental Gazette B 2464/3-12-2008).

Finally, a series of supplementary presidential decrees ministerial decisions were issued to enable the effective implementation of the laws. Though there is a large number of decisions and laws in order to be more effective and to increase sustainability, as it is described below, the result was not the expected one.

Especially for the Hellenic photovoltaic (PV) sector, this has started to take off in 2009 and was boosted in 2010 after two laws were introduced in 2006, aiming at improving FIT terms and softening bureaucratic conditions for old PV applications as well as new ones. The FIT introduced in 2006 seemed to be especially promising with respect to PV applications, as it was proven by more than 4,000 license applications submitted to the Regulation Authority for Energy within a year's time. However, the promising FIT could not offset a series of noneconomic hurdles, like the long and tedious licensing procedures, the considerations about the stability of the legislative framework and the limitations of the grid capacities, which lead to a more than 2 years' delay before the first investments began to be implemented EurObserv'ER 2011 and 2012.

Eventually, this development brought Greece into the seventh place of the EU PV market in terms of cumulative capacity in 2010. This position did not satisfy the internal market, as countries with similar FIT provisions and beneficial solar conditions (as located in the Mediterranean region), such as Spain and Italy, had a much faster growth, even when considering the difference in size. Furthermore, the cost of this policy was high, and became even more as the PV market took off (Papadopoulos et al. 2009). The legislation of 2006 had some further, "hidden" problems, like the fact that there were no cost-effective regional and time variations in the tariffs and no prediction for their evolution corresponding to the PV costs' decrease over time. Also, the annual revenues were guaranteed for a long period of time (20 years), thus resulting in a lack of drive toward state-of-the-art technologies or even innovation.

Most of those problems were dealt with, by means of the revision of the legislation in 2010, and as a result the spring of 2013 more than 1,850 MWs of PVs were operational (Operator of electricity market, 2013). With the electricity consumption having shrunk by almost 15 % with respect to 2009, the contribution of PVs has become proportionally more important; unfortunately so has the cost of their FIT. It was therefore not a big surprise, that the tariff for PV energy was amended in the fall of 2012, with reductions between 10 and 20 %, with respect to the size and the date of the investment EurObserv'ER 2012.

Similar conclusions can be drawn with respect to the utilization of wind power. Concerning the construction of a wind park the duration of the entire procedure can take more than 3 years. There are specific reasons for this, even after the revision of the legal framework, and these reasons can be divided into "hard" and "soft" ones. The former refers mainly to the insufficient capacities of the transmission network: areas with rich wind or solar potential, which are also available for big-scale exploitation, are as a rule remote, which unfortunately means that they lie off the routes of existing high capacity, high voltage networks. In extremis this is the case of the Aegean islands, but also the coastline of Thrace, in eastern Greece, or the north-eastern region of Portugal are regions with rather limited grid access. The "soft" problems concern more difficult to tackle, as they reach from public acceptance for renewables (which cannot always be considered as given) to the lack of a land registry and the deficiencies of the national land allocation plan, which lead to conflicts with local societies and disputes about the use of areas protected by international agreements like NATURA, RAMSAR, etc Kambezidis et al 2011.

Summarizing, the development of RES in Greece is an interesting example of how good political intentions can be offset by a complicated legal framework, by the inefficiencies of public administration, and by a lack of "soft" and "hard" infrastructure. More than a decade passed, with continual alterations and improvements of the legal framework and also with a quite high premium paid to investors in the form of really high FITs, before the RES-E market gained momentum.

This premium eventually lead to the much awaited success, with 1,850 MW of PVs and 1,650 MWs of wind power being operational in March 2013, while another 1,200 MWs of wind generators are expected to become operational by the end of 2013.

Still, the price for the final electricity end-user to pay, is high! All those delays and the high FITs created a financial burden that has easily been rolled to customers. Therefore, it remains to be seen whether the momentum will withstand the impact of the deep economic recession, which since 2009 is steadily reducing energy consumption as well as the ability and willingness of the consumer to pay, leading the market operator to struggle with debts to the RES-E producers of more than 450 mn Euros (Operator of electricity market, April 2013).

2.2 Portugal: A Story That Began Nicely, but Now Goes Slowly

Portugal has great prospects for RES investments, as it is characterized by high solar and wind potential. Although the Portuguese electricity system was fully liberalized in 2006 did not lead to a competitive electricity generation market. In fact, in 2010, apart from capacity payments to which conventional producers other than those benefiting from Long-Term Power Purchase Agreements (PPAs) or Guaranteed Compensation Energy Mechanism (CMEC) regimes are entitled since 2010, 83 % of installed electricity generation in Portugal mainland benefited from State guaranteed prices, independently of market conditions (Tolón-Becerra et al. 2011).

Unfortunately, despite the attractiveness of its FIT system, investors were discouraged and the country's PV market is far from mature, as Carvalho et al. (2011) reasonably support.

More specifically, the tariff was set according to the PVs' capacity and the stability criteria of the distribution gird, considering small-scale (mostly residential) systems. Installations benefiting from this tariff have to use a single RES technology and have a single-phase or three-phase load operating at a low voltage, with a capacity of no more than 3.68 kWp. If combined with renewable energy heating and/or cooling systems, then higher FITs apply for a 15 year period. In general, tariffs for PVs are guaranteed for 20 years and calculated based on a dependent formula according to systems' output and capacity. The most recent tariffs for small-scale systems, basically applicable to residential buildings, are presented in detail below (Table 1).

An interesting development was monitored in the field of wind energy: the country features a rich wind potential, particularly in the northern and north-eastern part of the country and ambitious plans were developed since the late 1990s for its utilization, aiming at more than 6,950 MW of installed wind power by 2020. By the end of 2010, the installed wind energy capacity reached almost

Table 1 Brief summary of the main incentive schemes provided particularly for small-scale (residential) PV systems in Portugal member states (Karteris and Papadopoulos 2012)

European member state	Support schemes	Amount	Eligibility period (years)	
Portugal	Tax incentives	Reduced VAT by 9 %	–	All PV systems
	Feed in tariff	0.45 €/kWh	20	
				<5
		0.32 €/kWh	20 First 8 years	>5–36
		0.40 €/kWh	Next 7 years	<3.68 kWp (micro units combined with renewable heating systems
		0.24 €/kWh		–

4,000 MW and the annually produced energy from some 9,000 TWh, corresponding to 17 % of total electricity production, providing the base for an optimistic outlook (EWEA 2011).

However, with the country plagued by the economic recession and the demand for electricity decreasing accordingly, the initial aims were met with skepticism, and the government revised its schedules, setting the goal at not higher than 5,300 MW, a figure which is only 960 MW higher than current installed power (Windpower Monthly 2013).

Summarizing, the Portuguese RES story is one of high ambitions, demonstrating a good start but also encountering significant difficulties.

2.3 Spain Attains a Dominant Position in the Mediterranean Region

For Spain, the growth rates of wind-installed capacity in the past decade are impressive. In the period of 1998–2005, wind capacity grew at an average annual rate of 42 %. 11,728 GW of wind power was installed with wind electricity generation reaching in 2006 22,941 GWh, accounting for approximately 10 % of the Spanish electricity demand (Gomez et al. 2011). This figure made Spain the second country in the world in terms of installed wind energy capacity, second only to Germany with 18 GW at the time (Miera et al. 2008). Great wind electricity potential, a favorable regulation, a high technological maturity and evolution, and a great manufacturing capacity of the domestic industry are factors behind this success. It is of interest to notice that more than 70 % of wind turbines installed by the end of 2005 were supplied by Spanish developers. Despite the economic recession starting from the year 2009, the electricity produced from wind energy continued to increase, reaching a capacity of 21.6 GW and a share of total electricity consumption of 15.9 % by the end of 2011 (EWEA 2011).

A similar success story can be noticed in the PV sector, where Spain is the second largest in EU, behind Germany. Most of this capacity was installed after 2008 when the country introduced an organized effort to promote this technology. Alone in 2008, close to 2,700 MWp were installed, taking advantage of a beneficial FIT mechanism.

The trend was so strong, that in the autumn of 2008 the Spanish Government introduced a cap on the yearly installations (Royal Decree No. 1578/2008 on the Payment for Electricity Generated by Photovoltaic Systems–RS 1578/2008), setting also considerably lower tariffs for new registered systems, and also introducing the provision that two-thirds of the systems will be applied in and on buildings (BIPVs and BAPVs). These changes resulted in a less progressive PV market (fifth in EU) in 2009 and 2010. Still, by the end of 2011, 4,214 GW had been installed, and at that same year a total of 7,912 TWh of electricity was produced [0] (EIA US 2013).

During 2012, due to the technological and market developments and demand and in order to keep up with the foreseen Decree, the tariff for PV energy was amended. Specifically, the FIT was reduced between 5 and 15 %, depending on the size and the date of the investment.

Synopsizing, Spain can rightly be considered a melancholic story considering the utilization of RES, but with a significant cost imposed on the Spanish energy system and eventually the consumer. In a country plagued by recession, it cannot be left unnoticed that more than 14 % of the households were suffering energy poverty in 2011 (Eurostat 2013), a figure which is unfortunately not expected to decrease.

On the other hand, one has to keep in mind that apart from sharing all Mediterranean natural benefits (climate and therefore potential for renewable energy increase), Spain is a much bigger economy with a robust industrial infrastructure. In that sense, and despite the country is now suffering as well from the Eurozone's recession, it stands with much better chances to overcome the problems and even strengthen its position in the European energy market. The development of companies like Acciona Energy, Gamesa, and Iberdrola, which have become global players within a decade, is a good example for how a country can benefit from a combination of national expertise, industrial capacities, and effective legislative framework.

3 Discussion

From the three countries analyzed (Greece, Portugal, Spain), it became evident that the government policies applied aimed to promote sustained and sustainable commercial development of renewable energy sources.

In Greece and Portugal major issues encountered, concerned the duration and complexity of the authorization procedure (including the licensing of the grid connection process). With respect to the average lead time for overall licensing procedure in Europe, Greece ranks last among the three countries. This is mainly because there is a rather complicated licensing procedure for RES. Further improvements in the legislation, as well as an effort to streamline the administration are expected to yield results in 2013/2014, but this remains to be proven.

A second observation can be made with respect to the FITs used to promote renewable: A successful FIT scheme should be a tool for policy makers in order to achieve sustainability in the energy sector at the minimum social cost. If that is too ambitious as a goal, it should at least enable a reasonable energy mix at optimized social benefits/privileges, with respect to other priorities set, like regional or sectoral development. In all countries mentioned, energy policies concerning RES-based electricity production began effectively to exist between 1997 and 1999. Since then significant progress has been made in wind power, while PV's remained a marginal application until the mid-2000s when on the one hand more appealing FITs were introduced and, on the other hand, the capital cost for PVs got significantly cheaper.

The potential of RES electricity market, especially of PVs and wind generators, has undoubtedly great development prospects in all countries, if an appropriate approach is applied. The tariff support schemes which are currently valid, certainly ensure the viability of investments, despite the administrative and bureaucratic barriers still existing. On the other hand, one has to keep in mind that maintaining a FIT scheme that favors strongly the investors, can lead to excessive costs for the consumers, as this is the case for PVs in Greece.

Given the fact that the development of the RES market unfortunately coincides with the deep recession of most Mediterranean countries economies, there are two, partly contradicting, demands which have to be met: The recession, which is not expected to be over soon, has led to an extreme unwillingness of the banking sector to fund almost any sort of medium- or long-term investments, even if they are otherwise considered to be mature and sound investments. On the other hand, there is the danger that with the shrinking economy and the reduced electricity demand, the cost of generously paid RES generated electricity may become disproportionately heavy for the consumer, leading to discussions about the viability of the whole scheme, and the reactions to the further penetration of RES.

The tariff scheme applied has therefore to achieve a delicate balance, providing incentives to keep the solar branch growing, as significant effort and resources have already been invested in this sector, keeping at the same time the overall cost for the final consumer, who is eventually asked to cover this cost, as reasonable as possible.

A continual reassessment of the FIT's efficiency and effectiveness, as discussed in this chapter, is therefore necessary, monitoring the impact of the volatile financial and economic environment; thus it is the evaluation of other parameters, especially of externalities like the social and environmental costs and benefits, which have also to be addressed.

4 Concluding Thoughts

Both policy makers and consumers have to keep in mind that the cost of an energy policy that ensures security of supply and sustainability at the same time cannot be anything but high, and becomes even more so as the RES-E market is capable to takeoff. This is initially due to the fairly high initial investment cost of RES systems: as those investments are new, their depreciation will burden the production cost for many years, leading to higher prices compared to the older plants which have been partly or entirely amortized. Furthermore, there is a demand for new grids, often connecting remote and isolated areas, which will also have to cope with the non dispatched and intermittent way of energy generation, which is so typical for RES.

Financing those investments in the liberalized European electricity market not only poses a problem for the states' budget, which is already difficult enough, but also turns into a problem of acceptability for RES by the final consumer, who is eventually asked to cover the majority of the costs required, while the investors are enjoying fairly safe revenues. Moreover, the current deep and lasting recession which is affecting mostly southern European economies cannot allow them at establishing a sustainable stable electricity market due to the existing contradiction: reduction of electricity demand and an increased contribution of the heavily subsidized wind and PV capacities of the past few years.

Still, even in those dire times one has also to keep some: The planning, licensing, constructing, and commissioning of a conventional power plant takes 4–5 years, while the planning, licensing, constructing, and commissioning a wind farm takes easily between 2 and 3 years. Covering the cost of peak loads, which are inevitable under the Mediterranean climatic conditions, is a costly exercise, as one has to rely on extremely capital intensive investments used over a short period of time. Integrated, efficient, and effective solutions are needed, in order to provide the much needed security of energy supply in a sustainable and feasible way.

The situation in the Mediterranean region cannot be compared to the challenges that northern European countries are currently facing, where nuclear power is gradually being phased out, mainly by means of wind power and natural gas. The cost for this process, which is certainly not small, is being born by a well doing and stable economy.

This is in stark contrast to Mediterranean region, where the cost for reaching the aims considering RES and their penetrations has to be borne by deeply suffering economy and society. *Sustainability is, in that sense, at least of the same importance as affordability, in order to re-establish economic growth and to combat the social problems generated by energy poverty.* This can be achieved in two ways:

(a) A realistic energy mixture is needed, optimizing environmental, economic, and technological parameters. Mistakes of the past few years have to be corrected, by reassessing and reestablishing FITs for each energy source, in order to achieve reasonably attractive conditions for the investors and affordable prices for the consumers, if need to be gradually reduce the contribution of certain renewable sources and increase the contribution of others.
(b) With the exception of Spain, where the utilization of renewable has lead to the development of a new industrial branch, namely the one of wind power, the implementation of RES *does not seem to have a significant impact on GDP and on employment.* It is absolutely necessary that the production of at least some components of RES systems has to take place in countries like Greece and Portugal, otherwise the benefits of RES propagation for the societies in those countries will be marginal.

It is in this line of thought that renewables have to be considered, as this was suggested by some unconventional pioneers like Amory Lovins in the 1970s, namely as a way to achieve sustainability, to foster economic growth, and to support employment on a regional level, while at the same time ensuring societal consensus and support.

Facing the energy problem remains a complex issue, which can only be done by effectively addressing problems that are deeper and more vital than simply meeting the goals of some energy policies foreseen by European or international agreements and imposed by the demands of an impersonal market.

References

BMU (2011) In: The Federal government's energy concept of 2010 and the transformation of the energy system of 2011. Available via DIALOG: http://www.bmu.de/fileadmin/bmu-import/files/english/pdf/application/pdf/energiekonzept_bundesregierung_en.pdf. Accessed 15 Jan 2013

Boemi SN, Papadopoulos AM, Karagiannidis A, Kontogianni M (2010) Barriers on the propagation of renewable energy sources and sustainable solid waste management practices in Greece. Waste Manag Res 31:967–976

Carvalho D, Wemans J, Lima J, Malico I (2011) Photovoltaic energy mini-generation: future perspectives for Portugal. Energy Policy 39:5465–5473

DEPOIR (2007) Developing practical guidelines for an observatory center on the progress of implementing projects on renewables and waste/wastewater, Project funded by the General Sectarteriat for Reaserch and Technology. Thessaloniki

Rio DP, Gual MA (2004) The promotion of green electricity in Europe: present and future. Euro Environ J 14:219–234

EIA US Energy Information Administration (2013) Today in energy. Available via DIALOG: http://www.eia.gov/todayinenergy/detail.cfm?id=4850. Accessed 20 Jan 2013

EurObserv'ER (2011) Photovoltaic barometer (pp. 1–22)

EurObserv'ER (2012) Photovoltaic barometer, April 2012

Eurostat (2013) News release. Flash estimate for the first quarter of 2013. Available via DIALOG: http://epp.eurostat.ec.europa.eu/cache/ity_public/2-15052013-ap/en/2-15052013-ap-en.pdf

EWEA (2011) Wind in power 2011 European statistics, February 2012, pp 4–11

Giatrakos GP, Tsoutsos TD, Zografakis N (2009) Sustainable power planning for the island of Crete. Energy Policy 37:1222–1238

Gómez A, Zubizarreta J, Dopazo C, Fuey N (2011) Spanish energy roadmap to 2020: Socioeconomic implications of renewable targets. Energy 36:1973–1985

IEA World Energy Outlook (WEO) (2009) Available via:http://www.worldenergyoutlook.org/media/weowebsite/2009/WEO2009.pdf. Accessed 15 Dec 2012

Jenner S, Groba F, Indvik J (2013) Assessing the strength and effectiveness of renewable electricity feed-in tariffs in European Union countries. Energy Policy 52:385–401

Kaldellis JK, Kapsali M, Katsanou Ev (2012) Renewable energy applications in Greece—What is the public attitude? Energy Policy 42:37–48 http://www.sciencedirect.com/science/journal/03014215/42

Kambezidis HD, Kasselouri B, Konidari P (2011) Evaluating policy options for increasing the RES-E penetration in Greece. Energy Policy 39:5388–5398

Karteris M, Papadopoulos AM (2012) Residential photovoltaic systems in Greece and in other European countries: a comparison and an overview. In: Advances in Building Energy Research, doi:10.1080/17512549.2012.672005

Lagie (2013) Special monthly bulletin account for RES & CHP–May 2013. Available via DIALOG: http://www.lagie.gr/fileadmin/groups/EDSHE/MiniaiaDeltiaEL/2013_05_Miniaio_Deltio_EL_APE_SITHYA.pdf

Michalena E, Hills JM (2012) Renewable energy issues and implementation of European energy policy: the missing generation? Energy Policy 45:201–216

Moreno F, Martinez-Val J (2011) Collateral effects of renewable energies deployment in Spain: Impact on thermal power plants performance and management. Energy Policy 39:7637–7657
Papadopoulos AM, Glinou GL, Papachristos DA (2008) Developments in the utilization of wind energy in Greece. Renew Energy 33:105–110
Papadopoulos AM, Karteris MM (2009) An assessment of the Greek incentive's scheme for photovoltaics. Energy Policy 37(5):1945–1952
de Miera GS, Gonzalez PR, Vizcaıno I (2008) Analyzing the impact of renewable electricity support schemes on power prices: the case of wind electricity in Spain. Energy Policy 36:3345–3359
The Washington Post (2013) Europe's endless recession, in one chart. Available via DIALOG: http://www.washingtonpost.com/blogs/wonkblog/wp/2013/05/15/europes-endless-recession-in-one-chart/
Tolón-Becerra A, Lastra-Bravo X, Bienvenido-Bárcena F (2011) Proposal for territorial distribution of the EU 2020 political renewable energy goal. Renew Energy 36:2067–2077
Tsoutsos T, Papadopoulou E, Katsiri A, Papadopoulos AM (2008) Promotional schemes for renewable energy sources and their impact on reducing the emissions of greenhouse gases in Greece. Renew Sustain Energy Rev 12:1767–1788
Windpower Monthly (2013) R. Wendling's Portugal slices 23% off 2020 wind target, Available via DIALOG: http://www.windpowermonthly.com/article/1128686/Portugal-slices-23-off-2020-wind-target. Accessed 20 Jan 2013

The Shadows Cast by Inadequate Energy Governance: Why More Sun Does Not Necessarily Mean More Photovoltaic Electricity

Sarah J. McCormack and Brian Norton

Abstract Many governments have used market stimulation programmes focusing on increasing awareness, regulatory interventions or financial incentives for photovoltaics (PV). Solar resource availability is thus not the primary determinant of economic viability of photovoltaic solar energy systems. Countries with a moderate annual solar radiation resource, such as Germany, have, via governmental interventions, become leaders in installed photovoltaic capacity. In other countries with abundant solar radiation, such as Spain, governmental interventions have led to a boom-bust effect in the PV industry. Governmental governance, is here defined as the combined effect of prevailing policy, regulatory, fiscal and legal environments and measures. In this chapter, international case-studies are presented where different types of governmental governance have led to policy formulations and whose implementation has either enabled and facilitated increased PV uptake or has created barriers to maintaining steady growth in the solar photovoltaic industry.

1 Introduction

There are four broad widely used groups of technologies that use the sun as the energy source; (i) supply-side (i.e. grid supplying), demand-side (i.e. local demand contributing) and mixed-mode, usually building-integrated, photovoltaics (PV); (ii) supply-side concentrating solar thermal power systems, that concentrate solar

S. J. McCormack (✉)
Department of Civil, Structural and Environmental Engineering, Trinity College Dublin, College Green, Dublin, Ireland
e-mail: mccorms1@tcd.ie

B. Norton
Dublin Energy Lab, Dublin Institute of Technology, Aungier Street, Dublin 6, Ireland
e-mail: president@dit.ie

E. Michalena and J. M. Hills (eds.), *Renewable Energy Governance*,
Lecture Notes in Energy 23, DOI: 10.1007/978-1-4471-5595-9_17,

radiation to create high temperature steam used to produce electrical power in relatively large-scale plants under direct insolation; (iii) domestic demand-side solar thermal and cooling which uses thermal energy directly from the sun to heat or cool domestic water or building spaces and (iv) a diverse range of specialist devices for cooling, refrigeration, drying and cooking (Norton 2012).

PV solar energy systems convert solar radiation directly into electricity. For PV systems, the amount of electricity produced is proportional to the incident solar radiation within normal operating conditions. The economic viability of PV is determined by many factors: climatic factors are solar radiation, and ambient temperatures. Site-specific factors are the area(s) in relation to latitude of surface(s) on which PV can be installed. Technical factors are the PV and inverter characteristics and the extent that electricity output matches the temporal pattern of the electrical load which in turn establishes whether battery storage is necessary. The costs of displaced electricity from other sources and life-cycle costs of the PV installation are also critical (Mondol et al. 2006). A brief introduction to PV solar energy systems is presented in the following section followed by a number of country case-studies highlighting the presence or otherwise of governmental interventions on the extent of installed PV capacity.

2 Photovoltaic Solar Energy Systems

Photovoltaics convert solar radiation directly to electricity. There are many variations of solar cells by material, design and method of manufacture. Crystalline silicon dominates the World market at 85–90 %; thin films account for a large proportion of the remaining 10–15 % with silicon also forming a portion of this sector (EurObserv'ER 2012a, b). Much of the industrial and research activity in PV technology has been guided towards lowering material and cell processing costs, whilst improving the operational performance of devices to near theoretically maximum efficiencies (Goetzberger and Hebling 2000). Research into the latter has been so successful that such theoretical limits have had to be carefully re-evaluated, to provide a more accurate target for the device designers based upon better known material properties, device/interface characteristics and additional or alternative attributes that are not subject to certain theoretical limits.

PV efficiency is defined as the ratio of electrical output energy to incident solar energy on the PV surface and the theoretical limit depends on the type and fabrication of the solar cell. A quantum well solar cell under isotropic blackbody radiation would achieve an optimal efficiency of 54.5 % (Honsberg et al. 1997), for homojunction solar cells, the limiting efficiency is 30 % (Shockley and Queisser 1961) where a 27.6 % module efficiency for a multicrystalline silicon solar cell has been reported (NREL 2012). Efficiencies vary depending on material from 4.4 % for organic cells up to 43.5 % for multijunction cells (NREL 2012).

As the incident solar energy increases so does the output electricity produced from a PV device array, but at a rate limited by temperature effects and mismatch losses within the device. For example crystalline silicon PV operating above 25 °C typically, shows a temperature-dependent power decrease with a coefficient of between 0.4 %/K (Weakliem and Redfield 1979; Krauter 1994) and 0.65 %/K (Raziemska 2003; Hasan et al. 2010). Even with this factor, PV output increases with incident solar radiation. PV produce direct current electricity. For many applications an inverter is required to convert this to alternate current electricity. In stand-alone applications such as lighting and vaccine refrigeration batteries may also be included for energy storage. These elements combine to form a PV system. In some systems, concentration of solar energy using mirrors or lenses is used to reduce the area of PV required per unit output.

Photovoltaic technology may therefore be expected to be a more economically viable option in those countries with higher incident solar radiation resource. However, when PV installed capacity is examined globally this is not the case that can be seen from Fig. 1. It is clear that there are other factors that have to be taken into consideration which have affected this economic viability. In developing countries that have a large solar resource (e.g. Sudan) there is often limited economic development and infrastructure in place to enable PV solar energy systems to become a large component of a decentralised and autonomous application-

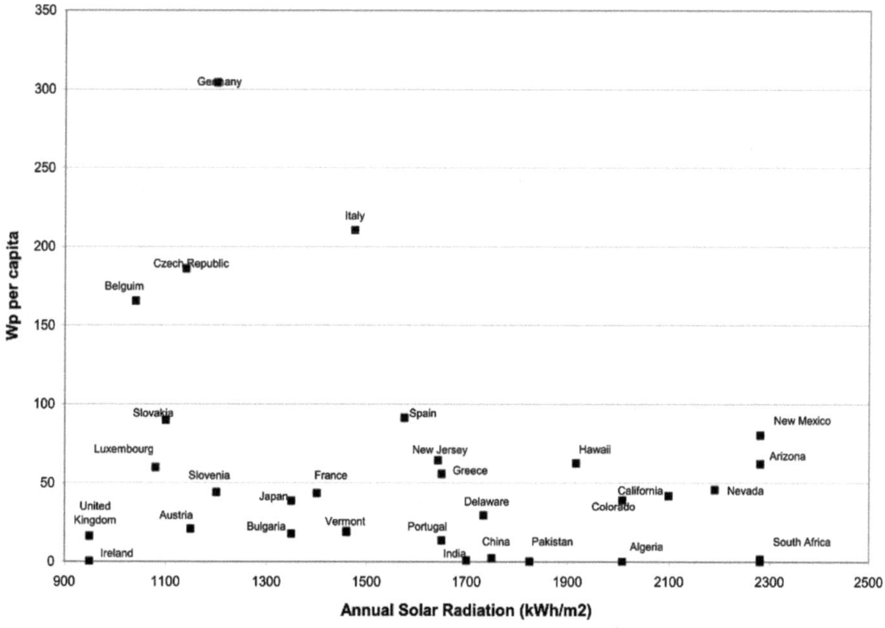

Fig. 1 Installed photovoltaic capacity (Wp) per capita with averaged annual solar radiation. Note Average solar radiation figures for smaller areas (such as European countries and North American States) are more representative than for larger countries (such as China and India)

specific (e.g. water pumps and vaccine refrigerators) electricity mix. Conversely, those countries leading the PV industry in terms of installed capacity (e.g. Germany) are not those with the highest solar radiation resource. Based on a long-term view of future economic growth, PV has been contrived in countries such as Germany to be a viable option to increase energy security, reduce reliance on finite fossil fuels, upskill a workforce and create export potential (BMU 2011).

A handful of countries that have strong policy regimes account for 80 % of global installed PV capacity (IEA 2009).

3 Diversity of Policy Options to Encourage Adoption of Photovoltaics (PV)

The International Energy Agency (IEA) PV Roadmap (2010) presented key actions for the development of PV industry. These included long-term targets underpinned by supporting policies to build confidence for investments in manufacturing capacity and deployment of PV systems. To accelerate learning and avoid duplicating efforts there is a need to expand international collaboration in PV research, development, capacity building and financing. Effective and cost-efficient PV incentive schemes are transitional and decrease over time so as to foster innovation and technological improvement. Increased R&D has both reduced costs and contributed to global commercialised readiness for rapid PV deployment, while also supporting longer term technology innovations.

Policy options to encourage adoption of PV, that are common to many countries include:

- Regulatory requirements on construction to include PV in building regulations and codes
- Incentivised PV capital cost reductions, such as grants, tax breaks, carbon taxes
- Utility obligations; to generate or buy an amount of PV generated electricity as a specified portion of overall electricity generation
- Feed-in-tariffs (FIT); an agreement to pay a guaranteed amount per kWh over a set period of time, of PV generated electricity (Cory et al. 2009).

These options are sequenced in timing depending on market maturity but can be used in combination. Regulatory requirements, include as examples building regulations, codes or the UK and Ireland's Building Energy Rating (BER). For buildings installing a renewable energy technology gives a high BER energy asset rating value. This encourages use of renewable in new builds, therefore creating an energy dimension to preferences in property markets. Incentivised capital cost reduction mechanisms are usually put in place to raise awareness of PV often in the context of full-scale trials where performance evaluation is undertaken in order to demonstrate the applicability both technically and economically. Such interventions can create an artificial demand, so that when removed it causes a collapse in the market. To sustain a long-term PV installation sector there is normally an obligation put on utilities to buy the electricity from the PV generator.

Feed-in-tariffs are one of the incentives used most widely for stimulating the development of renewable electricity, creating conditions that reduce risk and improve investment security. Globally, grid-connected solar PV have an average annual growth of 81 %, mainly driven by FIT schemes (REN21 2011). A variety of different strategies and tools adopted by EU states (Klein et al. 2010; Haas et al. 2011; Cansino et al. 2010; EPIA 2012; Duarte et al. 2011; del Rio and Mir-Artigues 2012; Timilsina et al. 2012; Furkan 2011; Solangi et al. 2011; Moosavian et al. 2013; Ranjit et al. 2012). It has been found that all interventions stimulate the PV market in the short term. However, the artificiality and/or impermanence of most measures has to differing extents led to inefficient market structures and supply chains, high initial profit taking and inappropriate application (in scale or end use) of PV. Successful longer term interventions have generally had a sector-wide remit to either the building or electricity sectors. For example, via regulating requirements to install PV on new buildings or an obligation on a utility, to have a specific proportion of PV generated electricity.

Here, we present case-studies from EU, China, Japan illustrating where differing interventions have had positive and negative influences on PV deployment.

3.1 The European Context

European global horizontal irradiation is presented in Fig. 2. Solar energy conversion contributes 1.4 % of the EU's electricity production, at ~44.8 TWh in 2011 (up 98 % on 2010) (EurObserv'ER 2012a, b).

Given the capacity installed by the end of 2011, photovoltaic electricity output should easily exceed 60 TWh in 2012. This will take the photovoltaic share close to 2 %. In 2011, it was approximately 3.6 % in Italy, 3.1 % in Germany and 2.6 % in Spain. Spain and Italy have high solar radiation, from 1,500 to 2,000 kWh/m^2, whereas Germany's resource is less, ranging from 900 to 1,200 kWh/m^2. In all three countries there are strong policies in place to incentivise PV installations. While the electricity output will vary between these regions due to the differences in solar resource, the financial gains (and therefore economic viability) are not dissimilar due to the nature of the policies in place.

To highlight the different PV governance policies, three EU countries are presented as case-studies in this chapter with different governance approaches to PV;

- Ireland with a low/moderate solar resource (800–1,100 kWh/m^2/year), where currently incentives for PV neither exist nor are likely to be prioritised in the near future;

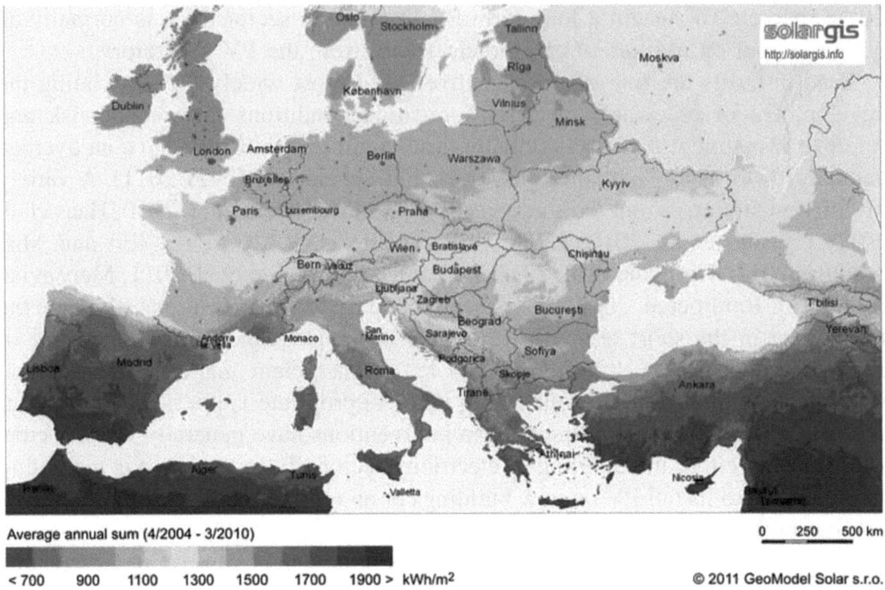

Fig. 2 Annual European global horizontal irradiation (SolarGIS © 2013 GeoModel Solar s.r.o)

- UK with a low/moderate solar resource (700–1,200 kWh/m^2/year), where current incentives are in place so PV form a small but growing part of electricity production;
- Spain with a moderate/high solar resource, with the strong PV supporting policies in place currently under revision.

3.1.1 Ireland

As can be seen from Fig. 2, Ireland has a moderate to lower solar resource (800–1,100 kWh/year) which has led to non-prioritisation of photovoltaic solar energy in targeted measures. A microgeneration trial, to assess the viability of microgeneration systems, including PV was held in 2009 by the Sustainable Energy Authority of Ireland. This consisted of 50 microgeneration sites, 15 of which were PV installations, amounting to almost 70 kWp (SEAI 2009). The payback period for the remotely monitored sites, based on interim data varied between 30 and 100 years. Such a high payback period was also found in a study by Ayompe et al. (2010) where a comparative analysis showed that current PV system costs are at least 50 % higher in Ireland than other jurisdictions with similar climatic conditions such as UK and Germany.

A feed in tariff of 0.19 c/kWh was available up until February 2012 from one specific operator which was state-owned and monopolised the market before liberalisation in 2005. The support package included free installation of an electricity import/export metre, a microgeneration payment (9 c/kWh) and a support payment (10 c/kWh) which applies to the first 3,000 kWh exported annually.

Ayompe et al. (2010) have suggested that the PV market was not operating optimally, possibly due to the small number of market actors or due to a lack of transparent market information such as system price and quality. Growing the number of installers would require market intervention such as sustained subsidies prior to the microgeneration trial (e.g. grants or low cost loans), higher feed-in tariffs or regulation. There is a need, however, to revisit the economic analysis of PV in an Irish context due to the substantial decreases in system costs globally. As these should have filtered through to the Irish market since previous studies, new studies may show a more favourable market.

3.1.2 United Kingdom

The UK has a similar solar resource to Ireland, but in contrast it has set up significant measures to increase PV electricity generation. The Renewables Obligation (RO) is currently the primary mechanism to support deployment of renewable electricity generation (Cherrington et al. 2013). The original design imposed an obligation on energy suppliers to provide an increasing proportion of their supply from renewable sources (Woodman and Mitchell 2011). Electricity suppliers can fulfil their obligation by delivering Renewable Obligation Certificates (ROCs) to the electricity regulator, paying the "buy-out" price or a combination of the two (Dusonchet and Telaretti 2010).

In 2009, the RO moved from a mechanism of single level support to a banded system varying by technology. Support levels for each technology are decided considering factors including their costs, relative maturity and potential for future deployment. In November 2011, the UK government published a banding review outlining the levels of banded support available for renewable electricity generation under the RO for the period 2013–2017 (DECC 2012). Solar PV installations between 50 kW and 5 MW are currently supported at 2 ROCs/MWh within the RO (DECC 2012). In September 2011, suppliers that fully complied with their renewable obligation received £14.32 for each ROC from the buy-out fund recycle payments (Ofgem 2011a). The money that the regulator Ofgem, collected in the buy-out fund has been recycled on a pro-rata basis to suppliers who presented ROCs minus the cost of Ofgem's administration of the RO. Suppliers that do not present ROCs pay into the buy-out fund at the buy-out price, but do not receive any portion of the recycled fund (UKGOV 2013).

Feed-in-Tarrifs (FITs), alongside the RO, were introduced by the UK government in April 2010 to promote the deployment of small-scale renewable and low-carbon electricity generation technologies as part of a national obligation to meet the Renewable Energy Directive (Ofgem 2011b). The main aim was to lower the cost of those technologies by providing potential investors with a secure financial return for the long-term financial commitment required.

The growth rate of solar PV installations was much faster than the government anticipated. Before the launch of the FIT, it was projected that the total capacity installed would reach 137 MW by the second year, i.e. in April 2012. However, this target was achieved in July 2011 (Muhammad-Sukki et al. 2013), and by the end of 2011, 899 MW of PV were installed in the UK bringing the cumulative installed capacity to just under 1 GW (IEA 2012). In order to reduce costs of the incentives, subsidy reductions were decided in June 2011, but they only applied to ground-based plants and bigger than 250 kWp roof-mounted plants.

As the market boomed, UK Department of Energy and Climate Change (DECC) announced on 1 November 2011 that it would reduce FITs, by applying the FIT reductions initially scheduled up until 1 April 2012 be brought forward to installations from 12 December 2011. The tariffs for small roof-mounted systems (<4 kWp) were more or less halved to € 0.2517/kWh (£0.22/kWh) (EurObserv'ER 2012a, b). However in December 2011, the UK High Court declared that this announcement was illegal, ruling that the government's schedule had sidestepped the mandatory consultation procedure for modifications to the incentive programme. This decision, which is subject to appeal, postponed the effective FIT reduction to March 2012. A further reduction has been planned for July 2013 linked to the installation level achieved in March and April 2013. The cost of purchasing and installing solar PV technology in UK has reduced dramatically since the introduction of the FIT. However, without the financial support from government, solar PV technology cannot currently compete with grid electricity generated from fossil fuels. The reductions in solar PV tariffs recommended by the UK government are expected to delay grid-parity due to the high cost of PV generated electricity extending the need for financial support of PV installations (Cherrington et al. 2013).

3.1.3 Spain

Solar PV growth in Spain took place under the 2007 FIT (Del Río and Mir-Artigues 2012).While in 2008, the introduction of a FIT in Spain achieved 2,708 MW of newly installed PV power from only 26 MW in 2005, 43 % more than Germany in 2008 (EPIA 2012). Although in Spain, FITs have been shown to be an effective instrument to promote PV, it has led to a boom-and-bust cycle, with negative consequences for, both, consumers and investors. During 2011, annual installed PV power was 345 MW with cumulative installed capacity reaching 4,260 MW (EurObserv'ER 2012a, b) with 99 % of PV installations grid-connected. To promote solar electricity, new procedures and technical regulations were published in 2011 that required: (a) connection to the electricity network of low-power embedded electrical energy production facilities (Royal Decree 1699/2011) including PV facilities of less than 100 kW; (b) self-consumption associated with PV installations (Royal Decree 1544/2011) and stipulating the technical and economic conditions relating to the consumption of electricity produced by the consumer. Royal Decree 01/2012 applied an immediate and indefinite moratorium (from January 2012), removing all financial incentives awarded to renewable energy-sourced power plants. This halted new installations. While the introduction of new support mechanisms is awaited, a law enforced in April 2012 will pay for the surplus electricity fed into the grid after self-consumption by <100-kWp installations (EurObserv'ER 2012a, b).

Spain does have further aspirations to meet 2020 targets where the share of renewables in Final Energy Consumption is to increase to 20 %, with 3.6 % of the Spanish electricity energy demand to be met by PV electricity.

3.2 China

From Fig. 3 it can be seen that there is a diverse range of solar radiation levels in China and therefore implementation of the same PV governance strategies will have very different outcomes for areas of high (>2,500 kWh/m^2) and low (<1,000 kWh/m^2) solar resource.

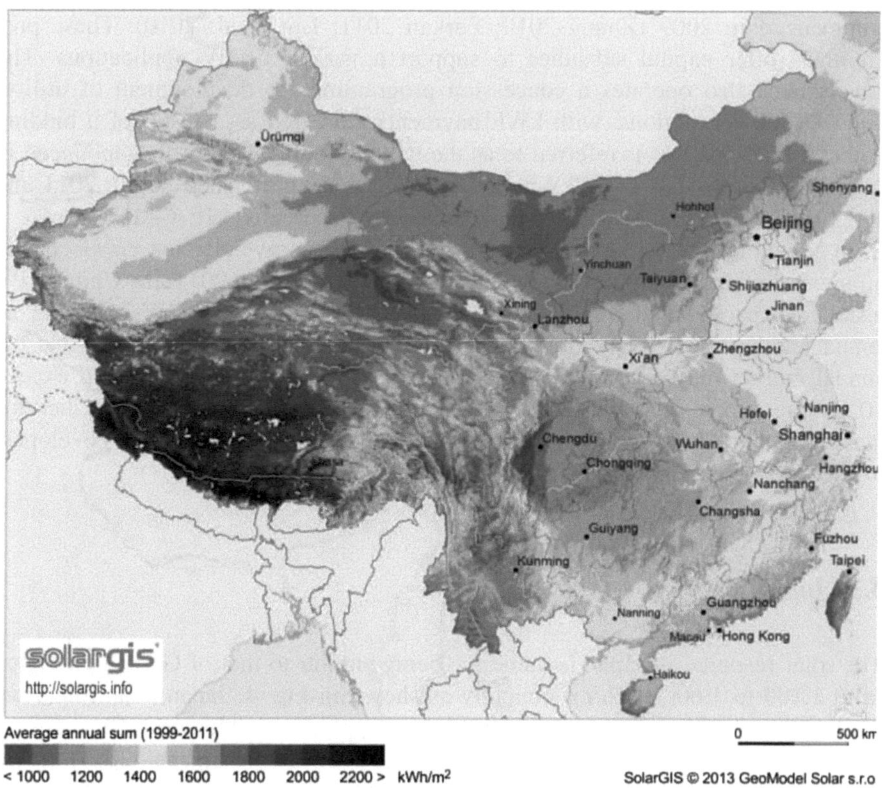

Fig. 3 Chinese global horizontal irradiation (SolarGIS © 2013 GeoModel Solar s.r.o)

According to the International Energy Agency (IEA) total installed capacity of PV reached almost 3,300 MW in China in 2011, a dramatic increase in 4 years from 100 MW. Internationally, there has been a huge growth in the PV industry with China producing over 57 % of PV cells in 2011 (Photon 2012) and with five companies in the top ten PV cell manufacturers (EurObserv'ER 2012a, b).

The Chinese Government is now providing strong support via incentive policies and financial measures to expand the domestic Chinese PV market, in an attempt to better balance domestic industrial production of PV with local PV market demand.

In 2011, 2,000 MW were large-scale grid-connected power stations, 480 MW were building-integrated (BIPV), or building-attached (BAPV) systems, and about 15 MW were installed for rural electrification and off-grid industrial applications. Since 2008, large-scale plants have been developed in China and are now clearly the dominant application, an example is presented in Fig. 3. In more recent years, building-integrated PV has been receiving strong support from the government and is playing an increasingly important role in the Chinese PV market (IEA 2012).

The "Solar PV Building Project" that focuses on BAPV and BIPV and the "Golden Sun Programme" (that focuses on rural electrification and building projects are the major market support demonstration programmes in China, both commenced in 2009 (Zhang 2010; Furkan 2011; Liu et al. 2010). These programmes offer capital subsidies to support a variety of PV applications. The government also operates a concession programme for development of utility-scale PV power stations, with kWh payments allocated as a result of a bidding process. In China this is referred to as the nationwide feed-in tariff (announced in July 2011), with 1.15 CNY/kWh allocated for projects finished before 2011 end and 1.0 CNY/kWh allocated for those completed thereafter (IEA 2012).

In November 2011, the surcharge for renewable energy collected via electricity bills was doubled to support renewable power generation. In 2011, the Chinese PV installed capacity goal for 2015 was updated from 10 to 15 GW, meaning an annual market of 3–4 GW of PV each year until 2015. It is anticipated that the annual market may reach 10 GW/year in the period up to 2020 and will exceed 20 GW/year thereafter (IEA 2012; EPIA 2012). The question remains whether this market outside China will be accessible on a wide scale to international developers of the next generation of PV technologies (EPIA 2012).

3.3 *Japan*

The solar resource in Japan is moderate being similar to that of Germany ranging from 1,100 to 1,600 kWh/m^2 annually as shown in Fig. 4. Japan is an important market player with respect to both the global supply, via its domestic PV industry, and the demand for PV, with a strong domestic market (Furkan et al. 2011). It was the global market leader up to the end of 2004 when it was surpassed by Germany; Germany has subsequently been overtaken by China. The Japanese PV market was

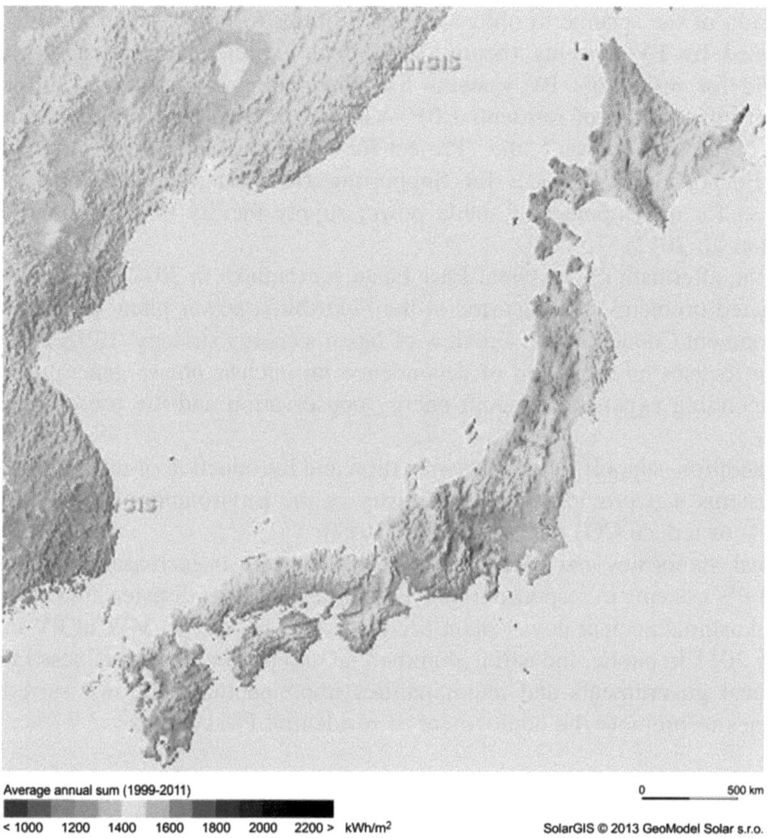

Fig. 4 Japanese global horizontal solar irradiation (SolarGIS © 2013 GeoModel Solar s.r.o.)

the second largest, with 287 MW of new installations in 2006, primarily using grid-connected residential systems under the Japanese PV residential programme. The cumulative installed PV capacity in Japan reached 1.71 GW in 2006 and has been steadily increasing (Furkan 2011), reaching 4.9 GW or 2.1 % of total national electricity generation capacity in 2011. In 2011, 1,295 MW of PV were installed in Japan, a 31 % increase from 2010 (IEA 2012). Most of these installations (over 1,245 MW) continued to be mainly residential, grid-connected distributed PV systems, with a further 45.9 MW comprising grid-connected centralised plants mainly installed by the electricity utilities. The PV market was led by a subsidy programme for residential PV systems and a programme to purchase surplus PV power at a preferential price from systems of less than 500 kW capacity.

In August 2011, the government enacted the Renewable Energy Law, under preparation since 2009. A Feed-in-Tariff programme came into force in July 2012. The Ministry of Economy, Trade and Industry (METI) resumed the Subsidy for Installation of Residential Photovoltaic Systems in 2009. In addition, with the extension of the scheme to oblige electric utilities to purchase surplus electricity generated by PV systems (below 10 kW) at a preferential price, the market demand for residential PV systems has been increasing steadily. Subsidies to support introduction of residential PV systems included the "Programme to Purchase Surplus PV power", the "Project for Promoting the Local Introduction of New Energy", the "Project for Supporting New Energy Operators" and the "Project for development of stable power supply facility for emergency cases" (Avril et al. 2012).

In the aftermath of the Great East Japan Earthquake in 2012 together with the associated problems that occurred at the Fukushima power plant, the Energy and Environment Council began a review of Japan's energy strategy (EPIA 2012). The review focuses on reduction of dependence on nuclear power generation with a corresponding expansion of both energy conservation and the use of renewable energy.

In addition, support for the dissemination and introduction of model projects for PV systems was provided by the Ministry of the Environment (MoE) as part of projects to reduce CO_2 emissions (IEA 2012).

Local authorities and industries in Japan showed an increased willingness to install PV systems in response to power shortages and as disaster prevention after the Fukushima nuclear power plant accident. This led to 140 MW of PV installed during 2011 in public, industrial, commercial and power sector business facilities. 875 local governments and municipalities implemented their own subsidy programmes to promote the deployment of residential PV systems.

3.4 United States of America

The size of USA leads to a large range in solar resource in different states as shown in Fig. 5 (range in annual resource 1,000–2,300 kWh/m^2). With high insolation, the greatest electricity demand per capita in the world, and ample available land for PV development, the U.S. presents an attractive long-term PV growth opportunity for developers, installers, financiers and other PV service providers (Furkan 2012).

From Fig. 5, the highest solar resource is received by the states of California, Nevada, Arizona, New Mexico, Texas, Utah and Colorado which matches somewhat with the states leading in installed PV capacity in 2011 (IREC 2012) as listed (1-10) below.

(1) California (538 MW)
(2) New Jersey (306 MW);
(3) Arizona (288 MW);
(4) New Mexico (122 MW);
(5) Pennsylvania (78 MW);
(6) Colorado (75 MW);

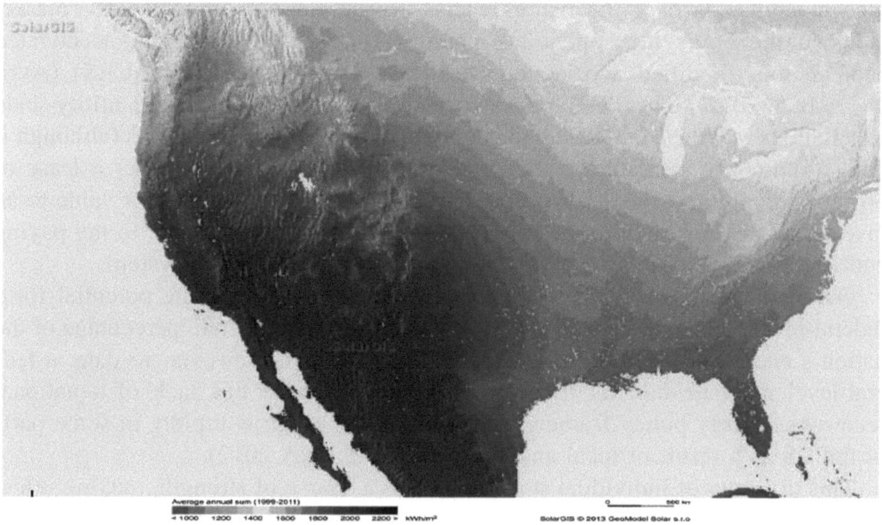

Fig. 5 Unites States of America global horizontal solar irradiation (SolarGIS © 2013 GeoModel Solar s.r.o.)

(7) New York (68 MW);
(8) Texas (51 MW);
(9) North Carolina (45 MW);
(10) Hawaii (40 MW).

Total PV capacity in the US increased by an estimated 1,867 MW in 2011 twice the growth in 2010. Cumulative installed capacity in the US reached 3,966 MW by the end of 2011 (IEA 2012). More than 60,000 PV systems were connected in 2011 for a 20 % growth in the number of grid-connected systems installed annually. By the end of 2011, there were approximately 214,000 distributed, grid-connected PV systems installed in the United States; the nation added 770 MW of utility-scale generation capacity that year alone (EPIA 2012).

The market developed evenly in the residential, commercial and utility-scale segments. The residential segment represented about 15 % of the new capacity. Its development is increasingly driven by third-party ownership instead of customer-owned systems. The commercial segment grew very rapidly in 2011 to reach nearly 45 % of the market mainly in California and New Jersey. Utility-scale projects helped to boost the market in 2011, representing 40 % of the new added capacity with PPAs driving the development of this segment (EPIA 2012).

PV capacity reached about 0.4 % of total national electricity generation capacity, with 9.4 % of new electricity generation capacity installed during the year being PV. The US PV market development is supported by financial incentives at the federal and some state levels. However, many policy drivers for renewable energy deployment remain at state and local levels with consequent high diversity. Two of the major federal drivers for growth in the PV market included the 30 % investment tax credit (ITC) and the five-year accelerated depreciation (modified accelerated cost recovery schedule or MACRS) (Avril 2012; IEA 2012). The ITC applies to residential, commercial and utility-scale installations and the MACRS applies only to commercial installations (although it is also indirectly available to the residential systems deployed under a lease or power purchase agreement). For commercial installations, the present value to an investor of the combination of these two incentives (only available to tax paying entities) amounts to about 56 % of the installed cost of the PV system.

Over the course of 2011, the federal government outlined the potential for a federal-level clean energy standard that would mandate a certain percentage of the nation's energy portfolio derived from 'clean' sources. However, to date, a federal-level mandate has yet to be implemented. Despite this lack of a national renewable energy policy framework, PV continues to grow rapidly in some parts of the US as a result of local and state initiatives (IEA 2012).

This diversity of individual state markets is a source of strength, making it less likely for the US to experience the boom-bust cycles seen in many other national PV markets (IEA 2012). California represented 29 % of new capacity installed during 2011 compared to 32 % in 2010, indicating stronger growth in other states. State incentives in the US have been driven in large part by the passage of renewable portfolio standards (RPS), also called renewable electricity standards (RES). As of December 31/2011, sixteen states and Washington D.C. had RPS policies with specific PV provisions (IEA 2012). Several other emergent policy and financing mechanisms have the potential to drive PV market expansion through the establishment of widespread local and utility programmes. Such policies include state-level FITs and time-of-use electricity tariff structures. Innovative financing programmes have also been a feature of the US PV market (EPIA 2012).

4 Discussion: Conclusions

In recent years, imperatives to promote local economic development and energy security whilst reducing the environmental impact of electricity production have all led to increasing need for national and local policymakers to develop renewable energy markets. There is a wide range of possible interventions that can increase market penetration of PV. Indeed, there are real success stories. Yet, among countries there are considerable differences in these strategies with respect to their economic efficiency or success in triggering a substantial number of new installations.

In the case of Ireland, a micro electricity generation trial was set up to increase the awareness of microgeneration technologies such as PV to the wider public. Through the scheme it was hoped to kick start the industry in Ireland by providing objective information about performance and cost. However, the prolonged waiting period for the introduction of the scheme caused the already small PV industry to stall as customers waited on the new trial to start that included a new FIT. In reality, only 15 PV installations proved to be qualified for the FIT under the scheme, the outcomes of which were to set PV at a payback period of between 30 and 100 years. So, while objective information on PV performance became available as a consequence of the scheme, it did little to encourage the industry in Ireland. Due to close proximity to the UK market's favourable incentive schemes, reduced PV costs are the beginning to filter through to Ireland.

In the UK and Spain, generous FITs were introduced that were predictable, continued for specified periods and decreased in amount over time. However, in Spain regulations should have included flexible (Srinivasan 2009) and adaptable revision mechanisms in order to avoid a short-term PV boom (Del Rio and Mir Artigues 2012). Similarly, the lack of such measures in the UK has led to legal proceedings, where specified periods were changed without consultation.

FITs have played an effective and instrumental role in PV development in Spain and UK. To reflect this emerging PV market maturity in UK, the FIT level is being reduced. It is, however, uncertain if the transition will produce expected results. Will the decrease in FIT, the primary basis for investors' confidence, drive investors away from solar energy markets?

A mix of policy portfolios that includes tax credits, subsidies and rebates, RPS or targets, net metering and renewable energy certificates (REC) have facilitated solar energy market growth in the United States (Timilsina et al. 2012), China and Japan. In the United States, upfront incentives are being shifted towards more FIT-like performance-based incentives with an aspiration to being the next large PV market (EPIA 2012).

Analysis of case-studies has shown that while solar resource is important to the economic viability of photovoltaic solar energy systems, the market intervention policies in place are more important. When market forces that take fully into account the externalities of fossil fuel electricity generation come fully into play, mass-manufactured PV will likely achieve grid-parity. PV will then be a rational economic choice for those countries with a higher solar resource; but until then, governance in the form of market and regulatory policies retain an invaluable role.

References

Avril S, Mansilla C, Busson M, Lemaire T (2012) Photovoltaic energy policy: financial estimation and performance comparison of the public support in five representative countries. Energy Policy 51:244–258, ISSN 0301–4215, doi: 10.1016/j.enpol.2012.07.050

Ayompe LM, Duffy A, McCormack SJ, Conlon M (2010) Projected costs of a grid-connected domestic PV system under different scenarios in Ireland, using measured data from a trial installation. Energy Policy 38(7):3731–3743, ISSN 0301–4215, http://dx.doi.org/10.1016/j.enpol.2010.02.051

BMU (2011) Development of renewable energy sources in Germany 2010. Based on statistical data from the working group on renewable energy sources—statistics (AGEE-Stat). http://www.bmu.de/files/english/pdf/application/pdf/ee_in_deutschland_graf_tab_en.pdf. Last accessed 20 Dec 2012

Cansino J, del Marıa M, Pablo-Romero P, Rocıo R, Rocıo Y (2010) Tax incentives to promote green electricity: an overview of EU-27countries. Energy Policy 38:6000–6008

Cherrington V, Goodship A, Longfield K, Kirwan K (2013) The feed-in tariff in the UK: a case study focus on domestic photovoltaic systems. Renew Energy 50:421–426, ISSN 0960–1481. doi:10.1016/j.renene.2012.06.055

Cory K, Couture, T, Kreycik C (2009) Feed-in tariff policy: design, implementation, and RPS policy interactions. Technical report NREL/TP-6A2-45549, National Renewable Energy Laboratory, Colorado, USA, 23 pp

DECC (2012) Consultation on proposals for the levels of banded support under the renewables obligation for the period 2013e17 and the renewables obligation order 2012. Department of Energy and Climate Change; 2011

del Río P, Mir-Artigues P (2012) Support for solar PV deployment in Spain: some policy lessons. Renew Sustain Energy Rev 16:5557–5566

Deshmukh R, Bharvirkar R, Gambhir A, Phadke A (2012) Changing sunshine: analyzing the dynamics of solar electricity policies in the global context. Renew Sustain Energy Rev 16:5188–5198

Dusonchet L, Telaretti E (2010) Economic analysis of different supporting policies for the production of electrical energy by solar photovoltaics in western European Union countries. Energy Policy 38:3297–3308

EPIA (2012) Global market outlook for photovoltaics until 2016. http://www.epia.org/index.php?eID=tx_nawsecuredl&u=0&file=/uploads/tx_epiapublications/Global-Market-Outlook-2016.pdf&t=1357819268&hash=75b2aa4df11bb4ec53e61f342d224b9a0017c2b3 Last accessed Jan 2013

EurObserv'ER (2012a) Solar thermal and concentrated solar power barometer. http://www.eurobserv-er.org/pdf/baro209-ST_H.pdf

EurObserv'ER (2012b) Photovoltaic barometer. http://www.eurobserv-er.org/pdf/photovoltaic_2012.pdf

Furkan D (2011) The analysis on photovoltaic electricity generation status, potential and policies of the leading countries in solar energy. Renew Sustain Energy Rev 15(1):713–720, ISSN 1364–0321. doi: 10.1016/j.rser.2010.09.026

Goetzberger A, Hebling C (2000) Photovoltaic materials, past, present, future, Solar Energy Materials and Solar Cells 62(1):1–19

Haas R, Panzer C, Resch G, Ragwitz M, Reece G, Held A (2011) A historical review of promotion strategies for electricity from renewable energy sources in EU countries. Renew Sustain Energy Rev 15(2):1003–1034 ISSN 1364-0321, http://dx.doi.org/10.1016/j.rser.2010.11.015

Hasan A, McCormack SJ, Huang MJ, Norton B (2010) Evaluation of phase change materials for thermal regulation enhancement of building integrated photovoltaics. Sol Energy 84(2010):1601–1612

Honsberg CB, Corkish RC, Bremner SP (1997) A new generalised detailed balance formulation to calculate solar cell efficiency limit, Proceedings of the 26th IEEE Photovoltaic Specialist Conference. Anaheim, California U.S.A.

IEA Photovoltaic Power Systems (IEA PVPS) (2009) Trends in photovoltaic applications, IEA PVPS http://www.iea-pvps.org/fileadmin/dam/public/report/statistics/tr_2009_neu.pdf

International Energy Agency (2010) Technology roadmap: solar photovoltaic energy. http://www.iea.org/publications/freepublications/publication/pv_roadmap.pdf

International Energy Agency (2012) Trends in photovoltaic applications: survey report of selected IEA countries between 1992 and 2011. http://www.iea-pvps.org/index.php?id=1&eID=dam_frontend_push&docID=1239

IREC (2012) U.S. solar market trends 2011. Interstate Renewable Energy Council (IREC). Retrieved Jan 2013 . http://www.irecusa.org/wp-content/uploads/IRECSolarMarketTrends-2012-web.pdf

Klein A, Merkel E, Pfluger B, Held A, Ragwitz M, Resch G, Busch S, Fraunhofer ISI (2010) Evaluation of different feed-in tariff design options—best practice paper for the international feed-in cooperation. 3rd edn. update by December 2010. http://www.feed-in-cooperation.org/wDefault_7/wDefault_7/download-files/research/best_practice_paper_2nd_edition_final.pdf Last access ed. Jan 2013

Krauter S (1994) Actual optical and thermal performance of PV-modules. In: Proceedings of the 1st world conference on photovoltaic energy conversion (joint congress of IEEE/PVSEC/EUPVC). Waikoloa, Hawaii (USA), Dec, vol 1, pp 734–737

Liu L-q, Wang Z-x, Zhang Y-c, Xue L-q (2010) Solar energy development in China—a review. Renew Sustain Energy Rev 14(1):301–311, ISSN 1364-0321. doi:10.1016/j.rser.2009.08.005

Mondol JD, Yohanis YG, Norton B (2006) Optimal sizing of array and inverter for grid connected photovoltaic systems. Sol Energy 80:1517–1539

Muhammad-Sukki F, Ramirez-Iniguez R, Munir AB, Hajar S, Yasin M, Hawa S, Abu-Bakar S, McMeekin SG, Stewart BG (2013) Revised feed-in tariff for solar photovoltaic in the United Kingdom: a cloudy future ahead? Energy Policy 52:832–838 ISSN 0301-4215. doi:10.1016/j.enpol.2012.09.062

NREL (2012) http://www.nrel.gov

Norton B (2012) Industrial and agricultural applications of solar heat. In: Sayigh A (ed) Comprehensive renewable energy, vol 101, no 3, Elsevier, Oxford, pp 367–594

Ofgem (2011a) The renewables obligation buy-out fund (2010–2011)

Ofgem (2011b) The feed-in tariff scheme E-Serve

Photon (2012) Photon Magazine. http://www.photon.de

Raziemska E (2003) The effect of temperature on the power drop in crystalline silicon solar cells. Renew Energy 28:1–12

Renewable Energy Policy Network for the 21st Century (REN21) (2011) Renewables global status report, REN21, Paris

Shockley W, Queisser HJ (1961) Detailed balance limit of efficiency of p-n junction solar cells. J Appl Phy 32(3):510–519

Solangi KH, Islam MR, Saidur R, Rahim NA, Fayaz H (2011) A review on global solar energy policy. Renew Sustain Energy Rev 15:2149–2163

SolarGIS © 2013 GeoModel Solar s.r.o. (www.solargis.info)

Srinivasan S (2009) Subsidy policy and the enlargement of choice. Renew Sustain Energy Rev 13:2728–2733

Timilsina GR, Kurdeglashvili L, Narbel PA (2012) Solar energy: markets, economics and policies. Renew Sustain Energy Rev 16:449–465

UK Government (2013) https://www.gov.uk/government/policies/increasing-the-use-of-low-carbon-technologies/supporting-pages/the-renewables-obligation-ro Last accessed July 2013

Weakliem HA, Redfield D (1979) Temperature dependence of the optical properties of silicon. J Appl Phys 50:1491–1493

Woodman B, Mitchell C (2011) Learning from experience? The development of the renewables obligation in England and Wales 2002–2010. Energy Policy 39:3914–3921

Zhang ZX (2010) China in the transition to a low-carbon economy. Energy Policy 38(11):6638–6653

Germany: Challenges of a Full Transition to Renewable Energy

Volkmar Lauber and Moritz Buschmann

Abstract The transformation of the German electricity sector to a near-total renewable supply ("Eneigiewende") by 2050 or earlier was embodied into law in 2012. This seemed to consecrate a development which began with the passage of the Renewable Energy Act of 2000. Naturally, such a major transformation needs considerable vision, and over time needs determination from political and social forces to overcome resistance from established ideas, interests, practices, and organizational arrangements. After a historical overview of the institutional politics of RES-E (renewable energy sourced electricity), this chapter will look at three major challenges that German politics and society was/is faced with: launching very rapid, indeed disruptive RES-E growth (disrupting carbon lock-in); building or maintaining political support and an actor network capable of supporting this change; and reshaping the electricity system to accommodate fluctuating generation by wind and solar power.

1 Introduction: The Political Context of RES-E Policy

The first Electricity Feed-in Law (*Stromeinspeisungsgesetz*) entered into force in 1991. It was only a first step, but it amounted to an important revision in the renewable energy policy of the Conservative (CDU/CSU)-Liberal government, which was in power from 1982 to 1998. This policy—formulated by the Ministry of Economic Affairs—limited support to renewable energy (RE) to R&D only,

V. Lauber (✉)
University of Salzburg, Rudolfskai 42 5020 Salzburg, Austria
e-mail: volkmar.lauber@sbg.ac.at

M. Buschmann
Energy Consultant, Bauerstraße 1 80796 Munich, Germany
e-mail: moritz.buschmann@stud.sbg.ac.at

E. Michalena and J. M. Hills (eds.), *Renewable Energy Governance*,
Lecture Notes in Energy 23, DOI: 10.1007/978-1-4471-5595-9_18,

supposedly in line with market principles. The Feed-in Law changed this by supporting market creation at least in a modest way, something that few governments did at that time. It introduced tariffs differentiated by technology and the obligation of utilities to purchase renewable electricity at a price set by law as a percentage of household prices (65–80 % for small hydro, 90 % for wind and solar power). Parliament—where the law was introduced as a private member's bill, against the preferences of the government—passed it without opposing votes (Jacobsson and Lauber 2006).

When the Social Democratic–Green Coalition came to power in 1998, it was determined to decisively accelerate the deployment of renewable energy as well as to phase out nuclear power by about 2020. The Renewable Energy Act of 2000 (EEG or *Erneuerbare Energien Gesetz*) was a law for supporting the transition to renewable energy, among other things by making it easy for citizens to become RES-E producers, thanks to a modest but highly predictable revenue stream for 20 years, and a strong position vis-à-vis electric utilities that had to accept all RES-E on offer. Other political goals of the Act included the replacement of fossil fuel imports by a domestic high-tech industry providing innovation and employment. The adoption of this act was surrounded by intense conflicts: Conservatives and Liberals opposed the legislation because it supposedly went against the principles of the market economy and the EU Treaty, and vowed to overturn it in case they should return to power. Similarly, intense conflicts prevailed at the first important amendment to the Act in 2004.

After the 2005 parliamentary elections, Conservatives and Social Democrats formed a grand coalition government. Given the close election results, the two parties agreed to continue with EEG and the nuclear phase-out. In fact, the main actors' positions came to be more alike (Buschmann 2011). In 2007–2008, EEG underwent a reasonably harmonious amendment process ("EEG 2009"). Preparatory steps were taken to allow self-marketing and—for solar generators—self-consumption in the name of greater reliance on market integration, reducing the scope of support by regulation.

The 2009 elections produced the first Conservative-Liberal coalition since 1998. The Liberal leadership (though not the majority of its voters) had opposed EEG all along while in opposition. Within the Conservatives, the economic wing[1] had maintained some scepticism but had become more moderate after 2005; the environmental wing wanted to maintain EEG on condition to further strengthen market integration. But in 2009, the two parties agreed on thoroughly revising RES-E policy and on postponing by about one decade (until around 2030) the nuclear phase-out decided in 2000. This led to several efforts to slow down RES-E deployment between 2009 and early 2013. The postponement of the nuclear phase-out in late 2010 was meant to be the key to reducing market space for renewable electricity.

[1] As opposed to the environmental wing of the party.

The Fukushima accident led to a policy U-turn on this point in the spring of 2011 (i.e., to reinstating the earlier nuclear phase-out) but the efforts to slow down RES-E continued nonetheless, even gaining in intensity in 2012 and 2013. It became clear that opposition against fast RES-E growth had intensified in the ranks of the Conservative–Liberal government; on the other hand, the chiefs of the *Länder* (territorial subunits) represented in the upper house of parliament strongly supported even faster growth and in 2012 forced the government to a compromise. In early 2013, a government proposal to cap the EEG surcharge which passes on the supposed extra cost of renewable electricity to consumers challenged basic elements of EEG; however, a decision was postponed until after the September 2013 elections.

2 The Challenges Unfolding Over Time

We view challenges as situations requiring commitment and/or major decisions which are crucial for a full transition to renewables but did not come easily to the actors of the political system caught in their own logic. These actors are political parties and state structures with their drive for power, their ideological and policy traditions and positions privileging a certain clientele; a great variety of economic actors (from electricity incumbents to highly differentiated new entrants into the field of renewable energy), citizens with their values and beliefs, environmental organizations, churches, etc.

The first challenge was to launch the rapid growth of RES-E deployment on a trajectory that would in principle allow the near-total replacement of fossil and nuclear sources in time to cope with peak oil and mitigate global warming, by installing a regulatory framework capable of setting the necessary incentives. This challenge was met in the period 2000–2004.

The second challenge was to build or maintain political support and an actor network capable of supporting this change. Political support for wholesale transition to renewables was already evident in public opinion surveys in the 1990s, and it remained so until early 2013. At the same time, it was important to convey the idea that renewable energy—and the equipment for generating it—could be produced domestically, even give rise to a whole new sector of industrial activities with new employment and exports. Such a sector would also be a natural ally of *Energiewende* against the attacks that could be expected from incumbents. Finally, costs needed to be plausibly justified and kept down to acceptable levels. This became a problem around 2009, a time when photovoltaic installations surged beyond all expectations and drove up costs and imports of modules from abroad, and led to intense political controversies in the subsequent years. But PV was not the only cause of the problem, as we shall show.

The third and most recent challenge which has only emerged in the present and has not been dealt with so far is to develop appropriate structures to deal with the problem of fluctuating generation by wind and solar power installations, and more generally to design a new electricity market scheme which—while no longer geared to the needs of fossil generation—will allow such generation to survive even as it is gradually phased out, a process likely to take some decades.

2.1 Challenge One: Launching "Disruptive" RES-E Growth with an Appropriate Regulatory Framework

2.1.1 Targets to Build Confidence and Commitment

The first challenge was to launch the rapid growth of RES-E deployment on a trajectory that would displace fossil and nuclear sources in time to cope with peak oil and mitigate global warming, by installing a regulatory framework capable of setting the necessary incentives.

This occurred in 2000 with the adoption of the Renewable Energy Sources Act or EEG and the more or less simultaneous decision to phase out nuclear power taken in 2000–2002. The explanatory memorandum attached to EEG (FME 2000) speaks of the urgency to shift to a sustainable energy supply to deal with the problems of environment and climate change, the problems of dependence on and costs of fossil fuel imports, the potential of an innovative industrial policy to create jobs and development, the need to compensate for external costs and subsidies of conventional generation and finally the need to break the vicious circle of high RES-E equipment prices and small production volumes by launching mass production. There was a target for 2010 (to at least double total renewable energy production) but the act clearly aimed further.

Soon, a pattern developed regarding RES-E targets and their fulfilment. RES-E deployment—all technologies taken together—regularly exceeded the targets laid down in the Renewable Energy Act and its 2004 and 2008 amendments; this was welcomed by the Ministry of Environment (which in 2002 took over the RES-E agenda from the Economic Affairs Ministry) and led to yet more ambitious targets. Thus, the goal of EEG 2004 was for renewable electricity to reach a 20 % share by 2020 (Section 1 of EEG 2004; see BMU 2004); actually this was reached in 2011. EEG 2008 proclaimed a 2020 target of 30 %. This was stepped up by the Conservative-Liberal government before it proclaimed the need to moderate deployment.

In its energy concept of 2010 (BMWT and BMU 2010), this government for the first time set long-term targets for RES-E growth (*see* Table 1). These targets were enshrined in EEG 2012 which was adopted in mid-2011 (EEG 2012a[2]).

In mid-2012, just a year after the unanimous adoption of that act, many *Länder* chiefs protested against the low level of ambition reflected in this target and the drastic plans of the government to impose regional caps on wind and solar growth,

[2] Another EEG amendment was adopted in mid-2012 (here referred to as EEG 2012b).

Table 1 RES-E targets in Energy Concept 2010 (BMU 2011) and in EEG 2012a, b

Target ("at least"), %	Deadline ("at the latest")
35	2020
50	2030
65	2040
80	2050

constraining many new *Länder* plans prepared after Fukushima. In 2012, the share of RES-E was about 23 %, up from 17 % just 2 years earlier. Environment Minister Altmaier proposed to slow down deployment, supposedly to avoid problems of energy balancing, grid stability and costs, by means of target corridors[3] on deployment on a regional basis, especially for wind and solar power. The heads of the *Länder,* determined to implement their recent deployment plans, disagreed strongly with this approach (Zeit online 2012). In June, however, in a compromise in the parliament's conciliation committee, they agreed to a 52 GW cap on PV (EEG 2012b). Once this limit is reached, EEG-style feed-in tariffs will no longer apply to PV; new support rules should be announced in time (PV capacity reached 30 GW in the summer of 2012). But the *Länder* chiefs opposed a similar cap, or regional allocation of new capacity, for other technologies, particularly for wind power. The federal government then started a negotiation process with the *Länder* chiefs to find a consensus during 2013, to be reflected in a new EEG amendment after the 2013 parliamentary election. This reflexion process was interrupted when the Liberals launched the conflict over the EEG surcharge described below.

2.1.2 Tariffs and Other Regulations to Stimulate Broad Deployment

The feed-in tariff system of the EEG is based on several principles which were designed to stimulate aggressive RES-E deployment: differentiated and highly predictable prices for RES-E based on a "full cost" model of an efficient installation; unlimited purchasing obligation and priority dispatch by grid operators; costs borne by electricity consumers, not taxpayers; and administrative simplicity. This was supplemented by low-cost credit and a highly predictable permitting system. The whole system was also meant to compensate RES-E generators for the fact that fossil and nuclear operators do not pay for substantial external costs and are/were subsidized on top of that (see Sect. 2.2).

The adoption of the feed-in tariff contained in the 2000 Act placed a clear emphasis on effective deployment. This tariff was highly differentiated by source (wind power, PV, biomass, etc.); size and vintage of the installation; quality of the site (for wind power). Tariffs were designed so that all well-built and well-

[3] Corridors are evolving targets with a minimum and a maximum value.

managed plants could expect to produce a reasonable return on investment (while not defined in any law, figures ranging around 6–9 % were discussed in this context). This meant that different technologies could be deployed simultaneously, not just the cheapest (such as waste combustion, which proved important under quota-cum-certificate systems), so they could all progress downward on the learning curve simultaneously. Windfall profits in the best locations were limited by differentiating for size (PV, hydro, biogas) or locational qualities (wind). Annual degression meant that the tariff paid for any given installation does not change for 20 years while subsequent vintages received somewhat less to accelerate learning by putting pressure on the RES-E equipment industry, which otherwise was expected to benefit from steadily expanding demand for its products. The fact that the tariff was laid down by a formal law and thus by parliament[4] offered a good guarantee against abrupt changes.

Unlimited purchasing obligation means that grid operators have to connect all RES-E installations in their territory and purchase all RES-E tendered to them at the tariff laid down in the law, without need for a contract. Grid operators in turn had to give priority dispatch status to RES-E, before taking on fossil or nuclear power.

The extra cost of the feed-in tariff to the grid operator is paid by electricity consumers in form of a surcharge, with no part of it coming from the state budget (something that conservative politicians frequently ignored in their criticism). Budget contributions would probably not have remained in place for so long, given the problems of public finances and of state aid.

All this created a simple, stable support system that made it possible for many people to become RES-E generators. Indeed, just more than half of all RES-E capacity is owned by private individuals and farmers (trend:research 2011, 45). With EEG in place, banks were perfectly happy to give loans for such a secure investment. A public development bank—KfW (Kreditanstalt für Wiederaufbau)—played an important role in providing low-cost loans to cover part of the necessary funds. Approximately 1.3 million people were engaged in renewable electricity installations (mostly PV) as owners or investors in late 2012 (Fig. 1).

Additional support came from the construction code which privileged RES-E generation installations, and from a regulation that required local communities to set aside zones appropriate for wind power ahead of construction requests. Tax regulations contributed to local support for RES-E projects since 70 % of the local industry tax on wind farms go to the community where the turbines are located, even if the owner resides elsewhere (Witt 2012a).

---

[4] In some countries support is set by administrative decrees which are changed more easily, something that tends to deter RES-E equipment producers. An attempt to introduce this in Germany just for PV was turned down by parliament in 2012.

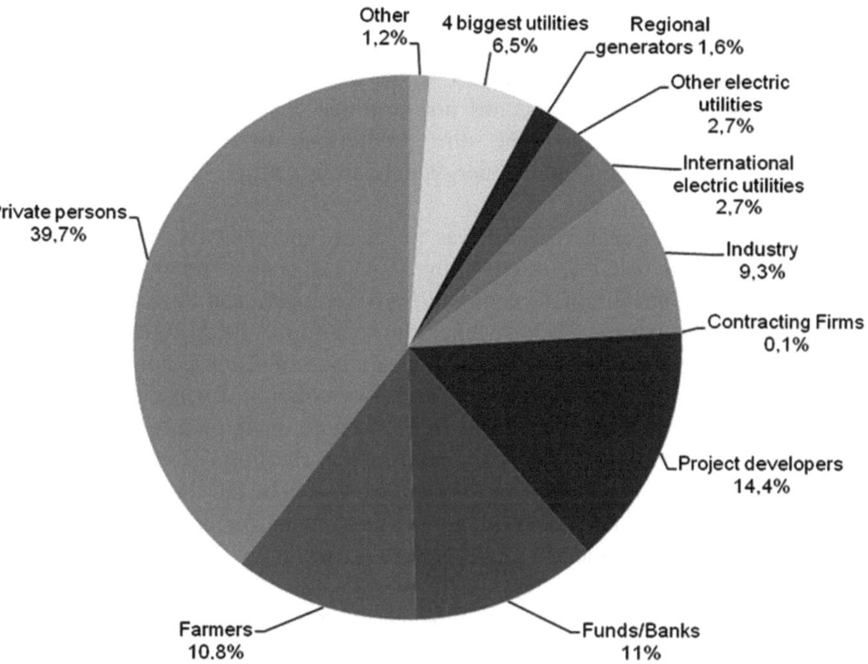

Fig. 1 Ownership structure in 2010 of renewable electricity installations in Germany (not including pumped storage) (Total installed capacity: 53.0 GW). *Source* Trend:research (2011)

2.2 Challenge Two: Ensure Political Support for a Long-term Deployment Framework

This challenge includes maintaining support by the public by making clear the larger goals and objectives at stake, keeping costs at acceptable levels, secure support by political parties and setting up networks that have an active interest in RES-E growth, particularly industries related to RES-E equipment production or installation, plus potential and actual owners of RES-E installations.

At the intellectual and emotional level, support for a transition to renewables was very strong in Germany even before the advent of the EEG; this is illustrated by a large number of surveys. Chernobyl, acid rain, and climate change made a deep impression in Germany and inspired various citizen movements. Many of them were looking for real-life alternatives to fossil and nuclear power. This was true for early groups working on biogas and wind energy back in the 1970s and 1980s; in the 1990s, similar groups paved the way for photovoltaic power by motivating municipal governments in a number of cities to require their electricity companies to provide special compensation for citizens producing electricity from PV installations on their roofs. Involvement with and support for renewable energy became remarkably widespread.

Of course, there were also critical stories in parts of the press of ugly wind farms and hugely expensive solar panels, but they were usually met with criticism and were not taken too seriously. Still, in the case of wind energy, it was felt in 2004 that onshore expansion could not continue much longer; this led to an increase for offshore wind. The dire predictions by electricity incumbents regarding high costs of balancing energy, dangerous grid problems, and enormous costs were soon disproven.

Once the Environment Ministry was in charge of renewable electricity (after 2002), it published valuable information about the cost of renewable power, its growth and its contribution to economy, environment, and society. During the years of the Conservative-Social Democratic coalition (2005–2009), the Conservatives seemed to arrive at a pragmatic acceptance of EEG although they maintained their commitment to a stronger market orientation. Even the Liberal Party leadership was instructed to support EEG by a vote of its membership at a party convention in May 2009, following a motion from the floor. However, it reneged on this commitment not very long afterwards. When in fall 2009, the Conservatives and Liberals formed the new government, their radical course on energy policy (postponing the nuclear phase-out, curtailing RES-E growth) came somewhat unexpected.

By that time, however, several new developments had changed the picture: the financial crisis of 2008 made for a new sense of financial fragility, and the PV surge in 2009 started to impact surcharge payments; worse, the share of imported modules (mostly from China, and quite possibly due to dumping or export subsidies incompatible with WTO rules) increased, while German cell and module producers entered a period of substantial difficulties; even some prominent firms (Solon, Q-Cells. Bosch) went bankrupt or left the business. This upset one of the political rationales underlying EEG, namely that EEG was supposed to replace imported fossil energy by domestic RES-E.[5] At the same time, the merit order combined with the new expansion of wind and solar electricity reduced prices on the electricity exchange and thereby drove up the EEG surcharge (which covers the difference between tariff payments and the sales value of EEG electricity on the exchange).

Finally, the Conservative-Liberal government generously increased exemptions from the surcharge for industry; this too increased the surcharge which is paid mostly by households. The net result was a steep rise of the surcharge for which the government, in rather simplistic but effective fashion, blamed the high extra cost of renewables, ignoring the other factors. In fall 2012 for the first time, a poll showed that a majority of respondents considered the surcharge to be excessive. This proved to be a window of opportunity for the energy conservatives of the governing coalition calling for a fundamental reorientation (*see* below).

[5] While German cell and module producers shrank, silicon and PV production equipment producers did well, partly due to strong Chinese demand; also, installers are usually local. Some observers estimate that close to two thirds of PV investments in Germany are still German-made.

Support for RES-E and more particularly EEG undoubtedly benefited from the ownership distribution of RES-E capacity (see Fig. 1). The "Big Four" oligopolists dominating electricity supply[6] rely on coal and nuclear generation and have been fighting EEG (and its predecessor law StrEG) for over two decades. But by 2010, all utilities taken together owned only 13.5 % of renewable capacity, while private persons and farmers owned about half of it. There were about 1.3 million RES-E generators in Germany in early 2013. They cannot be passed over easily. If RES-E deployment goes on, the Big Four will most likely fall behind even further (Becker 2011).

More support comes from the associations of business and labour in the areas of RES-E equipment and installation industry, farmer associations, and the big environmental and nature protection associations. Employment in the sector rose from some 66,000 jobs (AGEE 2011) to about 380,000 jobs in 2011 (FME 2012, p 49). After Fukushima, most *Länder* governments also became strong advocates of rapid RES-E deployment, taking the opposite course from that of the federal government.

2.2.1 Containing the Cost of EEG and Energy Transformation

From the beginning, the supposed extra cost of renewable energy deployment was one of the major arguments of its opponents. In the late 1980s, the Economic Affairs Ministry opposed RES-E market creation via feed-in tariffs with the argument that this would become a case of perpetual subsidies (Jacobsson and Lauber 2006). Most of the extra cost comes from feed-in tariffs and takes the form of a surcharge on electricity bills paid mostly by households and small business. Due to recent increases and also because of the Conservative-Liberal government's politicising this issue, acceptance of this surcharge became a significant problem for the first time in 2012.

Since 2010, the surcharge consists of the difference between the payments of grid operators to RES-E generators according to the EEG and the price they receive from the sale of EEG electricity on the exchange (where it is sold as gray, not as green electricity). From 2003 to 2009, it rose from 0.41 to 1.13 Eurocent/kWh. In the four subsequent years though it increased steeply to 5.3 ct in 2013, even faster than the RES-E share in total electricity generation (see Fig. 2).

Many critics—particularly from the Liberals or the business wing of the Conservatives—concluded that the cost of energy transformation was getting out of hand, supposedly due to rapid growth of PV installation. But the full story is quite different and considerably more complicated.

First, the (operational) costs of conventional generators used here do not include external costs. If those costs were included, only offshore wind and some

[6] The four oligopolists are E.on, RWE (Rheinisch-Westfälisches Elektrizitätswerk), Vattenfall Germany and EnBW (Energie Baden-Württemberg).

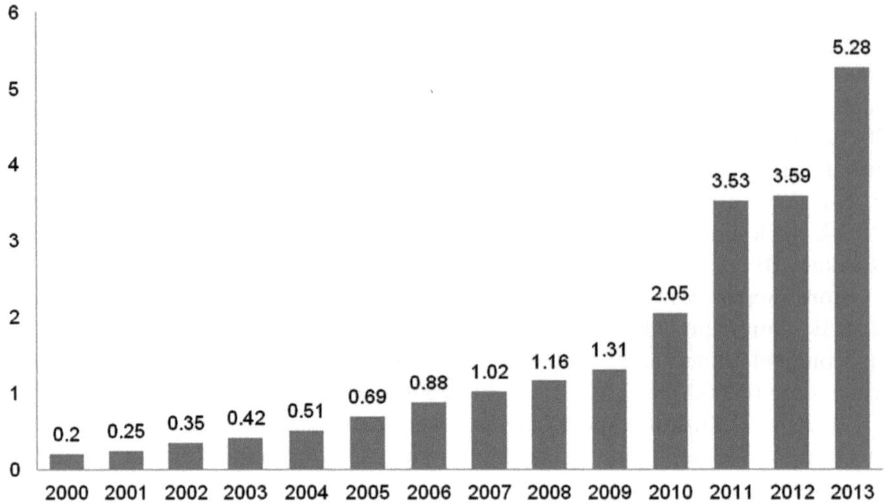

Fig. 2 Evolution of RES-E surcharge, 2000–2013 in ct/kWh. *Source* Ahmels (2012), p. 15

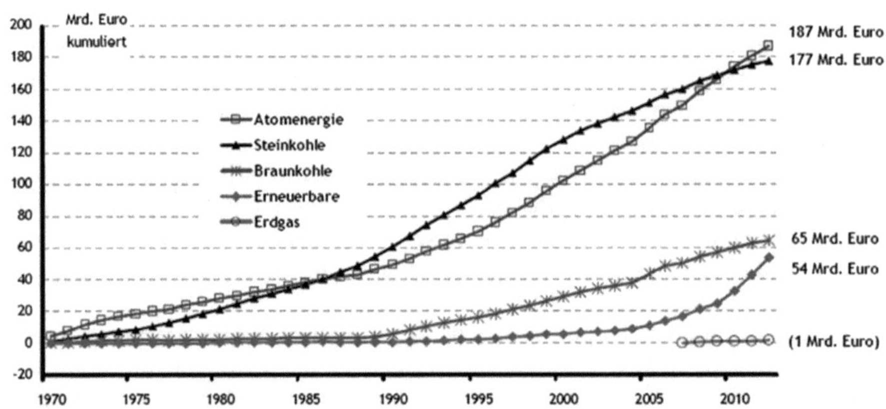

Fig. 3 Cumulated state subsidies to nuclear, hard coal, lignite, and natural gas generation and support for renewable generation including EEG support, 1970–2012. *Source* Küchler and Meyer (2012)

PV would still need EEG support in 2013. Küchler and Meyer (2012), relying on a study by Fraunhofer ISI (2012), report the following external costs: 8.9 Eurocent/kWh for electricity from hard coal, 10.7 ct/kWh from lignite, 4.9 ct/kWh from natural gas. For nuclear power, Küchler and Meyer (2012, p 25) assume a range of 10.7–34 ct. By comparison, the external costs for RES-E are: wind, 0.3 ct, hydro 0.2 ct, PV 1.2 ct. By January 2013, tariffs for bigger PV plants were below the full cost of coal.

Second, the operational costs for conventional generation do not include subsidies—i.e., extra cost of fossil and nuclear (direct financial support, tax breaks, other regulations with positive financial impacts such as free CO_2 certificates). Applied to electricity generation only, from 1970 to 2012, hard coal received 177 billion Euro, brown coal 65 billion and nuclear energy 187 billion in such subsidies, financed mostly via taxes from the state budget and—unlike the EEG surcharge—not communicated to the electricity consumer. Total support for new RES-E so far amounted to 54 billion, including payments outside the EEG such as government R&D funding or low-cost loans (Küchler and Meyer 2012, see Fig. 3). Clearly, those subsidies extending over several decades helped conventional generators to establish their market dominance. But in most public debates of *Energiewende*, only RES-E carries "extra costs".

Third, the calculation method of the EEG surcharge not only does not take the price reducing effects of renewable generation on electricity spot market prices into account ("merit order effect," *see* Sensfuß et al. 2008), but turns them into price increases—disproportionately so for small consumers. Under the merit order, generators are dispatched according to their marginal cost of operation. Given its marginal costs close to zero, wind and solar power push the more expensive thermal plants out of the market; this reduces overall prices at the exchange. This same price reduction then increases the difference between feed-in tariffs and spot prices and thus increases the surcharge by the same amount.

Fourth, a large number of industrial firms (theoretically "energy-intensive and exposed to international competition") are practically exempted from this surcharge. This exemption, originally introduced in 2003 by the red-green coalition for a limited number of firms, was expanded by later governments and finally doubled again by the current government in 2012 (Nestle und Reuster 2012, pp 14–15) and includes railroads, breweries, a fast food chain, etc. Small consumers paid already about 0.6 ct/kWh for this in 2011 and proportionately more in 2012. At the same time, industrial firms benefit from the lower prices resulting from the merit order system plus the lower costs of carbon emission certificates as RES-E reduces certificate demand from idling fossil fuel power plants. By contrast, household consumers do not benefit from the merit order effect as there is a lack of competition among suppliers. In fact, exempted industrial firms now pay lower electricity prices due to RES-E deployment, not higher ones. Yet the political rhetoric of the "energy conservatives" portrays industry as a victim of EEG.[7] On the whole, the surcharge no longer reflects the "extra cost" of RES-E growth accurately (see Fig. 4).

There is, however, one factor in RES-E supply that really did affect the surcharge in a major way, at least temporarily: The surge of PV in 2009 and 2010 when module price declined rapidly while tariffs were still governed by the relatively slow degression (about 9 % annually) of EEG 2009. New installations surged to a

---

[7] In late November 2012, the European Commission announced state aid proceedings against the EEG exemptions.

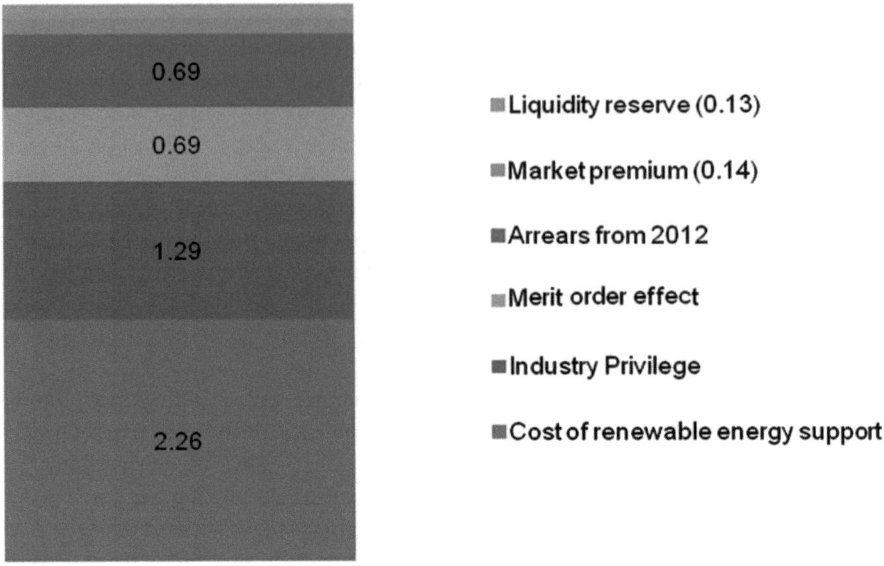

Fig. 4 Breakdown of renewable electricity surcharge paid by consumers in 2013. *Source* Ahmels (2012), p. 62

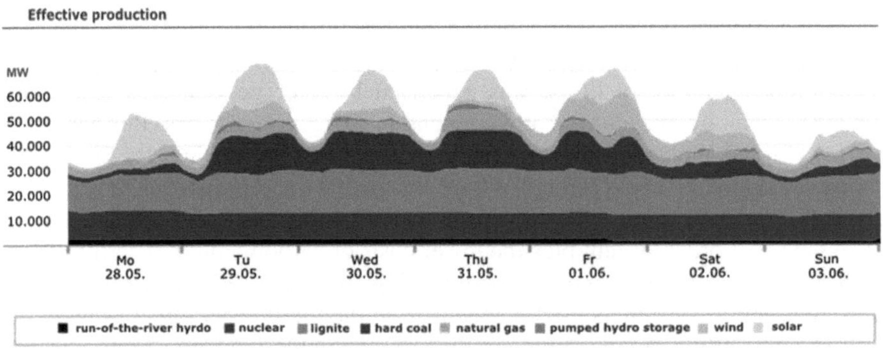

Fig. 5 Breakdown of electricity generation by source of daily loads during a sunny week in May/June 2012 (28 May through 3 June). *Source* Burger (2012), slide 192

level about six times higher than expected—7,400 MW in 2010. PV became the biggest cost item of the EEG surcharge increase in 2010. Many prominent members of the renewable energy research community feared that the legitimacy of this act might be seriously affected and in late 2010 RES-E wrote an open letter to the government to reduce this growth to acceptable limits (Figs. 5 and 6).

Between spring 2010 and summer 2012, the government responded to the PV surge by cutbacks of tariffs and by making degression strongly dependent on new build. The idea was to cut new annual installations to about 3 GW. Still, new

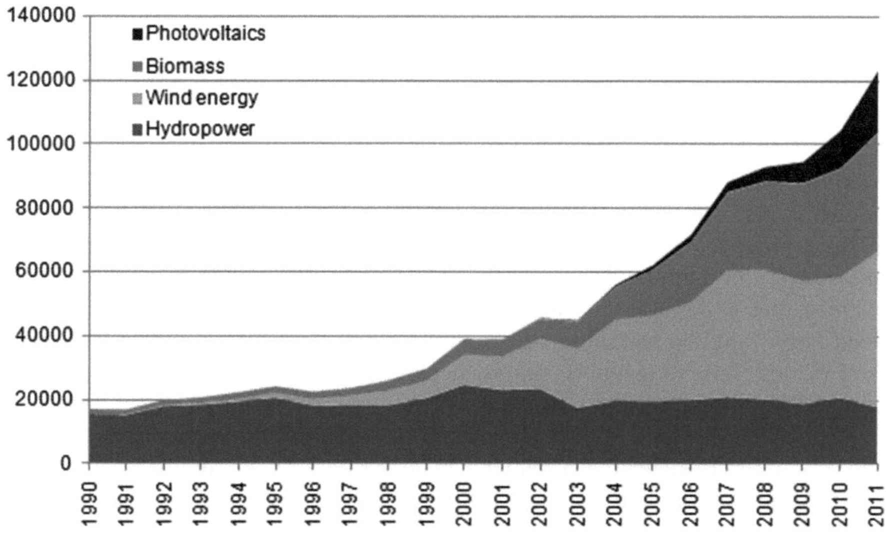

Fig. 6 Development of renewable-based electricity generation in Germany, 1990–2011, in GWh. *Source* FME (Federal Ministry for the Environment, Nature Conservation and Nuclear Safety) (2012)

installations in 2011, after a record installation of 3 GW in December, remained at the level of 2010. Even though new PV tariffs had already declined considerably by early 2012, reducing the impact of further PV installations on EEG costs, the economic wing of the Conservatives and many Liberal leaders[8]—most of them advocates of the life extension for nuclear power in 2010 and still regretting its phase-out—responded to the December figure with demands for more radical steps, in particular a firm cap of 0.5–1 GW on new annual installations and a more drastic reduction Liberal Economic Affairs Minister. Rösler at first had even demanded a 40 % cut in one step (Süddeutsche Zeitung 2012). This time Environment Minister Röttgen (author of the flexible cap regulation) had to yield. He now proposed reducing new PV build to 0.9–1.9 GW by 2017 (BMU 2012) as well as stepping up flexible degression which in the extreme case would have driven PV tariffs down to 2.9–3.16 ct/kWh in 2018 (Witt 2012b).

Eventually, the reform was made more moderate by the *Bundesrat*, the upper chamber of the German parliament in which *Länder* governments are represented and which voted against the bill by a two-thirds majority (a rare event[9]) which included Conservative-governed *Länder* from East Germany (chief location of the

[8] This included Michael Fuchs, one of the leaders of the Conservatives (CDU) in the lower house of parliament, and Rösler, Economic Affairs minister from the Liberal Party.

[9] In this case the *Bundestag* (lower chamber) also needs a two-thirds majority to override the *Bundesrat* (upper chamber) should the latter make a formal objection. The government could not have achieved such a majority.

solar cell and module industry) whose leaders protested against a policy that was bound to lead to deindustrialising their region. The compromise made tariff degression more moderate (*see* above), set an overall hard cap for EEG tariff support at 52 GW (higher than in the Bundestag bill; a new form of support would be developed in time), and obtained the assurance from the government that it would step up R&D for PV and launch a support program for batteries to store PV power. It also asked for (but did not obtain) a local content regulation to help German producers (Broer 2012a, b); but minister Altmaier subsequently came out in favor of anti-dumping proceedings.[10] Outraged at the compromise, the Liberals claimed that it was time to eliminate EEG principles such as priority feed-in or fixed rates; only the market could stop this systematic waste. As most PV critics, the Liberals chose to ignore its rapidly declining cost and hence the limited impact of new PV installations on consumers.

The surcharge crisis of 2012 intensified in January 2013 with renewed Liberal attacks on EEG. Repeating a demand by EU Energy Commissioner in summer 2012, Liberal Economic Affairs Minister Rösler called for a cap on the EEG surcharge; this had the potential of bringing deployment to a halt. Altmaier eventually gave in and called himself for a "stabilization" of the surcharge at its 2013 value. At a summit meeting in late March 2013, the *Länder* chiefs turned down the federal government's plans; EEG reform was postponed until after the parliamentary elections of September 2013. These plans had the potential of seriously undermining EEG implementation.

2.3 Challenge Three: Fluctuating Renewables and a New Electricity Market

Another challenge had emerged by this time: the clash between fluctuating renewables and inflexible nuclear and coal generation.

Before electricity liberalization, Germany's power plant portfolio consisted of two types of power plants. First, base load plants with almost no operating flexibility, combining high fixed costs with low marginal costs, such as nuclear or lignite plants. Second, power plants covering medium or peak load, not only more flexible but also more costly to operate, mainly fired with hard coal or natural gas. Planning and dispatch of power plants was in the hands of big energy utilities which at the same time controlled the electricity grid. RES-E was negligible.

Market liberalization did not really change the use of power plants. The daily load curve reflected marginal prices on the electricity exchange, leading to power plant operations according to merit order. Hours with low demand meant low prices at which only plants with low marginal costs could operate profitably. Hours

[10] Provisional anti-dumping duties were imposed by the European Commission in June 2013.

of high demand and high prices meant that power plants with high marginal costs could also achieve profits.

The recent rise of RES-E sources with fluctuating production and priority dispatch challenged this electricity market order dramatically. Daily load peaks are filled more and more with electricity from PV. Within a foreseeable future, demand for electricity from fossil fuels will no longer peak parallel to overall demand and high prices, thus reducing revenue for fossil plants; the peak for overall demand may even come to coincide with a trough for fossil demand and prices. The surging supply of renewable electricity with priority dispatch and marginal costs close to zero led to a significant wholesale price decrease on the electricity market (IZES 2012). Increasingly, natural gas and to some extent hard coal plants operate profitably only when high demand coincides with low production from wind and solar. As a result, the average operating times of these power plants—particularly at peak hours with their higher prices—fell dramatically in 2011 and 2012, bringing even written-off power plants into trouble. Thus, E.on, Germany's biggest utility, in November 2012 announced the closure of two big gas fired power plants[11] due to lack of profitability. So far, only hard coal and natural gas plants suffer from the decrease in operating hours. With further increases in RES-E generation, lignite power will come under pressure as well. The incumbents are bound to press for a solution that will remedy this problem from their perspective.

The post-Fukushima shutdown of old nuclear plants had led many to expect that Germany would face higher electricity prices, increase fossil generation, and import electricity to avoid blackouts. None of this happened after the first 2 months: prices went down, and so did fossil generation (strong reduction of natural gas use for power generation, slight increase for coal), while electricity exports reached an all-time high, quadrupling between 2011 and 2012 (IWR 2012c).

Apart from the profitability problem, there is a bigger systemic problem: shutting down fossil fuel plants permanently may threaten the security of supply. These power plants are still needed, e.g., on cold winter days with low wind speeds and little sun radiation but high electricity demand, at least until electricity storage becomes available on a large scale. In order to avoid blackouts in such situations, many politicians and experts call for a so-called "capacity market" under which operators of such plants are paid (e.g., via tenders) for standby power. As fluctuating production increases, more and more fossil capacity would need some financial support to remain on standby.

One possible way to deal with this situation—proposed, e.g., by the utilities' association BDEW—would require the following market design: In keeping with

[11] The plants Staudinger 4 and Irsching 3. In 2011, Staudinger 4 (622 MW) had only run for 65 h under full load, Irsching 3 (415 MW) 41 h. Strangely enough they were not even started up in December 2011 when one of the blocs of nearby nuclear plant Grundremmingen had an emergency shutdown; instead E.on—one of the four oligopolist, see fn. 9—imported electricity from a more distant and more expensive plant in Austria (IWR 2012a).

the current practice (electricity exchange based on merit order system relying only on marginal costs of production), RES-E installations with almost no marginal costs such as wind, solar, and hydro power are the first to be dispatched. Since this is likely to lead to regional over-investment in those technologies, regional caps on such sources could be envisioned, or reductions in compensation payments for RES-E which cannot be fed into the grid. To make sure that enough fossil fuel capacity is available without displacing electricity from renewable energies, power plants could be dispatched centrally by the grid operator (as currently for balancing energy). Generating technologies able to offer stand-by power—such as power plants fired with fossil fuels or biomass but also hydro storage plants—would draw additional revenue from auctions for maintaining standby power capacity.

The federal government favors a different solution: A decree of October 2012 requires fossil generators to announce their intention to retire a particular plant 12 months ahead of time; during this period, the Federal Grid Agency may reject such a retirement while providing for appropriate compensation for keeping the plant on standby (IWR 2012b). In both cases, the new design means eventually a phase-out of inflexible, base load-only power plants—such as nuclear and lignite power (Fraunhofer ISE 2012).

But the fluctuation problem can also be handled quite differently. A special market section not based on marginal costs could be set up for wind, solar, and hydro (Leprich and Hauser 2013). Fossil generators would still be subjected to marginal cost-based merit order, but at one point, shifting to 100 % RES-E will require renouncing fossil solutions unless carbon capture and storage enters the picture, which at present does not look likely in Germany. Various forms of power storage (from 6 h storage to seasonal storage, with storing an electricity surplus in summer and withdraw it in winter) and of other technologies capable of balancing demand could then deal with the fluctuation problem. This would require a significant expansion of storage capacity, which currently is limited in Germany. Hydro storage currently has a maximum capacity of about 7 GW and about 40 GWh (corresponding to less than two full load hours of total installed PV generation in mid-2012); this is about to be expanded by some 50 % over the next decade. Biomass power plants could add about 5.3 GW of capacity if equipped with the necessary storage facilities.

Some relief is available from hydro storage facilities in France, Switzerland, and Austria (Fraunhofer ISE 2012, pp 46–50). There is the idea of using Norway's vast hydro storage facilities—much bigger than storage facilities in Germany—for pumped storage to serve the German (and other European) markets. However, this would require not only installing pumping facilities but also vast power lines on land and across the sea (which could be coordinated with offshore wind build-up); a first cable linking Norway and Germany is scheduled for completion by 2018 (Fraunhofer ISE 2012, p 47).

Combined heat and power plants could serve both as generators and as sinks for RES-E electricity, transforming it into storable heat. Battery storage and pressurized air storage are further possibilities for short-term storage. Power-to-gas and power-to-liquids storage are at present more costly and less efficient, but could be acceptable if limited to supplementing highly efficient intra-day storage (Rasmussen et al. 2012; Welter 2012). They would draw on the surplus generation available during the summer from PV. The gas grid alone would allow storage of about 200 TWh (Welter 2012, p 50)—many times more than hydro storage can achieve. If capacity payments to standby fossil power plants are not to constrain progress towards these and other solutions (Fraunhofer ISE 2012), it will be necessary to design a new electricity market scheme which while no longer geared to the needs of fossil generation will allow a diminishing share of such generation to survive during its period of gradual phase-out (which may well last for several decades).

It seems likely that promising market designs will have to deal with the need to reform an electricity market based on marginal (operating) costs while fluctuating sources destined to supply the biggest share of electricity have mostly investment costs and nearly no operating costs. Integrating RES-E at its pure operating costs would drive prices toward zero and would not allow recuperating RES-E investment costs. Hence, the need to provide separate selection mechanisms (the counterpart to the current merit order) for RES-E, fossil generation, and storage systems (Leprich and Hauser 2013).

3 Discussion – Conclusions

Out of these three challenges, Germany has certainly met the first one very successfully (launching "disruptive" growth with an appropriate framework).

As regards the second challenge (securing acceptance and support), things seemed to go quite well until 2009 when the global overcapacity crisis of PV and Chinese market conquest made the EEG formula of replacing imported energy with domestic RES-E problematic. The Conservative-Liberal government coalition in power since 2009 reduced excessive PV tariffs (Bundesverband Solarwirtschaft 2012) though it kept treating PV as if it were still extraordinarily expensive. But deeper problems were also developed; falling exchange prices for electricity paradoxically drove up the surcharge, industry privileges did the same. Utilities and energy conservatives blamed it all on PV.

As to the third challenge (dealing with fluctuating generation and reforming the electricity market), it is too early to say how well Germany is doing on this account. It is struggling with the fossil standby power solution but has also launched a big research program for electricity storage. Success in our view would mean keeping the existing system viable (including the maintenance of sufficient fossil back-up capacity) while not impeding further growth of RES-E toward the 80–100 % target. Much of the transformation is a question of visions and imagination, technical, political, legal, and organizational. Though a huge project, it is far from being just a top–down process. In some ways, it resembles a huge citizen initiative, a bottom–up movement that is transmitting its energy to the political system and may actually be displacing an electricity system which combined industrial concentration, political connections, high profitability, and oligopolistic market power with a certain amount of insensitivity on social and environmental issues (Becker 2011). It is no wonder that the radical nature of its challenge would meet with opposition at some point. The year 2013 is likely to become a fateful year for *Energiewende*.

References

AGEE (Agentur für erneuerbare Energien) (2011) Arbeitsmarkt erneuerbare Energien. http://www.unendlich-viel-energie.de/de/wirtschaft/arbeitsplaetze-erneuerbare-karriere/arbeitsmarkt-erneuerbare-energien.html. Accessed 29 Dec 2012

Ahmels J (2012) Kostenentwicklung bei der Förderung der erneuerbaren Energien in Deutschland. Masterarbeit, Carl von Ossietzky Universität Oldenburg

Becker P (2011) Aufstieg und Krise der deutschen Stromkonzerne. Ponte Press, Bochum

BMU (2011) Erneuerbare Energien in Zahlen. Nationale und internationale Entwicklung, Berlin

BMU (2012) Energiewende Aktuell, 24 Feb 2012, listing the agreement between Röttgen and Rösler, Berlin, p 3. Accessed 5 July 2012 (since then removed from the server)

BMU (Bundesministerium für Umwelt, Naturschutz und Reaktorsicherheit) (2004) EEG Konsolidierte Fassung. http://www.clearingstelleeeg.de/files/private/active/0/EEG_2004_begruendung_konsolidiert.pdf. Accessed Oct 2012 (since then removed from the server)

BMWT (Bundesministerium für Wirtschaft und Technologie) and BMU (28 Sept 2010) Energy Concept for an environmentally sound, reliable and affordable energy supply. http://www.bmu.de/files/english/pdf/application/pdf/energiekonzept_bundesregierung_en.pdf. Accessed 12 Dec 2012

Broer G (2012a) EEG-Novelle: Vermittlungsausschuss ist am Zug. Solarthemen 376:8–9

Broer G (2012b) Europa als PV-Schutzzone: Local Content & Co, Solarthemen 377

Bundesverband Solarwirtschaft (2012) Kosten und Förderungen von Solarstromanlagen seit 2006. http://www.solarwirtschaft.de/pressegrafiken.html. Accessed 24 Jan 2012

Burger B (2012) Stromerzeugung aus Solar- und Windenergie im Jahr 2012, dated 17 Dec 2012. Fraunhofer-Institut für solare Energiesysteme ISE. http://www.ise.fraunhofer.de/de/downloads/pdf-files/aktuelles/stromproduktion-aus-solar-und-windenergie-2012.pdf. Accessed 26 Dec 2012

Buschmann M (2011) Erneuerbare Energien und das politische System. Der Ausbau von Wind- und Solarenergie als Resultat einer Interaktion von Politik und Technologie. Doctoral dissertation, University of Salzburg

EEG (2012a) Gesetz zur Neuregelung des Rechtsrahmens für die Förderung der Stromerzeugung aus erneuerbaren Energien, BGBl. 4 August 2011, Teil I, No. 42, p 1634

EEG (2012b) Gesetz zur Änderung des Rechtsrahmens für Strom aus solarer Strahlungsenergie und zu weiteren Änderungen im Recht der erneuerbaren Energien vom 17. August 2012 (BGBl. I S. 1754)

FME (Federal Ministry for the Environment, Nature Conservation and Nuclear Safety) (2012) Development of renewable energy sources in Germany in 2011. Graphics and tables version: July 2012. Based on statistical data from the Working Group on Renewable Energy Statistics (AGEE-Stat). http://www.bmu.de/files/english/pdf/application/pdf/ee_in_deutschland_graf_tab_en.pdf. Accessed 4 Dec 2012

FME (Federal Ministry of the Environment, Nature Conservation and Nuclear Safety) (2000) Act on Granting Priority to Renewable Energy Sources (Renewable Energy Sources Act). English version of the Act. http://www.climatelaw.org/laws/germany/germanyenergyact. Accessed 28 Dec 2012

Fraunhofer ISE (Fraunhofer- Institut für Solar Energiessyteme) (2012) Aktuelle Fakten zur Photovoltaik in Deutschland, version 19 Dec 2012. http://www.ise.fraunhofer.de/de/veroeffentlichungen/veroeffentlichungen-pdf-dateien/studien-und-konzeptpapiere/aktuelle-fakten-zur-photovoltaik-in-deutschland.pdf. Accessed 26 Dec 2012

Fraunhofer ISI (Fraunhofer Institut für System- und Innovationsforschung) (2012) Ermittlung vermiedener Umweltschäden—Hintergrundpapier zur Methodik—im Rahmen des Projekts "Wirkungen des Ausbaus erneuerbarer Energien". Untersuchung im Auftrag des BMU, http://www.erneuerbare-energien.de/files/pdfs/allgemein/application/pdf/hg_umweltschaeden_bf.pdf. Accessed 2 Dec 2012

IWR (2012a) Strom zu teuer – Warum E.On drei Gaskraftwerke wirklich abschalten will, 21 May 2012. http://www.iwr.de/news.php?id=21200. Accessed 29 Dec 2012

IWR (2012b) Bundeskabinett beschließt Maßnahmen gegen Strom-Blackout, 18 Oct 2012. http://www.iwr.de/news.php?id=22255. Accessed 20 Dec 2012

IWR (2012c) Stromexporte aus Deutschland auf Rekordniveau, 9 Nov 2012. http://www.iwr.de/news.php?id=22404. Accessed 20 Dec 2012

IZES (2012) http://www.solarserver.de/solar-magazin/nachrichten/aktuelles/2012/kw05/neue-studie-solarstrom-senkt-boersenstrompreise-um-bis-zu-10.html. Accessed 20 Dec 2012

Jacobsson S, Lauber V (2006) The politics and policy of energy system transformation—explaining the German diffusion of renewable energy technology. Energy Policy 34(3):256–276

Küchler S, Meyer B (2012) Was Strom wirklich kostet. Vergleich der staatlichen Förderungen und gesamtgesellschaftlicher Kosten konventioneller und erneuerbarer Energien. Langfassung, überarbeitete und aktualisierte Auflage. FÖS, Berlin. http://www.foes.de/pdf/2012-08-Was_Strom_wirklich_kostet_lang.pdf. Accessed 15 Dec 2012

Leprich U, Hauser E (2013) Drei Säulen für ein neues EEG. Klimaretter, 6 June. http://www.klimaretter.info/politik/hintergrund/13175-drei-saeulen-fuer-ein-neues-eeg. Accessed 6 June 2013

Nestle U, Reuster L (2012) Ausweisung der EEG-Umlage: eine kritische Analyse. Forum Ökologisch-Soziale Marktwirtschaft, Berlin. http://www.foes.de/pdf/2012-11-14-EEG-Umlage-kritisch-analysiert.pdf. Accessed 20 Dec 2012

Rasmussen MG, Andresen GB, Greiner M (2012) Storage and balancing synergies in a fully or highly renewable pan-European power system. Energy Policy 51:642–651

Sensfuß F, Ragwitz M, Genoese M (2008) The merit-order effect: a detailed analysis of the price effect of renewable electricity generation on spot prices in Germany. Energy Policy 36(8):3076–3084

Süddeutsche Zeitung (2012) Länder blockieren Röttgens Solarkürzungen, 11 May. http://www.sueddeutsche.de/politik/bundesrat-kuerzungen-der-solarfoerderung-gestoppt-vorerst-1.1354530. Accessed 29 Dec 2012

Trend:research (2011) Marktakteure Erneuerbare-Energien-Anlagen in der Stromerzeugung. http://www.kni.de/pages/posts/neue-studie-bdquomarktakteure-erneuerbare-energien-anlagen-in-der-stromerzeugungldquo-32.php. Accessed 15 Dec 2012

Welter P (2012) Like this, Minister Altmaier! Photon International, pp 30–45

Witt A (2012a) Gewerbesteuer für Solarstandorte. Solarthemen 387:2

Witt A (2012b) EEG: Koalition schraubt PV-Vergütung runter. Solarthemen 373:8–9

Zeit online (2012) Erneuerbare Energien: Bundesländer wehren sich gegen Altmaiers Reformpläne, 16 Oct, http://www.zeit.de/wirtschaft/2012-10/altmaier-oekostrom-reform-bundeslaender. Accessed 29 Dec 2012

Part V
Hopes and Fears: Considerations for Future Governance

Green Electricity Certificates in Flanders: The Gradual Extension of a Market-Based Mechanism and Doubts Over its Cost-Efficiency

Ils Moorkens, Yoko Dams, Luc Van Wortswinkel and Gerrit Jan Schaeffer

Abstract At the time of its introduction in 2002, the Flemish system to support renewables was a pure market-based green electricity certificate system. Starting in 2004 a string of changes to the system, up till the current year 2013, culminated in the addition of a minimum allowance for green electricity certificates. This minimum allowance presents a minimum market price for green electricity certificates. The gradual transformation of the system took place as a consequence of a series of policy responses to imperfections of the system as perceived by policy makers. In our analysis, we investigate whether the system has been effective and efficient in reaching its goals. We focus on four consequences of the system's structure: three related to the aspect of minimum allowances (the time-lagged nature, the technological orientation and the differentiated rights for technologies) and one related to the market-based green certificate aspect (the short-term target setting by limited annual quota increases). At present Flemish renewable targets have been reached, thus the system seems to have been effective but there are doubts about its efficiency. Whether the current form of the system will still be effective in the future, so that Flemish renewable energy targets can be met, and whether the system will turn out to be efficient, is as yet undetermined.

I. Moorkens (✉) · Y. Dams · L. Van Wortswinkel · G. J. Schaeffer
Flemish Institute for Technological Research (VITO), Boeretang 200, 2400 Mol, Belgium
e-mail: ils.moorkens@vito.be

Y. Dams
e-mail: yoko.dams@vito.be

L. Van Wortswinkel
e-mail: luc.vanwortswinkel@vito.be

G. J. Schaeffer
e-mail: gerritjan.schaeffer@vito.be

G. J. Schaeffer
University of Leuven (KU Leuven), Kasteelpark Arenberg 10 - bus 2445, 3001 Leuven, Belgium

G. J. Schaeffer
EnergyVille, Dennenstraat 7, 3600 Genk, Belgium

E. Michalena and J. M. Hills (eds.), *Renewable Energy Governance*, Lecture Notes in Energy 23, DOI: 10.1007/978-1-4471-5595-9_19,

1 Introduction

In Belgium the legislative power in the field of renewable energy is delegated to the three regions: Flanders, the Walloon region and the Brussels Capital Region. In support of renewable electricity generation Flanders implemented a Flemish green electricity certificate system (abbreviated as the FGECS) that assigns green electricity certificates to producers of renewable electricity and obliges electricity suppliers to hand in a number of certificates (quota) yearly. The FGECS was revised multiple times and evolved into a market-based certificate system with the addition of technology multipliers (banding factor) and bottom prices for certificates (minimum allowance)..

We will present an analysis of the FGECS and the direct and indirect effects of its implementation.

The central question of this chapter is: "Is the Flemish implementation of a technology-based hybrid certificate system an effective and efficient tool to support renewable electricity deployment and does it help to achieve the renewable energy target?"

The consequences of the FGECS's implementation on which our analysis is based, are:

(1) The time-lagged nature of the system (related to the minimum allowance aspect)
(2) The pure technological orientation of the system (related to the minimum allowance aspect)
(3) The way in which the implementation of green electricity goals influences investment decisions
(4) Differentiated rights for technologies (related to the minimum allowance aspect).

In order to substantiate some of the consequences of the system's implementation, we will make use of a tailor-made cost benefit analysis (financial gap analysis) of renewable electricity technologies compared to reference technologies. This methodology is based on financial concepts such as the income statement and resulting cash flow on a project basis and plays a major role in the FGECS.

2 Setting the Scene: The Flemish Green Electricity Certificate System

The Flemish green electricity certificate system has evolved throughout several iterations since its adoption in 2000 up to the latest changes taking effect in 2013.

2.1 Why a Certificate System?

The FGECS was registered in the Electricity Decree on 17 July 2000 (Belgian government 2000) and became effective in 2002. The introduction of the Decree referred to a draft Directive of the European Commission on harmonising support systems for renewable energy within the European Union. This Directive aimed at a Member State obligation to introduce market-oriented systems. Member States could choose between two alternatives: the introduction of green electricity certificates to impose minimum market shares for renewable energy, or initiating tendering procedures to stimulate construction of new green electricity production installations. In 2001 the European Commission considered introducing a harmonised system throughout the European Union and there were strong indications that a system of tradable green certificates would be preferred. In anticipation of an international green electricity certificate system in Europe the Flemish government choose to introduce quota controlled by green electricity certificates as the central instrument in its renewable electricity policy (Bollen et al. 2011). Afterwards, the European Commission decided to defer its decision about a harmonised system.

2.2 General Functioning of the Flemish Green Electricity Certificate System

The Flemish green electricity certificate system consists of rights and obligations.

The system assigns a green electricity certificate to producers (installation owner) of renewable electricity for each MWh_e of renewable electricity produced. The system obliges electricity suppliers to hand in, to the regulator, a fixed amount of green electricity certificates each year. In case an electricity supplier does not hand in sufficient green electricity certificates he will incur a fine per missing certificate. The systems rights and obligations are integrated by means of a certificate market among renewable electricity producers and electricity suppliers. The Flemish energy market regulator (VREG) assigns the certificates and facilitates the certificate market. The certificates are tracked by means of an online database (see Fig. 1, Initial functioning of the system). Besides income generated by the selling of certificates, producers also generate income by selling green electricity on the wholesale market.

The certificate database is no trading platform and initially trade took place via bilateral contracts which were reported to the regulator. This system had the disadvantage that some parties exerted substantial market power caused by the combination of many sellers of certificates (renewable electricity producers) and one dominant buyer of certificates (electricity supplier). The situation was to be rectified by the introduction of an electronic exchange for the trade of green electricity certificates in 2010. This electronic exchange was realised by the

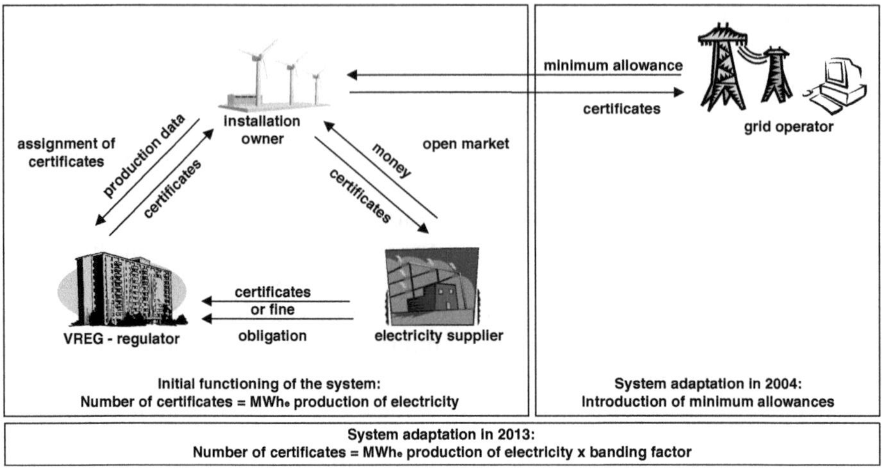

Fig. 1 Functioning of the FGECS: initial functioning, system adaptations in 2004 and 2013

coupling of the certificate database to the Green Certificate Exchange (GCE) of the Belgian Power Exchange (Belpex).

The number of transactions via the GCE is rather limited so far, among others because of the economic crisis of 2008–2009 has led to a decrease of electricity demand and, subsequently a decrease in the number of certificates to be handed in. Moreover there are serious doubts that the GCE resolves the market power of the dominant buyer. Anonymity in the transactions does not reduce the market power of the dominant buyer (Bollen et al. 2011).

The price of green electricity certificates is determined by the balance between supply and demand, where the demand level is a consequence of the obligation of the electricity suppliers. However, the price in this market is limited by the level of the fine, since in the case of prices being higher than the fine, the obligated parties would rather pay the fine than purchase certificates. Figure 2 shows that the evolution of the market price follows the evolution of the fine, but from 2010 on there was a decline of the market price compared to the fine because of the supply surplus of green electricity certificates at that time. This surplus was caused by the massive investments in solar energy due to the high level of the minimum allowance for this technology at that time (see Sect. 2.3). In direct response to this threat of market failure, the Flemish government decided to raise the green electricity certificate quota in 2012 and to further adapt the system, introducing regular evaluation and adaptation of the minimum allowance (see Sect. 2.3).

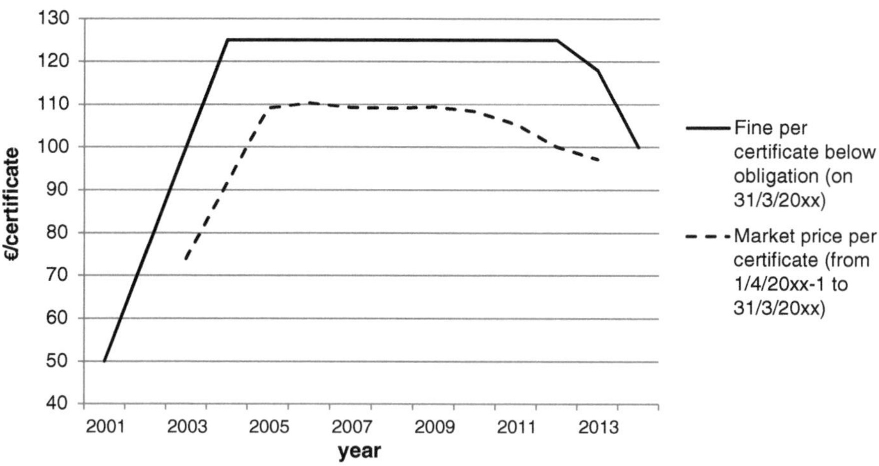

Fig. 2 Evolution of the fine and the market price per green electricity certificate

2.3 Key Innovations of the System: Minimum Allowance and Banding Factor

Besides the trade mechanism described above, the system also contains the concept of a *minimum allowance*, which was introduced in 2004. The minimum allowance is granted to renewable electricity producers by grid operators who have the obligation to buy certificates at the price of the minimum allowance (see Fig. 1, System adaptation in 2004). If the market price is less than the minimum allowance then electricity producers prefer to sell their certificates to the grid operator at the price of the minimum allowance. Grid operators can resell these certificates on the certificate market in order to, at least partially, recuperate the cost of their obligation. The remaining part of the cost (the positive difference between the minimum allowance and the market price) is charged to electricity consumers by means of distribution tariffs imposed by the grid operators. Since this mechanism of cost recuperation leads to different costs for different grid operators, and therefore different costs for their consumers, it was decided in 2010 to consolidate all costs among the grid operators.

After its introduction in 2004 the height of the minimum allowance has been changed several times but no systematic evaluation has been performed because there was no legal binding obligation to do so. Instead an ad hoc evaluation took place which was triggered by signals from the energy market and societal actors. The method of the financial gap (FG) (see Sect. 3) was used several times to determine the optimal level of the minimum allowance per technology. In theory the concept of minimum allowances was introduced to guarantee financial stability to green electricity producers, but in practice the level of the minimum allowances was adapted slowly and in an ad hoc fashion which in turn leads to windfall profits

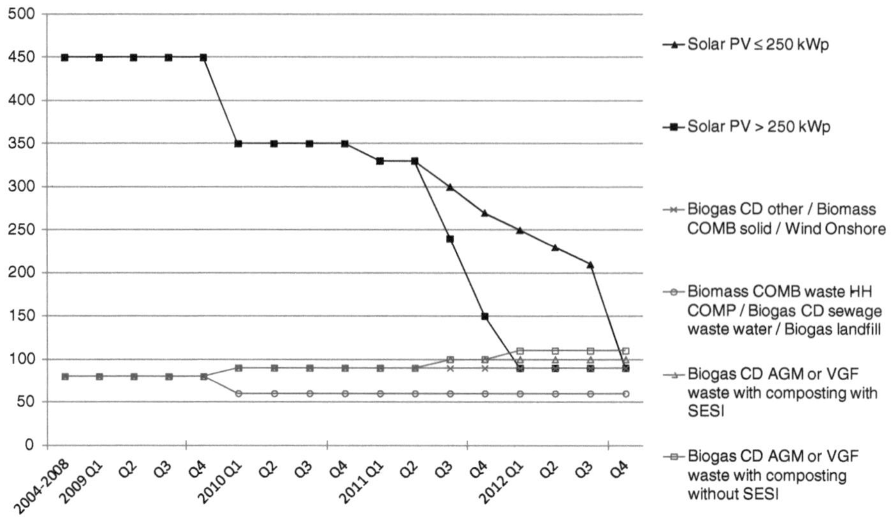

Fig. 3 Evolution of the height of the minimum allowance for new renewable electricity installations[1]

(if the minimum allowance is too high) and losses (if the minimum allowance is too low). This is demonstrated by Fig. 3 which depicts the evolution of the height of the minimum allowance for the different technologies. For solar PV the minimum allowance amounted to 450 €/MWh$_e$ until the end of 2010 while the financial gap for installations amounted to maximum 300 €/MWh$_e$ at the end of 2010. Installations which entered into production in the course of 2010 gained considerable windfall profits. Hence, the concept of minimum allowance and the linked ad hoc evaluation did not succeed in its endeavour to guarantee a smooth course of the FGECS.

To overcome these problems, the Flemish government introduced drastic changes to the FGECS in 2012 (see Fig. 1, System adaptation in 2013). The concept of the *banding factor* is a central pillar in the adapted system. With its application each technology receives a number of green electricity certificates per MWh$_e$ equal to its banding factor. This banding factor differs per technology and it is calculated by dividing the financial gap (see Sect. 3) by the expected certificate value (also called banding divider), which is set at 97 €/MWh$_e$ from 2013 on. For instance: a project with a financial gap of 48.5 €/MWh$_e$ will have a banding factor of 48.5/97 = 0.5 and thus receives a green electricity certificate for each 2 MWh$_e$ of renewable electricity produced. The Decree of 13 July 2012 (Belgian government 2012), in which also the value for the banding divider is set, stipulates that the Minister of Energy can determine the maximum authorised banding factor for

[1] Legend: COMB = combustion; HH = households; COMP = companies; CD = co-digestion; AGM = agrarian and/or manure; VGF = vegetable, fruit and garden; SESI = subsidy for ecologically sound investments.

new projects from 2014 on in an annual cycle and that in any case the banding factor is capped at 1.25. For 2013 the banding factor is capped at 1.

Besides the introduction of the banding factor the adapted system comprises an annual update of the financial gap for different installation types. To this end a new team called Monitoring and Evaluation has been established within the governmental administration. The methodology and parameters to be used in the financial gap model are established by the Flemish government in a transparent way to guarantee a clear investment framework. Parameters are based as much as possible on publicly available data, such as for example exchange indicators for electricity and (fossil) fuels. Only when no references are available for a parameter does the Monitoring and Evaluation team make assumptions based on expert judgment which are in turn objectively legitimised as much as possible.

The ultimate result of all changes to the FGECS is to assign a different number of certificates to each technology in order to realise a more adjusted support, which in turn should correspond to the real cost price. *In effect these changes have turned the tradable green electricity certificate system into system very similar to a feed-in premium system with only a minor trade component.*

The principal aim of the changes implemented in 2012 and 2013 is to increase the cost efficiency of the system, to guarantee an equitable distribution of the costs over the different societal actors and to provide a stable and beneficial investment climate for green electricity producers.

The future will reveal whether these changes are sufficient to be able to cope with the diverse requirements for a green electricity support system.

3 A Methodology for Analysis: The Financial Gap Analysis

The methodology to determine the minimum allowance per renewable energy technology is the so called "Financial Gap Analysis." The methodology was developed and detailed by the Dutch research institute ECN in 2003 and was used to calculate the production subsidy levels for renewable energy projects in the Netherlands. In 2005, the method was revised and adapted in order to match the specific needs/requirements of the Flemish context.

For the purpose of this chapter, we define the financial gap (FG) as the deficit in a project's revenue that is needed for the net present value (NPV[2]) of the project to equal zero based on a discount rate[3] set equal to a predefined Internal Rate of Return (IRR).[4] The financial gap is expressed per unit of electricity produced:

---

[2] In finance, the net present value (NPV) of a time series of cash flows, both incoming and outgoing, is defined as the sum of the present values of the individual cash flows of the same entity.

[3] The discount rate is the rate used in discounting future cash flows.

[4] The Internal Rate of Return (IRR) is used in capital budgeting to measure and compare the profitability of investments.

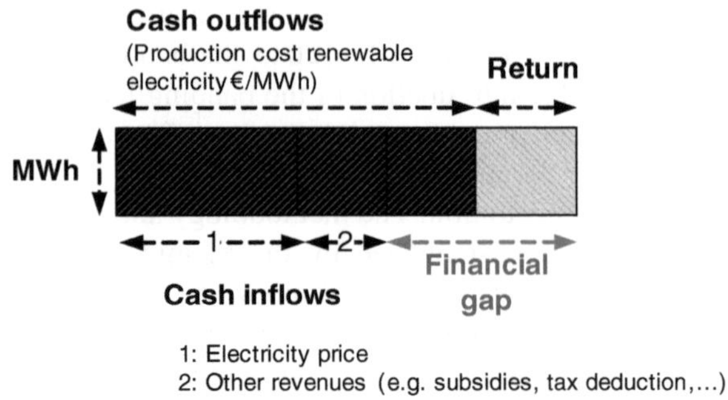

Fig. 4 The construction of the financial gap: a cash flow approach

$$NPV(FG) = 0 \, with\, discount\, rate\, set\, to\, target\, IRR$$

A cash flow approach is used to gain insight in the cash inflows and outflows over the projects life time. However, there is an important difference with a conventional cash flow approach. Here, the overall return of the project is not the output of the calculation. Instead, the FG is computed as a function of the cash flows and a presupposed return (IRR) of the project. Hence, the value of the IRR is determined in advance and serves as an input. The IRR percentages can be found in the Decision of 2012 (Flemish Regulator for Energy and Gas 2012) and differ per type of installation. The values vary between 5 and 12 %. These values are used in the financial gap calculations.

In order to determine the net present value of the project, the *cash flows* which will accrue during the lifetime, are estimated. In Fig. 4 the cash outflows include production and operating costs such as fuel, maintenance, insurance contracts, etc. The cash inflows include the electricity price and other possible revenues. At the cash inflow the value of the generated green electricity is determined. Sometimes the investor can apply for a subsidy for ecologically beneficial investments or a tax deduction for investments in renewable energy production. In these cases this additional cash inflow is also taken into account.

Investments in renewable electricity projects require financial incentives. Investors[5] do not only look for a cost-neutral project, they also request a higher *return* on their investment to compensate for additional project risk.

As illustrated in Fig. 4, the financial gap is the area needed to fill the gap between expenditures, revenue and requested return. Although the FG analysis is straightforward, the selection of parameters and assumptions are always prone to criticism.

In order to perform an FG analysis, a hypothesis with respect to the timing of the cash flows is required. The FG assumes the capital cost is incurred right at the start in year 0 after which the construction of the green electricity project may take

[5] The term 'Investor' refers to all organisations or individuals who contribute capital and/or resources to the development of renewable electricity projects, small or large scale, and who anticipate a financial benefit.

some time. After the construction period the investor receives periodic public financial support for a period equal to the depreciation period. Any assumed bank loans are assumed to have the same duration. This approach however is not feasible for all technologies. In some cases the economic life time of an installation is longer than the policy period (support period) which is used in the FG analysis. In these cases, the model assumes that there is financial support for a shorter period than is actually true and hence the FG will be overestimated. It is because of this issue that, in the Decree of 2012 (Belgian government 2012), the FGECS is extended with a possibility to level down the FG.

In terms of funding an FG analysis evaluates each project in a vacuum. As a result possible tax benefits only influence the project budget, and if needed they can be transferred to the following year(s). This does not allow for fiscal optimization where profit and loss of multiple projects are combined.

While the FG model processes revenue and expenditure, it does not take into account future price adjustments. This might lead to somewhat defective estimated cash flows as energy related prices do fluctuate over time and they, for their part, affect expenditure and revenue. Because of this, the Flemish government decided to adjust the methodology in July 2012. Since then the model is extended with the expected average annual change in the value of electricity and other fuels.

4 Flanders Green Electricity Certificate System: Consequences of Implementation

To answer our central question we will address a number of consequences of the FGECS's implementation which merit an in-depth discussion. In this section we present an analysis of four "issues" which we identified.

4.1 A Time-Lagged System

Time lag plays an important role in the effectiveness of economic policy. Today's information, which is translated into parameter values of the FG analysis, is used to reach a future goal. Aiming at an attractive investment climate, the financial gap analysis should reflect the actual market situation, year after year.

In order to calculate the financial gap, a lot of techno-economic (e.g. investment, maintenance, operating time, electrical/thermal return,...) and financial-economic (e.g. ratio equity/debt capital, electricity and fuel prices,...) assumptions are made. On top of the uncertainty resulting from parameter estimates, there is an additional uncertainty that the parameter values will change substantially over time.

After the system reform of 2012, the team Monitoring and Evaluation evaluates the financial gaps and banding factors of different technologies on an annual basis, with an exception for photovoltaic which is evaluated every 6 months. Additionally, the FG model is extended with an assumed pathway for some parameters (as discussed in Sect. 2).

Before these changes in 2012 the FGECS was rigid: Once the minimum allowance of a project was determined, it was fixed for the entire policy period (20 years).

Because the minimum allowance remained constant for such a long period, and the model did not include dynamic pathways for the economic parameters (technology learning), the FGECS casted doubt upon the accuracy of the support level. It is obvious that outdated calculations can cause incorrect levels of financial support and economic parameters change rapidly. Probably, the decrease of PV system costs is the most spectacular and unpredicted change. In 2006, the FG was calculated with an estimated investment cost of 7,000 €/kW$_e$ (residential installation). During the reassessment in 2010 the cost was scaled down to 4,235 €/kW$_e$. In the latest update of 2012 the cost is again reduced to 1,702 €/kW$_e$. This large and unforeseen price erosion gave the market an additional stimulus to invest in the technology.

In 2011 the economic and financial impact (Meynaerts et al. 2011) of the FGECS was calculated. The authors of the study assessed per technique a FG pathway by incorporating potential evolutions for some parameters. It became clear that the FG of most green electricity techniques decreased over time.

From a sensitivity analysis which provides a basic understanding of how strongly the FG reacts to a modification of a parameters value, we learned that in some cases the change in evolution of a parameter over time has a more profound impact than the introduction of variation in its absolute value. In a similar fashion, by introducing pathways for some core parameters and by evaluating the FG annual, the policymakers tried to tackle the time-lag problem. In the current iteration of the FGECS anno 2013, the problem is not solved, e.g. the banding factor is still fixed for a year.

4.2 A Technology-Oriented System

The Flemish government defines typical project categories for which a financial gap and banding factor are calculated. These categories are differentiated towards technology (solar PV, wind onshore, co-digestion of biomass and biomass combustion) and within the different technologies towards capacity range and type of biomass.

For the typical project categories, average parameters for the FG calculation have been established. The choice of categories implies that for particular technologies or situations, basic parameters can vary widely and to a large extent influence the FG and resulting banding factor.

To demonstrate the effect of variation in parameters for a particular technology, we take a closer look at the following example: the variation of the number of full load hours in the category of onshore wind turbines. The reference installation for calculation of the FG for 2013 is a turbine of 2.3 MW_e with 2,000 full load hours a year and a life time of 15 years. By using the average number of full load hours the difference in the number of full load hours between windy locations and less windy locations was not taken into consideration. Based on 2,000 full load hours the financial gap amounts to 80 €/MWh_e. A wind turbine in the more windy locations of the Flemish region demonstrates 2,500–3,000 full load hours a year. The FG of a wind turbine with 2,700 full load hours amounts to 47 €/MWh_e (no variation in other parameters to calculate this FG). On the other hand, turbines in the less windy locations demonstrate to have 1,700–1,800 full load hours a year. The FG of a turbine with 1,750 full load hours is about 98 €/MWh_e. Using the banding divider of 97 €/MWh_e the corresponding banding factors are: 0.48 (windy); 0.83 (reference) and 1.00 which is the maximum banding factor for 2013 (less windy). According to the current system, the banding factor of 0.83 applies to all wind turbines that will be installed in the year under consideration. As a result, wind turbines in more windy locations receive more total subsidy than turbines in the less windy locations.

Another clear example to demonstrate the importance of varying parameters for a particular technology is the price of biomass in the category of solid biomass combustion up to 20 MW_e. The reference installation (10 kW_e, 7,900 full load hours) is assumed to combust wood chips of recycled wood. The price of recycled wood is assumed to be rather low compared to the price for clean wood chips and wood pellets. In general, price information for solid biomass is rather limited, especially for recycled flows. Best estimates are based on limited information from recycled wood flows in Flanders and market information from the German market for clean wood chips. One of the questions to be asked here: *is the assumed price high enough to stimulate the building of new combustion installations?* The FG with the price for recycled wood chips (0.0141 €/kWh) amounts to 81 €/MWh_e (corresponding banding factor (BF) = 0.84). The FG with the price for clean wood chips (0.0228 €/kWh) amounts to 117 €/MWh_e (corresponding BF = 1 = max BF).

The consequence: if investors in new installations cannot obtain recycled wood flows then there will be no new investments in such installations. In reality the existing recycled wood flows are indeed limited in the Flemish region. Moreover, these flows are wanted as a raw material by the wood and paper industry which is specialised in processing recycled material because of the small wood areal in the Flemish region. The wood and paper sector requested sufficient allocation of material (recycled and fresh wood) to its use as a raw material. Since the Flemish region has a strong tradition of recycling, the Flemish government granted the wood and paper sector its request and stipulated that only short rotation wood and wood flows, which cannot be used as an industrial raw material, should be considered to receive green electricity certificates.

Another effect that plays a significant role at the technology level is the application of a cap on the banding factor, set at a maximum of 1.25 by the Flemish Government (for 2013 there is a maximum banding factor of 1 set by ministerial decision). For 2013 we observe that the BF reaches its cap for the following technologies: all biogas technologies, except co-digestion of sewage water sludge and biogas from landfills, and all fluid biomass combustion. This cap implies that investments in these technologies are discouraged and will probably be slowed down for some years to come. A possible consequence of this choice is that the accrued knowledge on these technologies (e.g. for co-digestion of manure and agrarian flows) is at stake. On the other hand this can be a conscious choice to stimulate other technologies like solar PV, wind and solid biomass combustion. For the first two this might seem reasonable because of fuel independence. On the other hand these sources demonstrate a highly intermittent supply profile and hence require a more intensive mitigation of—and adaptation to intermittency. Biomass combustion can be used as a base load source and could hence has a distinct appeal to be included in the (renewable) electricity mix. While the above arguments seem reasonable and logical it is by no means clear that these are the real reasons behind the cap on the banding factor.

The fact that the FGECS is a technology-oriented support system is amplified by establishing typical project categories for calculating the FG and BF. The examples illustrate the necessity to carefully make choices. It is important to introduce sufficient differentiation in order to stimulate investments in desirable technologies and to limit support for other costly technologies. On the other hand a higher number of different categories considered necessitates more information gathering on underlying parameters and hence higher administrative costs. The choice between many or few typical technology categories should be based on a conscious weighing of the pros and cons for both approaches.

A global renewable energy vision for the next decades, in which the importance of the different technologies is estimated, could be a valuable foundation for these choices.

4.3 Green Electricity Goals and Investment Decisions

Another important consequence of the Flemish green electricity certificate system's implementation is the way in which it implements its green electricity certificate goals. The FGECS obliges electricity suppliers to turn in a fixed amount of green electricity certificates each year and allows electricity suppliers to buy those certificates on a certificate market should their own green electricity production be insufficient to cover their obligation. In such a market mechanism, the value of a certificate is based on total supply and demand of these certificates. By means of quota, the FGECS fixes the demand level along the lines of an annual incremental path starting at 14 % in 2013 up to 20.5 % green electricity in 2020.[6]

---

[6] The incremental demand path has already been changed a few times in the past. E.g. from 01/01/2011 to 29/07/2012 the 2013 target was 8 % and the 2020 target was 12.5 %.

The supply of green electricity certificates is determined by technology specific certificate rights per MWh_e of green electricity produced. Given *the nature of the demand and supply side of the green electricity certificate market, it is defined by strong market power at the demand side.* The final quantity of green electricity certificate demand is largely[7] determined by the Flemish government and one could argue this market resembles a market with one final consumer with a price-inelastic demand.[8] For this comparison the demand of the Flemish government acts as the one final consumer and its demand level is fixed by law at a predefined level and hence is largely price-inelastic. As such the green electricity certificate market exhibits traits of a consumer monopoly. There are two traits of this demand level setting that we will discuss further: the small annual increments and the fact that the path is fixed for the short to medium-term future.

A first question we pose ourselves is: *if and how small annual demand level increments impact green electricity investment?* If we translate the small annual increment in green electricity certificate demand to an energy quantity it approximates somewhere around 1.8 PJ[9] of green electricity for each percentage increase in green electricity certificate demand.[10] While this is a significant amount of green electricity, it is still small compared to the green electricity output of one big green electricity plant. For illustration take the E.ON coal plant of Langerlo in Genk which will be converted to a biomass plant by 2014. After conversion this plant would generate somewhere around 9 PJ of green electricity which matches approximately 5 % of total green electricity demand in 2011. Obviously, the addition of one or more of these large installations would severely impact the green electricity certificate value and hence the profitability of all operational green electricity installations. If such an installation is implemented too early, the green electricity certificate value might plummet, and if it is implemented too late, the green electricity certificate value might soar. Based on this observation, we would state that the green electricity certificate demand level setting *promotes small to medium installations* and entails a *risk on expected returns should one implement a large scale installation.* The previous statement holds true in the context of a green certificate market on the limited Flemish

[7] In addition to final demand for green electricity certificates as set by the Flemish government, the possibility to bank green electricity certificates for 10 years allows for a temporal increase or decrease in demand. (banking of certificates = storing obtained certificates for later usage) The decision to bank green electricity certificates can be based on an expected future certificate price increase.

[8] A price-inelastic demand translates to a demand level which remains constant regardless of an increase or decrease in price.

[9] PJ = Petajoule = 10^{15} Joule: a derived metric for energy where one joule is equivalent to the work required to produce one watt of power for one second.

[10] Final electricity consumption in Flanders was 181.7 PJ in 2011. Flemish Energy Balance 2011, http://www.emis.vito.be/sites/default/files/pages/1332/2012/balans_2011_versie_nov_2012_correctie_0.xlsx.

regional scale. In case a similar system is applied to a larger region the above statement becomes less of an issue.

A second question we would like to discuss is: *how the fixed path demand level*[11] *influences green electricity investment decisions?* The fixed path demand level seems to have an upside and a downside. An upside is a certain level of *market stability* and *market information.* Information on the current green electricity generation capacity is publicly available and, in combination with information on the fixed green electricity certificate demand path, this provides good market information for investment decisions. Since every market participant has access to this market information, it is safe to assume a relatively stable market which in turn adds to a good investment climate. A downside to the fixed path demand level is the lack of incentive to go beyond the set green electricity certificate target in a given year. Investing in overcapacity for green electricity production seems undesirable since such an investment would incur diminishing returns based on a lower green electricity certificate value. Additionally, past investments would also incur the diminishing returns based on the lower certificate value.

As shown by the previous discussion, there are multiple consequences tied to the way in which the green electricity certificate goal is set. The choice for an annual incremental fixed path demand level leads to a relatively stable market. On the other hand, it does not promote large ambitious projects nor does it offer a financial incentive to go beyond the preset green electricity target. Whether or not this trade-off is desirable depends on the point of view. The fixed target almost guarantees that renewable electricity production will be sufficient to cover Flemish international obligations and, in that way, is a safe choice. If on the other hand a longer term, more ambitious goal has been envisioned, this way of demand level setting might not provide an optimal incentive to step up to far reaching green electricity targets.

4.4 Differentiated Technology Rights

The Flemish green electricity certificates system, with the addition of a technology dependent banding factor, confers differentiated rights to the different technologies. In the text below we will explain the concept of *differentiated rights,* the consequences of its application and the reasoning behind its importance as a consequence of the systems implementation.

[11] With the expression "fixed demand path level" we refer to the mechanism where the Flemish government defines a fixed number of green electricity certificates to be handed in each year and hence defines the demand for green electricity certificates. The quantity to be handed in each year is known years in advance. So the demand for certificates is known as well as the yearly evolution of this demand. We call this a "fixed" (by law) "demand path" (yearly evolution) "level" (known quantity).

Differentiated rights, what do we mean by this term? To understand its meaning we will take a closer look at the financial gap analysis and the resulting banding factor for a given technology. The financial gap analysis asserts the gap amongst revenue and expenditure in order to achieve a set Internal Rate of Return (IRR) over a projects operational lifetime. Assuming the targeted IRR is the same for different technologies, the resulting financial gap might differ based on technology specific operational parameters. In these cases we state them having *differentiated rights* because the FGECS grants them a different number of certificates albeit with the same IRR.

To illustrate this concept with an example assume:

- two different technologies A and B both have a lifetime cost of 10,000 €
- a green electricity certificate has a value of 50 €
- technology A its IRR is 13 % and technology B its IRR is 14 % without green electricity certificates
- the targeted IRR for both technologies is 15 %
- hence technology A will receive 4 certificates (10,000 * (15−13 %)/50) and technology B 2 certificates (10,000 * (15−14 %)/50). They both have different rights.

Differentiated rights, what are the consequences of its application? The consequences of the application of differentiated rights are numerous but they all resolve around the way the FGECS influence the free choice of potential investors and hence define the short-term and medium-term future technology mix for electricity generation.

Assuming the targeted IRR is the same for multiple technologies, a potential investor has no financial incentive in terms of IRR to choose one technology above another. Hence the financial incentive *does not provide directional guidance* to green electricity investments. For the investor's part, he/she will still perceive advantages and disadvantage which are, amongst other, project risk, project payback period, capitalization rate, feasible financial leverage and sometimes genuine irrational human perception. Whether this aspect is seen as positive or negative depends on whether one adheres a policy of being reluctant to picking winners or to a policy that is guided by a shared vision on the energy mix needed in the future.

Another possible consequence of differentiated rights is the *non-optimal use of financial community resources*. As different technologies present a different financial gap to obtain the same IRR, the community resources used to support them will also differ. One could argue that an optimal use of community financial resources can only be obtained if financial incentives are granted solely to the best performing technologies and only to a level where total expected green electricity production would match the predefined desired quantity of green electricity.

Furthermore differentiated rights allow for *more competition*. By means of differentiated financial incentives, different technologies become economically feasible with a resulting market entrance. Since multiple technologies strive against each other for market share there is a strong incentive to lower costs and increase performance. If only one technology would be economically feasible there would be less incentive to improve product performance and competition might shift from cost lowering and performance increases towards better service to gain a competitive advantage. While this is valuable for investors, it does nothing to alleviate the future community cost of green electricity support.

Another interesting effect of differentiated rights is their contribution to market diversity and as such the *guarding against future technology lock-ins.*[12] As differentiated rights make multiple technologies economically feasible, potential investors will most likely implement a more or less differentiated mix of technologies based on perceived advantages and disadvantages. As argued above this could be labelled as non-optimal use of financial community resources. On the other hand this same technology mix avoids overcommitting to one single technology. While such over commitment would be a good thing should this technology indeed turn out to be the optimal choice, there is always a chance that the financial gap analysis turns out faulty. This possibility exists because a financial gap analysis requires various assumptions about future costs and operational parameters and in real life there is no such thing as prescience in these matters. Hence differentiated rights guard against a technology lock-in with what in hindsight might proof to be a sub-optimal technology.

Differentiated rights, is it any good? With the upsides and downsides of differentiated rights discussed above, one cannot draw a clear conclusion whether it is a concept worth keeping or an aspect of the FGECS which should be subject to change. It largely comes down to personal belief in terms of forecasting ability and ones view on the role of the policy maker. If one strictly believes in the ability to accurately assess the financial gap of technologies now and in the future, one could opt for an optimal strategy based on perfect foresight and as such revise the current system of differentiated rights. If one questions the ability to make this assessment with reliable and robust results, one would indeed be better off with the current concept of differentiate rights and the resulting technology differentiation.

5 Discussion—Conclusions: The Flemish Green Electricity System: Did it Deliver?

At the onset of this chapter we formulated the research question: "Is the Flemish implementation of a technology-based hybrid certificate system an effective and efficient tool to support renewable electricity deployment and does it help achieve

[12] A technology lock-in is a situation where past investment decisions limit the number of present technology investment options. If an investor has committed a large amount of capital to a new installation it is unlikely that he will end its exploitation before the end of its technical life. Hence he will not make new investment decisions during this time.

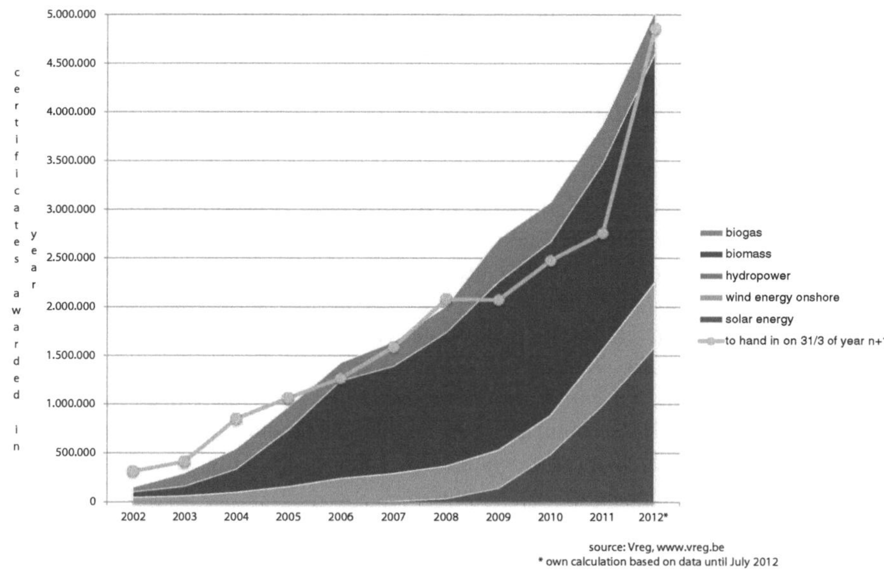

Fig. 5 Evolution of the number of awarded certificates and the certificate obligation

the renewable energy target?" Throughout the chapter we discussed several consequences of the system's implementation and their effect on green electricity production within Flanders. To conclude the chapter we will present the demonstrated results of the FGECS and discuss its effectiveness as well as its efficiency.

The evolution in the number of awarded certificates versus the certificate obligation shows a shortage during the period 2002–2005, a market balance, after the introduction of minimal allowances, during the period 2006–2008, and a surplus during the period 2009–2011 (Fig. 5).

Based on the above supply and demand figures one can conclude that the FGECS demonstrates a good efficacy,[13] or in other words, the Flemish green electricity targets are met.

Still, while these supply and demand figures inform us of the FGECS efficacy, they do not inform us on its effectiveness[14] or efficiency.[15] Based on currently available data it is not possible to draw definite conclusions on effectiveness and efficiency of the FGECS. In 2011 the Social and Economic Council of Flanders published the book Energy for a Green Economy in which this topic was also discussed for the FGECS before its revision in 2012 (Bollen et al. 2011). The

---

[13] Efficacy measures in how far goals are met.

[14] Effectiveness measures the causal relation between an action and goal satisfaction. Has the action resulted in the goal satisfaction or would the goal also be obtained without the action?

[15] Efficiency measures the balance between effort and result. Is the amount of effort proportional to the obtained result?

writers noted the lack of data but based on what was available they drew some general conclusions: the cost of the FGECS was increasing more rapidly than the green electricity certificate goal during the period 2002–2009, some case-specific technologies generated significant windfall profits, some technologies got more support than the financial gap suggested, and future electricity consumers could become heavily penalised for the FGECS past liabilities. While some of these conclusions might be invalidated by the FGECS's 2012 revision, they still serve as a cautionary tale of what the consequences can be of a green certificate system that started out as a pure market-based system, but was later on expanded with minimum allowances.

The Flemish green electricity certificate system can best be described *as a market-based certificate system with the addition of technology multipliers (banding factor) and bottom prices for certificates (minimum allowance).* The resulting system exhibits certain advantages and disadvantages compared to the commonly used pure feed-in tariff systems. In conclusion we give an overview of some of these advantages and disadvantages.

(+) The market-based character ensures least-cost implementation of green electricity production by price competition and allows for free choice whether or not to deploy renewable technologies.

(+) Another advantage of a market-based system is that it is less prone to a financial backfire. A pure feed-in tariff gives the same tariff when suddenly huge amounts of new capacity would be installed. A market-based system its certificate value would decrease and hence it is not opportune to overinvest in renewable technologies.

(+) The addition of minimum allowances ensures a minimum level of certainty for investors. In case the certificate price would plummet, the minimum allowance functions almost like a feed-in tariff, the only exception being that the certificates can be stored for later use.

(+) The addition of technology multipliers (banding factors) enables a level playing field for a diverse set of technologies so that all can be cost-effective investments while at the same time preventing windfall profits by over subsidising.

(−) The disadvantage of a market-based system is an increased risk for investors when compared to feed-in tariffs. The value of a certificate is dependent on the market and hence on the actions of all market players. In a pure feed-in tariff system the level of financial support is known in advance and not a resultant of a certificate market.

(−) There is an additional administrative cost related to the support of the certificate system. Certificates have to be awarded, a market needs to be established and certificates need to be recalled as per predefined quota. The cost of establishing the multiplier is comparable to the cost of establishing the level of the feed-in tariff and as such does not present an extra administrative cost.

References

Belgian government (2000) Decreet houdende de organisatie van de elektriciteitsmarkt. Belgian Off J. Available via http://www.ejustice.just.fgov.be/cgi/article_body.pl?language=nl&caller=summary&pub_date=00-09-22&numac=2000035898

Belgian government (2012) Decreet houdende wijziging van het Energiedecreet van 8 mei 2009, wat betreft de milieuvriendelijke energieproductie. Belgian Off J Available via http://www.ejustice.just.fgov.be/cgi/api2.pl?lg=nl&pd=2012-07-20&numac=2012035855

Bollen A, Van Humbeeck P, Lamote A (2011) Energie voor een groene economie. In: Hernieuwbare energie: beleid en evaluatie. Boekdeel 2. SERV. Gent Academia Press

Flemish Regulator for Energy and Gas (2012) Principiële goedkeuring Vlaamse Regering Ontwerpbesluit tot wijziging van het Energiebesluit, wat betreft de groenestroom- en warmtekrachtcertificaten en de garanties van oorsprong. Available via http://www.vreg.be/ontwerpregelgeving

Meynaerts E, Moorkens I, Cornelis E (2011) Doorrekeningen ter ondersteuning van evaluatie GSC en WKC-systeem. VITO, Mol

Building on Norway's Energy Goldmine: Policies for Expertise, Export, and Market Efficiencies

Tomas Moe Skjølsvold, Marianne Ryghaug and Jon Dugstad

Abstract This chapter deals with the governance of renewable energy in Norway. The Norwegian case is a peculiar one. Norway has been blessed with an abundance of waterfalls, which for the past 100 years have been used to produce cheap and renewable electricity. In addition, Norway has vast oil resources which have been exported to generate national wealth. How does Norway's odd position of on the one hand being one of the most renewable nations in Europe, while fuelling the world with hydrocarbons affect the governance of 'new' renewables? Does the fact that Norway is already one of Europe's largest producers of renewable energy make a transition toward more renewable policy and production, easier or harder? What cultural, political, and financial factors seem to influence Norway's strategies in the area of renewable energy production? Historically, energy in Norway has been both cheap and profitable, and Norwegian energy policies have traditionally been geared toward this goal: energy use and production should first and foremost be cost-effective. This situation has been challenging for the implementation of new renewable energy technologies. In the liberalized Norwegian electricity market, the governance of renewable energy has largely been left in the hands of the market participants. The low electricity prices over the past years have not attracted investment in renewables. In an attempt to mitigate this, Norway recently introduced electricity certificates in a joint market with Sweden, thereby creating a new class of incentives for investment in renewable energy generation. Further, to increase Norwegian renewable energy deployment, the Norwegian government has funded research and development projects and a number of large

T. M. Skjølsvold (✉) · M. Ryghaug
Department of Interdisciplinary Studies of Culture, Norwegian University of Science and Technology, Trondheim, Norway
e-mail: tomas.skjolsvold@ntnu.no

M. Ryghaug
e-mail: marianne.ryghaug@ntnu.no

J. Dugstad
Norwegian Renewable Energy Partners—INTPOW, Oslo, Norway
e-mail: jon@intpow.com

E. Michalena and J. M. Hills (eds.), *Renewable Energy Governance*, Lecture Notes in Energy 23, DOI: 10.1007/978-1-4471-5595-9_20,

research centers for environmentally friendly technologies. However, there appears to be challenges in transferring the R&D activities to commercial products that could reach the market.

1 Introduction

In a world where the term 'energy crisis' has entered everyday language, Norway stands out. Norway has been blessed with an abundance of waterfalls. Historically, this has been used to produce hydropower which has allowed for the establishment of an energy intensive industry with access to cheap electricity. Similarly, householders in this cold country have lived comfortably, at least energy-wise. In Norway, access to cheap electricity is taken for granted, so much that it has been vital in shaping a 'comfort culture' (e.g., Aune 2007; Aune et al. 2011). Here, hydropower is the 'gold standard' of energy production (Sørensen 2007).

If Norway struck gold with its hydropower, it found more on the day before Christmas in 1969 when oil was found off its shore (Tamnes 1997). This oil became the economic basis for the nation now ranked as number one on the UN development index (UNDP 2011). Thus, Norwegian energy has allowed for wealth through export, but it has also been cheap while catering for industrial development and comfort domestically.

Hence, the Norwegian identity has developed as an 'energy nation'. Its position as such is anchored in its oil and gas endeavors. However, climate change has caused some friction. How can oil production and climate concerns be reconciled? While Norwegian electricity production onshore is largely 'clean', its oil and gas adventure is not. Not only does it fuel the world's reliance on fossil fuels. GHG emissions from Norway's oil production are substantial, but not included in official national emission statistics. The same can be said of its transport sector where fossil fuels still dominate. Many advocate electrification of the offshore oil installations. However, this will require massive investments both in transmission lines and expanded electricity production capacity. Currently, no one is willing to make such investments. And, if they were: what power plants should be built? There are not many large waterfalls left in Norway, and using the remaining few for electricity production would be extremely controversial. Realization of Carbon Capture and Storage (CCS) in gas power plants is one option that Norway has invested heavily in, with increased production and transportation of liquid gas as a goal. In addition, Norwegian geography allows for the best wind resources in Europe, both on- and offshore. The theoretical potential for new renewable energy is significant.

Another much discussed idea has been to offer Norwegian hydropower an alternative for a transition toward European low carbon societies. Norway is the largest European hydropower producer, and number six worldwide. The storage capacity of this hydropower system could be used to refine intermittent renewable energy due to its regulating capacity. Again, however, this would require investments both in cables to Europe, and increase the need for more electricity production in Norway. Who will bear the costs of such expansions in the Norwegian and European energy systems?

In this chapter we discuss the Norwegian efforts for renewable energy governance. We want to highlight the importance of considering the existing energy system; its physical structures, political landscape, and the way it relates to Norwegian society in ways that sometimes helps, sometimes hinders the implementation of renewable energy. While much rhetoric focuses on ideas like "transitions", we argue that the existing Norwegian energy regime-based oil and hydropower combined with the current policies of governance is a challenge for the implementation of new renewable energy technologies.

2 A Policy of Energy Economization and a Culture of Comfort: The Market Participants as Shadow Actors

Resource abundance has been important for the shaping of Norwegian energy policy. While comparable countries like neighboring Sweden have focused on energy savings and security, saving energy has not been a goal in itself in Norway. Rather, Norwegian policy has sought to use and produce energy more profitably (Ryghaug and Sørensen 2009). These policies have their own Norwegian name: ENØK—"energy economization". One practical consequence of ENØK policies have been that costs are expected to drop whenever one energy carrier substitutes another. The result has been complete dominance on behalf of hydropower both in terms of production capacity and end use of energy. This is true even in cases where it would be more energy efficient to use other energy carriers, such as for example hot water for space heating. Other renewables have had a hard time, and technology subsidies have not been a relevant policy tool.

In 1990, the Norwegian electricity market was one of the first in the world to become fully liberalized, and the availability of electricity and associated tools of production have since been seen as products of supply and demand. If electricity prices are high the rationale is that electricity producers will build new production capacity (See Karlstrøm 2012a, for more). This, of course, is also applicable to renewable energy. Thus, much renewable energy governance is delegated to market participants. However, the Norwegian public still largely views provision of electricity as a political task. Meanwhile the reliability of supply has been stable with power outages being extremely rare. The electricity prices have remained low, so very few investments have been made in new production capacity.

While ENØK is still important in Norway, the world is different today than when ENØK was conceived in the 1970s. Climate change has taken the center stage and opened the door to new discussions and policies—also in Norway. In 2006, the government set a goal of adding 30 TWh of new renewable energy production from 2001 levels before 2016 (OED 2006, p. 1). Since then, goals and strategies for renewable energy have been elaborated on in various settings. One example is the Official Norwegian Report on climate policy from 2006. Here, power stations based on both coal and gas combined with CCS were seen as important, but so were public support schemes for renewable energy (NOU 2006, p. 114).

Another development is the integration of European energy markets. This has influenced Norwegian renewable energy governance in at least two ways. First, Norway has made the European renewable energy directive part of its legislation. The Norwegian target is a total of 67.5 % renewable energy by 2020 for all energy consumption, including transportation. While the authorities see this as highly ambitious, we should remember that Norway's electricity production is already almost fully renewable. Thus, the increase in share of renewable energy must come from transport or through replacing the few sources of heat production that are based on fossil fuels. This has led to a focus on electric vehicles, and Norway is now the largest market for many e-car producers. However, if vast amounts of Norwegians were to start driving electric cars there would be a substantial demand for more electricity production.

Second, in 2012 Norway joined the Swedish market for electricity certificates; a market-based support system for renewable energy production. Sweden has had this system since 2003, and it is often cited as a success. The Norwegian participation in this market is seen as a step toward tighter electricity market integration, and it is also an attempt of reaping benefits from economies of scale rather than having two relatively small and similar markets side-by-side. The goal of the scheme is to increase the investments in production capacity based on 'new' renewable energy technologies. The scheme is technology neutral, which means that all renewable electricity production is eligible. However, production of renewable energy in the form of heat is not included. This means that many producers of bioenergy, such as district heating facilities and small-scale heating facilities falls outside the scheme. Through the implementation of the quota system, the authorities would like to see 26.4 TWh of new Norwegian and Swedish renewable electricity generation by 2020.

The scheme is set up as follows: producers of renewable power receive a certificate per produced MWh. The certificates are sold to energy suppliers, required to have certificates for roughly 3 % of annual production in 2012, gradually increasing to between 15 and 20 % in 2020. The costs are covered by consumers as an addition to the electricity bill. The addition to the total bill will probably be marginal. For an average household with an annual consumption of around 20,000 kWh, current estimates suggest an addition of 150 Norwegian kroner, or approximately 20 € a year. While this number will increase toward 2020, little suggests that consumption will be affected substantially. The electricity certificate scheme represents a new element in Norwegian energy governance, since it is in principle a subsidy. However, it is based on market pricing and does not favor specific technologies over others. Thus, *the scheme keeps focus on technologies that deliver in terms of economic efficiency*. Since this is a new tool of governance, its long-term effects are still unknown. However, given its market pricing, many speculate that it will favor an expansion of the hydroelectric base in Norway through constructions of small hydroelectric power stations (that would have been profitable also without the certificates). Some also think it might serve the development of wind power. Others believe that structural, tax and cost differences between Norway and Sweden will favor investments in Sweden instead of Norway. So far, reports suggest that the scheme has not elevated prices enough to attract investors. One small wind power plant with two 0.8 megawatt turbines has so far received certificates. The Norwegian regulator asserts that many wind power licenses are issued, but that few of these materialize in investments (Adomaitis 2012). How things will pan out, however, is still an open question of the future. None-the-less, electricity certificates represent a new governance tool to increase the use of renewable energy in Norway.

Norwegian renewable energy governance strategy is mainly based on a market philosophy and the market is left to regulate the pace of implementation. Lately, though, the electricity certificates, while derived from the same market philosophy, are taking Norwegian governance a step further. In addition to the market-based measures, Norwegian renewable energy governance has since 2001 been reliant on Enova SF, a public enterprise, to drive the transition to an environmentally friendly energy regime through direct investment incentives. Its goal was twofold. On the one hand, it was to promote efficient energy consumption, both to lower the need for investments in increased production capacity and to the climate impact of industries and households. Second, since the possibilities of expanding the hydroelectric base in Norway are limited, it was to increase the production of "new" renewable energy through targeted investment programs in areas where the greatest effect could be documented. From 2001 to the end of 2010, they supported projects with an output of 2 TWh renewable electricity and projects generating an output of 6 TWh worth of heat.

While this suggests that the climate issue has affected Norwegian energy governance, we should ask whether or not the basic principles of governance have changed over the past three decades. So far, research suggests that it has not. Norway promotes renewable energy through a focus on cost-efficiency and technology-neutrality (Hanson 2011). This means that for state-funded Enova to support a project, *cost-efficiency must be documented.* Support is given in an investment phase, day-to-day subsidies of operations are not provided. Once operative, the projects must be competitive on their own. The electricity certificates *might* represent a shift, but again; current price levels might not facilitate increased investment.

Many argue that the focus on cost-efficiency and reliance on markets to resolve energy issues have removed much of the leeway for political governance and intervention. Most political parties agree on the need for some sort of energy transition. However, *it is in principle the market participants that decide when a transition is due, based on the information available.*

In an attempt to change some of the market dynamics, an important governance strategy for renewable energy in Norway has been research, development, and innovation. Over the past years, many research activities have been initiated. The goal has been to make new technologies competitive based on the assumption that the R&D activities over time will lower the costs of investing in new renewable energy production in Norway. Thus, R&D is seen as a route toward modernizing and expanding the Norwegian electricity system. On the other hand, this focus on R&D and innovation is an attempt to create internationally competitive Norwegian products and businesses. Both ideas were observable in the so-called Norwegian "climate settlement", a 2008 agreement, signed by the Norwegian Parliament. As a result of this agreement, 11 national research centers focused on 'environmental-friendly' energy technology was established. It has been noted that an outcome of the "climate settlement" was to transform Norwegian climate policy into a set of innovation policies (Kasa 2011). Thus, on the one hand, funding R&D is a central part of Norwegian renewable energy governance while, on the other, the implementation of new technologies, their acceptance, and commercial viability is left to the market to decide.

3 Norwegian Renewable Energy Strategies in Practice: Two Technologies

The Norwegian renewable energy governance strategy is twofold. First, implementing production of new renewable energy is mainly left to the market, despite the newly introduced electricity certificates. Supply is seen as governed by demand. Second, R&D funding is a core policy, expecting that new energy technology derived will become competitive and commercially viable.

How do the Norwegian governance efforts interact with and affect specific renewable energy industries? In the following we will consider two examples: offshore wind and bioenergy.

3.1 Offshore Wind

Many have pointed to offshore wind energy as strong opportunity for Norway. On paper, Norway sounds like an ideal place to establish an offshore wind industry. The country has a long costal line with good wind conditions. The Norwegian industrial structure has long traditions in marine operations and offshore petroleum, and many see this as a competitive advantage. It is expected that much of the competence in the petroleum industry can be applied in a new offshore wind industry.

However, political opinion suggests that there is no current governance efforts supporting offshore wind deployment in Norway. At one level this has to do with the material and cultural configurations already present in a nation so heavily reliant on the oil and gas industry. First, this is obviously related to money. The Norwegian state relies on income from oil and gas, as does much of the Norwegian industry. Thus, it is not clear that sharing the pool of resources with oil and gas is an advantage for offshore wind. The two industries are often in direct competition for investments and for the same bright heads. In comparison, where oil and gas can promise mature markets, huge and safe investment payoffs, and steady jobs, offshore wind can promise none of these benefits. This is still predominantly uncharted territory, most suited for entrepreneurs willing and capable of pursuing risks. So far, policies have offered little risk mitigation.

The Norwegian approach to offshore wind has not differed much from its general approach to energy since the 1970s. Proof of cost-efficiency before implementation is considered key. The sentiment is that cost-efficiency will lead to competitiveness and investments in production capacity. However, offshore wind costs are high compared to hydropower and current electricity certificate incentives are too low to attract investors. Thus, while the first license to operate a Norwegian offshore wind park was granted in June, 2009, it has yet to materialize. Specific subsidies, however, have been regarded as unwanted market intervention and an irrelevant tool to speed up the process of implementation. Thus, Norwegian offshore wind policies avoid market intervention. Instead, Norway has focused on R&D to create new innovative technologies. Two national research centers on offshore wind energy have been partly funded by the government. In one of these centers vision statement we can read that the goal is "[...] to provide innovative and cost efficient solutions and technology for large water depths and harsh offshore environments" and to build "strong clusters on offshore wind energy in Norway by developing new knowledge and by providing skilled persons for the industry" (NORCOWE 2012).

While this is highly welcome among research institutions, it is not clear that the R&D experiences gained in these settings will be directly transferable to the Norwegian electricity production regime. Currently, offshore wind does not deliver the returns on capital required by the investors to be a viable source of energy production in Norway without additional substantial public financial support. While the international cost reduction targets of at least 30 % might be met through new technology developments and R&D, deployment of offshore wind, and hence new technology, in Norwegian waters would still need substantial public support over and above the current technology neutral electricity certificates.

Nevertheless, since the market for offshore wind technology is global, the Norwegian governance could be seen as a pure technology and industry development project. That implies that the research centers are to create technologies that can compete internationally, benefiting Norwegian offshore wind industry participants with offerings to offshore wind installations abroad.

This interpretation is well supported in the Norwegian offshore wind industry. However, the sentiment is that the lack of a domestic market as an arena for trial and error is a major drawback in international competition (Hansen and Steen 2011).

3.2 Bioenergy

Another set of frequently discussed renewable energy technologies in Norway comes in the form of bioenergy. While many possibilities (such as biofuels for transport and biogas) are on the table, the most commonly discussed and applied technology is heat production based on some sort of biomass from wood. The potential is great; the country has vast forest resources. As in the case of offshore wind, an important governance strategy has been to develop bioenergy and related industries through R&D. A national bioenergy innovation center has been funded by the authorities with a goal "to develop a sustainable, cost-effective bioenergy industry in Norway" (CenBio 2012).

However, bioenergy policies are not only anchored in R&D, but are to a larger extent directed at energy production than in the case of offshore wind. While it is unlikely that we will see much electricity delivered from offshore wind any time soon, Norway aims to use bioenergy domestically, first and foremost as a source of space heating. In the Norwegian governments 'bioenergy strategy', targeted bioenergy production is projected to be 14 TWh by 2020 (OED 2008).

Today, almost all space heating in Norway is done by electrical panel heaters, fuelled by "green" hydropower. In addition, many households have traditional wood heating. To use energy more efficiently, the current goal is to replace as much as possible of this heating with water carried district heating, primarily based on bioenergy and waste incinerators. To reach the target, two sets of policies have been implemented. The first is the opportunity for municipalities to enforce obligatory connection to district heating grids for new property developments (KRD 2008). This has been described as important to secure a market for the high cost investment that a district heating plant is. The second policy tool is administered by the public agency Enova who manages monetary support for the establishment of heating facilities based on district heating and support for establishing smaller heating stations based on waste, biomass, or other renewables. These policies have been quite successful: Statistics Norway reports that since 1999 the consumption of district heating in Norway has increased from around 1.5–4.3 TWh in 2010. This represents around 1.6 % of net Norwegian domestic energy consumption (SSB 2011).

One explanation for this success is found in an interesting interplay with a governance tool aimed at energy and climate issues at the municipal level. Currently, all municipalities in Norway are required to make climate and energy plans. Rygg (2012) has studied the content of these plans and how they have been part of shaping a set of quite different new renewable energy activities throughout the country. The plans, she finds, have been of particular importance for bioenergy. Typically, the plans combine climate and environmental concerns with a quest for new ways of using resources, creating jobs, and developing new markets locally. Many municipalities see heating based on biomass as an opportunity. They are often actively involved in establishing the new production arrangements, either through funding, ownership of production facilities, resources, or a combination.

They could own forest, they own public buildings, and they often struggle with a lack of industrial and commercial activities (Skjølsvold 2012).

Thus, the production of municipal climate and energy plans are part of a governance regime related to the climate change and to energy transitions. However, they can also be considered as a tool of governance in a separate domain; the historically important Norwegian "district policies". For decades this has been a central doctrine stressing a distributed population pattern of the entire country. The general demographic trend, however, sees people moving out of rural areas and into cities. Bioenergy is often framed as a new way to generate activity in struggling areas.

The success of the Norwegian efforts to govern toward implementation of bioenergy could be contested. The numbers for district heating indicate that there has been an increase in the production of this form of energy over the past years. However, *research on the practice of individual companies suggests that they very seldom are competitive.* It is very difficult to thrive as a Norwegian bioenergy entrepreneur. Some argue that this is related to the strong emphasis on R&D in Norwegian policy. While R&D provides opportunities for developing technology, implementation still lags due to relatively high cost. Ole I. Gjerald (2012) have studied the links between government support schemes and renewable energy producers. Many of the energy producers who work with bioenergy observe a mismatch between policies aimed at increasing energy production and policies aimed at innovation. Energy producers find it difficult to invest in new energy generation technologies in competition with the strong hydropower returns. As a result, most bioenergy companies are struggling. Those who manage to remain in business tend to highlight that they do so in spite of government policies rather than *because* of them, or that they do so because they are able to work on an idealistic nonprofit basis (see, e.g., Skjølsvold 2012; Gjerald 2012). Another issue is that because producing electricity from biomass is noncompetitive compared to hydropower in the Norwegian setting, many bioenergy producers only produce heat, and not electricity. Thus, their renewable energy production is not eligible for electricity certificates. Many see this as discrimination of bioenergy technology.

4 Norwegian Governance Strategies Toward Renewable Energy Technologies

Our discussion suggests that Norwegian renewable energy governance is multi-faceted. It is influenced by the Norwegian situation in terms of access to natural resources and social organization. However, two trends are clear. The governance of renewable energy rests on two principles: technology development and a strong market belief.

We have argued that Norwegian governance of renewable energy technologies can be linked to:

- *Energy economization.* Has been the most salient line in Norwegian energy policy. It implies economically optimal energy production and consumption. In Norway this means using hydropower to produce electricity on the domestic market, while exporting oil and gas to foreign countries.
- *R&D policy.* Efforts to tackle climate change issues have been seen as a technology challenge, with the funding of research centers as the main tool to meet the challenge and cost targets.
- *Industry development.* The Norwegian market for new renewable energy has been small. Norwegian governance is emphasizing the commercialization of Norwegian technology and industry in international markets.
- *District policy.* Renewable energy is seen as a tool to mitigate changing demographic and economic trends in rural Norway.
- *Energy policy.* Some energy policies are geared toward energy transitions. The district heating policies are one example, the implementation of an electricity certificate market another.
- *Climate policy.* The link between Norwegian renewable energy policies and Norwegian climate policies are largely rhetorical. Why should renewables be added in already "green" Norway? However, the climate profile of Norwegian renewable energy efforts can become clearer if the Norwegian energy market becomes more tightly integrated with the Swedish and the European markets, if these markets are able to maintain renewable energy growth by utilizing Norwegian hydropower as the balancing battery to mitigate the negative grid effects of intermittent solar and wind power generation.

The implementation of renewable energy governance across sectors and different issues might be problematic, but is also symptomatic of support being offered indirectly to create commercial opportunities to fit current markets. While many actors see a potential in renewable energy, we observe that a grand narrative regarding what role renewable energy really should play in Norway is missing. Are there any signs of such a narrative emerging?

The implementation of new intermittent renewable energy production in Europe creates a demand for balancing power so that the system can provide security of supply. Some argue that Norway, with its hydropower reservoirs, could take a role as a "green battery" for Europe. Hydropower can be switched on and off quickly. The basic idea, then, is that the Norwegian hydropower reserves can provide balancing power for Europe. When the wind calms down in Germany, Norwegian hydropower could be the needed renewable backup (See Karlstrøm 2012b). Norwegian importers might benefit from cheap power during a surplus of wind and solar power in Europe. The realization of such an idea will require massive investments in transmission cables and domestic Norwegian cabling. There are currently two new cables in planning, both to be financed by Statnett and eventually with trading profits from the transport duties.

5 Discussion and Concluding Remarks

In this chapter we have studied the Norwegian governance toward an energy system where new renewable energy might hold a stronger position than it does today. These efforts are anchored in two strategies. Norwegian energy policy has historically been rooted in ideas from economic theory. "Energy economization" is an example of this—energy should be produced and used in the most profitable way. With the early liberalization of the Norwegian electricity market, the integration of economy and energy became strengthened. Today, the commercial market is expected to bear the brunt of the transition. A problem, though, is that the prices of electricity in Norway are low. The implementation of a market for electricity certificates aims to mitigate this problem. However, we do not know at this point whether the certificates will be strong enough to attract investors.

The funding of R&D to improve technologies might benefit the research communities, but it is still uncertain to what degree the R&D effort will actually lead to development, commercialization, and implementation of more renewable energy generation in Norway or if the R&D will prove helpful in establishing an industry.

So far, combining the commercial view on generation additions with R&D funding has not led to much further renewable energy generation capacity. Some claim that Norwegian renewable energy governance is characterized by a lack of tools for policy intervention. From decades of literature on innovation we know that linking R&D to commercialization is difficult (e.g., Rosenberg 1994). While Norwegian R&D policies hint at a linear relationship, the same innovation literature warns us that a strong R&D sector is no guarantee for market penetration of products and technologies. Thus, the major challenge today is related to commercialization and implementation of new renewable energy technologies.

How can the Norwegian renewable energy governance be strengthened? We believe that further public support of added generation would be of benefit, in particularly in those technologies that are supported by the R&D efforts. Inspiration for how to do this could be found in literature on the multilevel perspective (MLP, see e.g., Geels 2004), advocating for strategies of establishing niches, and/or arenas for learning. Such arenas have been called for by Norwegian entrepreneurs, for instance in the offshore wind industry. While this industry largely aims to cater for markets abroad, such as in the UK, many emphasize that there is a need for a domestic market that gives room for practical learning.

In addition to creating arenas and spaces to mitigate the market risks, there is evidence that successful transitions presuppose positive public engagement with the new technologies. Large-scale transformations of the energy infrastructure affect the public in new ways, and often create new types of political friction that can be difficult to bypass without an active strategy of public engagement. Typically, such challenges emerge around questions of placement of specific developments, debates over land use, the relationship between technology and ecological degradation as well as the development of price. Generally "green" technologies tend to be packed with controversy (see e.g., Skjølsvold 2012). Thus, governance strategies for implementing new renewable energy should also consider the need for positive public engagement with the new technologies. This, in turn relies on public evaluation of new renewable energy production. Public acceptance is vital in shaping the widespread implementation of renewable energy and the achievement of energy policy targets.

An added effect could be an improvement of renewable energy standings in markets. If outstanding and sublime perceptions of the technologies could be achieved, some studies indicate that market participants could be willing to pay a premium, thus enabling competitiveness (Karlstrøm 2010). While the importance of public acceptance, public engagement, and public understanding has long been recognized in the scholarly literature (e.g., Devine-Wright 2009), we see little reflecting this in Norwegian renewable energy governance.

Today, the Norwegian public participates in the transition processes, *but only in its capacity as market participants*. As market participants, they are—not surprisingly—most interested in the price of the final product. In the Norwegian setting the result has so far been a nontransition. So far this has not been very problematic, since Norway has been well catered for through hydropower. Now, however, as Norway intends to expand its production capacity as well as to make its transportation sector more renewable, the image looks different.

References

Adomaitis N (2012) Norway's ambitious plans to boost wind power output to cut greenhouse gas emissions and diversify supplies are at risk because of low power prices and an underfunded subsidy scheme, industry sources say. Reuters

Aune M (2007) Energy comes home. Energy Policy 35(11):5457–5465. doi:10.1016/j.enpol.2007.05.007

Aune M, Ryghaug M, Godbolt ÅL (2011) Comfort, consciousness and costs—transitions in Norwegian energy culture 1991–2010. Energy efficiency first: the foundation of a low-carbon society. In: Proceedings of the ECEEE 2011 Summer Study

CenBio (2012) Cenbio—bioenergy innovation centre. Available via http://www.sintef.no/Projectweb/CENBIO/

Devine-Wright P (2009) Rethinking NIMBYism: the role of place attachment and place identity in explaining place-protective action. J Comm Appl Soc Psychol 19(6):426–441. doi:10.1002/casp.1004

Geels FW (2004) From sectoral systems of innovation to socio-technical systems: insights about dynamics and change from sociology and institutional theory. Res Policy 33(6–7):897–920. doi:10.1016/j.respol.2004.01.015

Gjerald OI (2012) Fornybare strategiar? Innovasjon, implementering og kommersialisering for ny fornybar energi. Vestlandsforskning, Sogndal

Hansen GH, Steen M (2011) Vindkraft til havs. Teknologi-og industriutvikling fra et norsk bedriftsperspektiv. Trondheim, CenSes

Hanson J (2011) Innmatingstariffer—marked som læringsarena. In: Hanson J, Kasa, S, Wicken O (eds) Energirikdommens paradokser. Innovasjon som klimapolitikk og læringsutvikling. Oslo, Universitetsforlaget

Karlstrøm H (2010) Den deregularte forbruker. Norske strømforbrukeres holdninger til strømmarkedet, fornybar energi og bærekraftig kraftproduksjon. NTNU, institutt for tverrfaglige, kulturstudier

Karlstrøm H (2012a) Empowering markets?: the construction and maintenance of a deregulated market for electricity in Norway, vol 79. Trondheim, Norges teknisk-naturvitenskapelige universitet

Karlstrøm H (2012b) Norge—Europas Grønne Batteri. Manifest tidsskrift

Kasa S (2011) Klimakamp blir innovasjonspolitikk. In: Hanson J, Kasa, S, Wicken O (eds) Energirikdommens paradokser. Innovasjon som klimapolitikk og læringsutvikling. Universitetsforlaget, Oslo

KRD (2008) Ot.prp. nr. 45 Om lov om planlegging og byggesaksbehandling (plan- og bygningsloven) (byggesaksdelen). Ed. Ministry of local and regional development

NORCOWE Vision (2012) Available via http://norcowe.no/index.cfm?id=229099

NOU (2006) Norges offentlige Utredninger 2006: 18. Et klimavennlig Norge

OED (2006) St.meld. nr. 11 (2006–2007). Om støtteordningen for elektrisitetsproduksjon fra fornybare energikilder (fornybar elektrisitet). Ed. Ministry of petroleum and energy

OED (2008) Strategi for økt utbygging av bioenergi. Olje- og Energidepartementet

Rosenberg N (1994) Exploring the black box. Technology, economics and history. Cambridge University Press, New York

Rygg B (2012) Local governments' engagement with bioenergy. Paper given at Innovation and learning in energy policy: new pathways to sustainability. Trondheim, December 12–14, 2012

Ryghaug M, Sørensen KH (2009) How energy efficiency fails in the building industry. Energy Policy 37(3):984–991. doi:10.1016/j.enpol.2008.11.001

Skjølsvold TM (2012) Towards a new sociology of innovation: the case of bioenergy in Norway and Sweden, Norwegian University of Science and Technology, Faculty of Humanities, Department of Interdisciplinary Studies of Culture, Trondheim

SSB (2011) Fjernvarmestatistikk 2010. Sterk økning i fjernvarmeforbruket. Statistics Norway, Oslo. Available via http://www.ssb.no/fjernvarme/

Sørensen KH (2007) Fra "hvite kull" til grønn varme? Utfordringer for energi. In: Sørensen KH, Aune M (eds) Mellom Klima og komfort. Utfordringer for en bærekraftig teknologiutvikling. Tapir Akademisk Forlag, Trondheim

Tamnes R (1997) Oljealder. 1965–1995, Norsk utenrikspolitikks historie. Universitetsforlaget, Oslo

UNDP (2011) Human development report 2011. sustainability and equity: a better future for All

The Significance of the Environmental Communication for the Renewable Energy Governance Scenario: Who Decides for Whom?

Constantina Skanavis, Christos Giannoulis and Vassilis Skanavis

Abstract This chapter focuses on the power that citizens hold, when it comes to the decision making on wind turbines planning and implementation, in their hometown area. Basically, it covers the theoretical concepts related to the governance of renewable energy projects from the local communities' perspective and involvement. Also, it illustrates the similarities and differences in two Greek geographic locations when it comes to citizen participation in the decision-making process. The results from the two selected areas in Greece are discussed and the conclusions reveal the empowerment that environmental communication can attribute to the local element of the sites where wind turbines are to be located.

1 Overview

Industrialization and economic development have required a rapid increase in energy demand (Hernandez-Escobedo et al. 2010a, b; Tolon-Becerra et al. 2011; Manzano-Agugliaro and Carrillo-Valle 2011). The renewable energy sources (RES) are currently enjoying the majority of attention due to the environmental problems, like climate change, that have been associated with the fossil fuels (Manzano-Agugliaro et al. 2012, 2013; Hernandez-Escobedo et al. 2010a, b).

C. Skanavis (✉)
University of the Aegean, University Hill, 81100 Mytilene, Greece
e-mail: cskanav@aegean.gr

C. Giannoulis
Environmental Education and Communication Research Unit, University of the Aegean, Aegean, Greece
e-mail: cgiannoul@gmail.com

V. Skanavis
Technical University of Crete, 73100 Chania, Greece
e-mail: billskan@gmail.com

E. Michalena and J. M. Hills (eds.), *Renewable Energy Governance*, Lecture Notes in Energy 23, DOI: 10.1007/978-1-4471-5595-9_21,

Keeping the reduction in greenhouse emissions and climate change in mind, the dependence on fossil fuels has been less promising in recent times (Charters 2001; Jagoda et al. 2011; Cansuno et al. 2010; Dijkman and Benders 2010).

Several authors define renewable energy sources (RESs) as unpolluted supplies of energy. Efficient utilization of these resources reduces impacts on the environment, generates least amount of secondary wastes, and can respond efficiently to both present and future economic needs (Panwar et al. 2011). Fourteen percent of the global energy needs is derived from RES, but still many RESs are presently under-utilized due to social factors. One growing trend is for consumers to make arrangements to directly control their energy systems (instead of the conventional management) at the level of the individual customer or by regional/national utilities (St. Denis and Parker 2009). As a result, the influence of education and awareness can play a role in establishing a wide community acceptance of RE technologies and their successful and efficient use.

The extent to which state guidelines have been fruitful in implementing renewable energy technologies differs, with wind energy probably emerging as the most impressive development in certain countries. However, with the number of wind turbines increasing, it has been realized that the potential obstacle to the accomplishment of set goals is the social recognition of the value of RE. This issue was largely ignored in the 1980s when policy programs began. Studies, however, were already indicating that stakeholder and community support cannot be taken for granted (Wustenhagena et al. 2007).

This is the point where the NIMBY-ism (Not In My Back Yard phenomenon) develops, with people debating the disparity between universal recognition of the value of RE and opposition to particular schemes. This is attributed to the fact that people back renewable energy technologies but only if they are not installed on their own patch. Others interpret this reaction as a generalization of people's hidden agendas. Factors that influence the community's acceptance have to do with distributional justice (how is profit and loss split?), procedural justice (do all important stakeholders have a chance to partake?) and (does the regional society have faith in the information and motives of shareholders and players beyond the region?) (Gross 2007; Huijts et al. 2007).

Having faith in the objectives, approach, and the capability of the shareholders, and facility owners becomes an issue, particularly when they do not belong to the community. Trust between citizens, investors of RE and representatives of local governance is fragile. It takes a long time to be established and it can be broken quickly if the RE investment process deviates from the announced plan (Slovic 1993). The trust of the locals must be earned through specific approaches from potential developers of wind power projects (Aitken 2010).

2 The Headache of Wind Power

Electricity generation using wind energy is a mature, viable, and nearly pollution-free process, commonly used in various parts of the world (Balat 2009). Electricity production from wind is intermittent and is supposed to be a zero-cost fuel. Several billion barrels of oil can be saved and many million tons of carbon and other discharges can be avoided by electricity generation through this energy source (Thomas and Urquhart 1996). According to opinions, energy services derived from RESs emit practically no air pollutants and greenhouse gases, improve energy supply dependability and organic fuel economy, resolve issues of indigenous energy and water supply, increase living standards and employment rates of the local inhabitants, and fulfill agreements on environmental protection (Zakhidov 2008). But how are all these put down in practice in the local community's agenda? How can those statements become the basis for reasonable decisions?

In many countries the development of wind energy infrastructure has been an uphill battle, including Greece (Wolsink 2007). Among the most important but quite debatable public fears around a wind power plan are the concerns about the scenic values and the visual impact which can affect the landscape values. One study revealed that the reactions of the respondents were influenced more by the physical characteristics of the terrain and wind turbines than the sociodemographic and attitudinal factors. The respondents were extremely sensitive to wind turbines being installed at picturesque locations. Their acceptance was better when the number of turbines were restricted within an area and kept away from inspection points, such as communes, transportation infrastructure, and viewpoints (Molnarova et al. 2012; Kaldellis 2006; Frantal and Kunc 2010; Bishop and Miller 2005).

Respondents living near wind turbines, and educated people tended to be more judicious of the installation of wind turbines in scenic locations. (Molnarova et al. 2012). Detractors protest at the alteration of natural landscapes into "landscapes of power production." If people see more than five on-land turbines, they are less positive (Ladenburg and Jens-Olav Dahlgaard 2012). The site of the wind power development creates resistance as well, especially for offshore wind farms (Ladenburg, 2008, 2010; Kaldellis 2005; Gee 2010). Critics further assert that the landscape impacts of wind farms will hurt tourism, result in noise pollution, influence property prices, and harm flora and fauna. Communities are scared with the thought that wind power schemes could reduce local incomes by having a negative impact on tourism. Another reported consideration has been the light emissions from aircraft obstruction markings, which could produce discomfort or even considerable irritation to dwellers near wind farms (Fisher 2005).

Then the opposite of NIMBY, known as PIMBY (Please in my Backyard) has been identified as a possible reaction. PIMBY springs out when wind turbines are cited as a cause of income (Joberta et al. 2007).

It has been suggested that offshore wind parks are generally favored over onshore parks and have less chance of causing public disagreement (Soderholm, and Pettersen 2007).

Some planners of wind power believe that citizens' opposition is based on misconceptions. Therefore, according to their hypothesis, people oppose because they are ignorant or misinformed or that opposition to wind power is deviant (Aitken 2010). Having this in mind, public response toward wind farm developments could take a turn for the positive, if a change in development models toward community ownership occurred (Pasqualetti et al. 2002).

3 Citizen Participation in the Wind Power Decision Making

3.1 The Public and the Wind Turbines

Residents willingly participate in a societal movement when they observe that their way of life is endangered (Skanavis et al. 2005a). Social acceptance is based on variables that deal with geography, visual impact, ownership, local economy (role of tourism), information, and participation.

The concept related to faith and justice is not simple. The manner in which it may possibly be approached is not explicit. Justice can be distinguished in relation to either procedure or results and this can create further confusion. The advantages are usually not distributed among locals. The advantages are experienced by the general public on the promotion of an improved environmental quality, exploitation of a variety of energy resources, and conformity of international treaties. The price though is paid locally (Devlin 2005).

The fears of the public with regard to visual impacts of wind turbines on the landscape, and planning information pertaining to visual effects of future construction developments are particularly significant in the case of wind farm development. A question that needs to be answered is: *What would it look like?*

If residents oppose to the wind power installation then this results in crippling planning delays and significant stress for both the local communities and the developers. The planning process is facilitated if citizen involvement is increased and information is made readily accessible, which subsequently results in better rates of participation and improved decision making (Berry and Higgs 2011).

3.2 The Environmental Communication as a Tool Toward Social Acceptance

The main goal of suitable environmental communication is that residents build a conscientious environmental approach. Environmental consciousness is a significant factor in the formation of a responsible environmental behavior. The mass media can fortify the extent of environmental consciousness of the community via release of environmental information (Skanavis and Sakellari 2007).

There is real promise for the use of inventive digital visualization and communication methods to augment current practices of visual information distribution and public involvement. Public involvement can be improved significantly through Web-based GIS. Bad communication can always result in issues, but the salient issue is always why there is bad communication. It should not be assumed that opposition to wind power is irrational. It is of extreme importance to understand the social context if we do wish renewable energy to be established. Nevertheless, debates on the role of "community" in local wind power development and governance remain a key factor to successful implementation of renewable energy resources. Concerns must be accommodated for conditional supporters and skeptics not to become antiwind. *Visual examination of the influence of the wind power seems to be a powerful tool.*

Direct impacts on local residents must be assessed. Contingent valuation (CV) is a commonly employed procedure within cost benefit study and it produces approximations of either an individual's willingness to pay (WTP) for a betterment in the quality or quantity of a certain environmental product or to avert environmental decline or of their willingness to accept (WTA) rewards for a decline of environmental quality or to give up an environmental improvement (Hanley and Nevin 1999).

Also a point to be considered by local communities is the safeguarding of the value of their area in time to come. Popular tourist destinations, due to increased energy demands, are vulnerable to environmental degradation resulting from overuse of fossil fuels, and this in turn can affect their potential to remain a favorite destination (Skanavis and Sakellari 2011).

4 Greece and Wind Energy

Greece objectives' of 2020 (i.e., 20 % of the national gross energy use and 20 % of the national gross electricity use should be covered by RESs) demand a large-scale incorporation of RES appliances. However, accessibility of appropriate sites for the installation of those technologies is questioned. The fact that most of the sites which are able to support RES infrastructure have already been allocated does not seem to be the only problem as the community resistance frequently poses further stumbling blocks (Kaldellis et al. 2012).

The development of civilization in the Greek islands was based on the power of the wind. The local populations had developed ways to enable their everyday life demands based on the advantages the powerful winds provided them with. A common application that was taking advantage of the exploitation of wind power was the traditional windmills. In the Aegean islands the sea wind intensities of 8–10 Beauforts are a familiar fact of life. Greece, particularly in Aegean Archipelago and the mainland shores, possesses some of the best wind potential sites in the world. Also it should be noted that over 50 % of the CO_2 emissions in the country come from electricity production, which underlines the importance of use of wind power (Fragoulis 1994).

Furthermore, the reliance on imported fuel (nearly 70 % of its domestic energy use is imported), results in substantial exchange loss, particularly with countries outside the E.U. (Kaldellis 2005).

The rapid installation of a large number of huge wind turbines in a few areas resulted not only in stern backlash from the locals but also led to termination of new RE installation projects in other places A good example is Southern Euboea with wind farms which pushed the operational capacity of the area to nearly 50 % of the total wind capacity in Greece. Euboea was perceived as an example to be avoided, because of the development of a skeptical outlook toward new wind parks, largely as a result of extraordinary wind power concentration in a condense time frame. This scenario sent damaging implications to all over the country. On top it has been seen in several cases, in Greece, that investors and individuals trying to introduce wind energy in a particular area were met with unwelcoming and aggressive responses from the locals who were not well informed in advance, or were not invited to participate in the project planning or they were feeling that they were giving their land for free. There were also reported instances where locals were motivated by local authorities (Kaldellis 2005). It has also happened that sometimes a small group would stoutly oppose to wind turbines installation, for reasons of personal financial interest.

A few years ago, research was carried out in various areas of Greece to assess the local approval of wind parks (Kaldellis 2006). This research indicated that the 70 % of the locals approve the existing wind parks, while 20 % of them disapprove, largely due to visual effects and the noise. The visual effects differ according to various attributes of the wind parks such as the size, color, and the type of the wind turbines, the distance from residences, the type of landscape, the time passed from their opening date, etc. (Tsoutsos et al. 2009).

Another problem faced in Greece is that there is a lack of individuals who are specifically trained to interpret the environmental and technological information, in a way that could be fully understood by the average citizen. Environmental interpretation analysis is just starting to develop as an academic field. There are no nature manuals or particular maintenance objectives, and no expert preparation for nonformal environmental educators and/or analysts. The art of environmental analysis as a type of environmental teaching, incorporated in planned communication in regions with touchy issues for the local environment plays a tremendous role for the successful implementation of innovative projects in Greece (Skanavis and Giannoulis 2009).

5 The Cases of Skyros Island and Pindos Region

Skyros, located in Central Aegean Sea, is the biggest island of the Northern Sporades complex. It is though considered a small Greek island which appeals to investors for wind energy appliances due to the admirable wind potential available.

The island peaks at almost 4 MW the year in energy demand and shows a yearly energy use of 15 GWh (Kaldelis 2011). Pindos is an inland mountain range that is located in northern Greece, which has also excellent wind conditions that make this area desirable for wind power installation as well.

Domestic opposition to large-scale wind power ventures has been intense in Skyros and quite strong in Pindos area as well. Particular attention on the issue has been given in Skyros, where the local and state media has reported the matter extensively (Giannarou 2007; Dimitropoulos and Kontoleon 2009).

The two regions were chosen because they have regional heterogeneity, exhibit geographical and cultural differences, and of course they both have strong wind potential.

Skyros is quite well known for the endangered species, the Skyrian horse. Pindos has unique forests of black pine and beech, and in the higher parts of the area Bosnian Pine (*Pinus leucodermis*) is widespread. The region is a refuge for bears, wild cats, and lynxes. Both areas have substantial biodiversity value, a natural breathtaking beauty, and landscapes that are characterized as protected environments.

In an attempt to investigate citizens' willingness to accept wind power installation, a search of what has been written in local and national news has been done. Also, a visit at the town halls revealed what has been experienced, when residents reacted to the power energy potential in their geographic area. The similarities were striking: Independently from which area they were living in, residents, considered wind power as a pollution-free energy source for electricity production, with minimal effects on the surroundings, nature and human life. The more serious impacts considered, were those on the landscape and the ones that concerned noise emanation. They definitely stressed that impacts on humans and environment caused by traditional electricity generation methods, like thermal or nuclear power plants, were by far more serious when compared to the ones resulting from wind parks. Still, it was stated that the wind turbines are visible and therefore the influence on the landscape is not questionable.

Addressing all the above reactions, the question still remains-*whether wind power installation can be acknowledged or not.* The answer to this question from some researchers seems to be that it is *rather biased*, for it relies on how the issue of power energy is confronted (Thayer and Freeman 1978). Wind turbines which do not function in low wind speeds, they do not produce any sound, whereas in the case of wind speeds in excess of 8 m/s, the noise emanating is overshadowed by the sounds produced by the wind itself (tree leaves, etc.) (Pedersen and Larsman 2008). Therefore, noise emission is not a problem unless someone wants to underline it as one. Locals are likely to be against proposed projects if they think they are not sufficiently informed and knowledgeable about wind energy and wind parks. Skyros has experienced by far more reactions probably because this power energy potential has been on the local agenda for few years, while in Pindos it is a quite recent topic of discussion for the reasons below.

In Skyros, some years ago, two power turbines were installed at a quite visible landscape location and then were left broken and abandoned for many years after they have stopped operating. This has left the residents quite bitter and resistant to accept another source of ugliness. Further, people at Skyros demonstrated a higher willingness to accept compensation for the installation of wind power (Dimitropoulos and Kontoleon 2009). Until today, the Skyrian community is alerted against wind energy installations and does still split into two groups with the greater part of residents perceiving the wind power scheme as an unjust one.

In Pindos there is mistrust as far as the wind power installation is concerned, but this does not seem to concern locals to such a level in order to actively participate against it.

Contribution of the residents is of paramount importance in environmental decision making, if a good quality of life is to be acquired. Local groups usually know all alternatives for suitable management of their region. The results of a study show that participants would like to have access to an appropriate non-formal/informal environmental education that will help them in increasing their knowledge and obtaining skills for critical thinking and vigorous contribution (Skanavis et al. 2005b).

6 Discussion and Conclusions

The promotion of renewable energy technologies among municipalities requires a special attention to be placed on local dynamics, as local dynamics are extremely important in such innovative projects (Michalena and Angeon 2009). A lot of effort should be directed on how local societies could best accept the implementation of wind power projects. The dissemination of the results from success stories implemented elsewhere should be high in the communication agenda between governmental/local authorities and the communities. Through this effort, it is important that “hidden barriers,” problems, and promising perspectives surrounding the use of wind energy should be directly addressed.

Toward this end, embedding accountability in every phase of citizens’ participation should be a top ranked priority. Residents need also to be introduced to tactics related to avoidance of insufficient maintenance or dealing with the negligence to replace the wind turbines after their lifetime ends. These issues, once discussed, would empower local communities and give them the confidence that they are in control of their environmental destiny. Surprisingly enough, although it seems that these barriers are well known, the RE planners seem to act as those issues are not worthy to be addressed and somehow the barriers would disappear…The saddest story is the one, where the professionals decide to proceed to the implementation of a wind project ignoring the lessons that research on civic education have delivered.

Negative local feelings that have to do with unfairness or threats on the area’s biodiversity and the endangered species are sensed and lead to negative consequences. This is especially true for countries that face economic struggle and where the major industries that could bring possible development are the tourism and maritime sectors. On the other hand, development on these business sectors would require increased electricity demand and this route mandates careful consideration for the most appropriate energy source. Not surprisingly, communities with limited vision of clear and specific short-, medium-, and long-term expectations of realistic outcomes *(what we wish to achieve)* and impacts (*what we aim to change*) are unwilling to accept technological innovations and can postpone projects for years and for no good reason, just because of inadequate information or because of prior misconceptions (unsuccessful removal of old wind turbines….).

Citizens want RESULTS, and those RESULTS should be measurable.

Highlighting achievements that lead to local upgrading must be shared with residents in order to challenge them to try to invest in wind energy.

What is only sure is that negative argumentation from the local communities would definitely create an agony to wind power investors and the inevitable instability in everyone's decisions would determine the final relationship of wind energy installations in desirable Greek locations. Prediction, in this case, is a difficult task. Taking into consideration past experiences and as many lessons the global literature has offered is the most prudent reaction RESs investors can depend on. Unfortunately, developers do not base their plans on experiences and literature. They base their acts on whether the mayor will accept their money or not and, also, on how much competition they might encounter.

Still, there are alternative ways to preserve local environments: We firmly believe that that the promotion of environmentally resilient societies is a solution and toward this end, environmental communication and relevant programs should be offered. But these programs, that aim to focus on responsible environmental behavior of citizens, *MUST* concentrate on strengthening the participants' decision making skills and critical analysis abilities. The communication programs should be tailored to the needs of each specific community, taking into consideration their unique characteristics, their values, the hierarchy of their priorities and mostly the residents' relationship with the environment.

The present educational system has failed to give away a basic understanding of energy supply options and their impact on society and environment. Locals have been caught unprepared. Education plays a central role in helping increased understanding of the technology among the masses and the formation of user belief in the technology (Jennings 2009).

The extent of public know how in relation to wind energy application, the public understanding regarding environmental effects of wind energy and the public outlook toward current and proposed wind parks, in line with the NIMBY syndrome are concerns that need to be identified *before* an environmental communication program is launched. Also, before an education program is launched, it should be identified why the locals disapprove the wind power implementation and which specific issues this resistance is associated with. Only when these findings are fully analyzed, one could reveal the reasons that frequently motivate residents to accept (or not) wind plans in their area. If all these addressed points are incorporated in an educational program, it is very probable that a mutual agreement would be reached.

Finally, bringing environmental and gender concerns together is a relatively recent phenomenon in the United Nation's history. Sustainable development can only be achieved if women participate completely in environmental management and development. This does not often happen in rural areas, and this should be taken into consideration when the educational programs are designed (Skanavis and Sakellari 2012). However, there is still a long way from society's *truthiness*[1] to actual implementation of embracing gender equality in environmental decision making procedures.

Finally, it is recommended that the educational and communication responsibility is based on scientific/academic research and placed under the care/auspices of professional environmental interpretation experts, like the ones in the Environmental Education and Communication Research Unit of the University of Aegean. The environmental interpreters can back up the role of administrative managers in regions where wind energy projects are to be implemented. Keeping in mind the particular social peculiarities and the uniqueness of each biophysical environment, the experts would make the wind power approach in Greece and around EU a success story.

References

Aitken M (2010) Why we still don't understand the social aspects of wind power: a critique of key assumptions within the literature. Energy Policy 38:1834–1844

Balat M (2009) Review of modern wind turbine technology. Energy Sources Part A Recovery Utilization Environ Eff 31(17):1561–1572

Berry R, Higgs G (2011) Web-based GIS approaches to enhance public participation in wind farm planning. Trans GIS 15(2):147–172

Bishop ID, Miller DR (2005) Visual assessment of off-shore wind turbines: the influence of distance, contrast, movement and social variables. Renew Energy 32:814–831

Cansino JM, Pablo-Romero MP, Roma′n R, Iñiguez R (2010) Tax incentives to promote green electricity: an overview of EU-27 countries. Energy Policy 38:6000–6008

Charters WWS (2001) Developing markets for renewable energy technologies. Renew Energy 22:217–222

Devlin E (2005) Factors affecting public acceptance of wind turbines in Sweden. Wind Eng 29(6):503–511

Dijkman TJ, Benders RMJ (2010) Comparison of renewable fuels based on their land use using energy densities. Renew Sustain Energy Rev 14(9):3148–3155

[1] Truthiness is a quality characterizing a "truth" that a person claims to know intuitively "from the gut" or because it "feels right" without regard to evidence, logic, intellectual examination, or facts. Meyer, Dick (December 12, 2006). "The Truth of Truthiness." CBS News. Retrieved December 14, 2006. Truthiness voted 2005 Word of the Year by American Dialect Society and for 2006 by Merriam-Webster. Truthiness (2006). In Wikipedia. Retrieved January 18, 2013, from http://en.wikipedia.org/wiki/Truthiness#cite_note-dialectsoc-7

Dimitropoulos A, Kontoleon A (2009) Assessing the determinants of local acceptability of wind-farm investment: a choice experiment in the Greek Aegean Islands. Energy Policy 37:1842–1854

Fisher RS, Harding G, Erba G, Barkley GL, Wilkins A (2005) Photic- and pattern-induced seizures: a review for the Epilepsy foundation of America working group. Epilepsia 46:1426–1441

Fragoulis AN (1994) Wind energy development and future perspectives. Renew Energy 5(I):642–649

Frantal B, Kunc J (2010) Wind turbines in tourism landscapes, Czech experience. Ann Tourism Res 38:499–519

Gee K (2010) Offshore wind power development as affected by seascape values on the German North Sea coast. Land Use Policy 27:185–194

Giannarou L (2007) Series rejects the giant wind farm. Kathimerini 04/04/2007

Gross C (2007) Community perspectives of wind energy in Australia: the application of a justice and community fairness framework to increase social acceptance. Energy Policy 35(5): 2727–2736

Hanley N, Nevin C (1999) Appraising renewable energy developments in remote communities: the case of the North Assynt Estate, Scotland. Energy Policy 27:527–547

Hernandez-Escobedo Q, Manzano-Agugliaro F, Zapata-Sierra A (2010a) The wind power of Mexico. Renew Sustain Energy Rev 14(9):2830–2840

Hernandez-Escobedo Q, Manzano-Agugliaro F, Gazquez-Parra JA, Zapata- Sierra A (2010) Is the wind a periodical phenomenon? The case of Mexico. Renew Sustain Energy Rev 15(1): 721–728

Huijts NMA, Midden, CJH, Meijnders AL (2007) Social acceptance of carbon dioxide storage. Energy Policy 35(5):2780–2789

Jagoda K, Lonseth R, Lonseth A, Jackman T (2011) Development and commercialization of renewable energy technologies in Canada: an innovation system perspective. Renew Energy 36:1266–1271

Jennings P (2009) New directions in renewable energy education. Renew Energy 34:435–439

Joberta A, Laborgneb P, Mimler S (2007) Local acceptance of wind energy: factors of success identified in French and German case studies. Energy Policy 35:2751–2760

Kaldellis JK (2005) Social attitude towards wind energy applications in Greece. Energy Policy 33(5):595–602

Kaldellis JK (2006) Evaluation of Greek wind parks visual impact: public attitude and experts' opinion. Fresen Environ Bull 15(11):1419–1426

Kaldelis JK (2011) Critical evaluation of financial supporting schemes for wind-based projects: case study Greece. Energy Policy 39:2490–2500

Kaldellis JK, Kapsali M, Katsanou Ev (2012) Renewable energy applications in Greece—what is the public attitude? Energy Policy 42:37–48

Ladenburg J (2008) Attitudes towards on-land and offshore wind power development in Denmark; choice of development strategy. Renew Energy 33:111–118

Ladenburg J (2010) Attitudes towards offshore wind farms—the role of beach visits on attitude and demographic and attitude relations. Energy Policy 38:1297–1304

Ladenburg J, Jens-Olav Dahlgaard JO (2012) Attitudes, threshold levels and cumulative effects of the daily wind-turbine encounters. Appl Energy 98:40–46

Manzano-Agugliaro F, Carrillo-Valle J (2011) Location of air in-leakage in power plants condensers by helium test. DYNA 86(2):173–181

Manzano-Agugliaro F, Sanchez-Muros MJ, Barroso FG, Martı′nez-Sanchez A, Rojo S, Pe′ rez-Baño′n C (2012) Insects for biodiesel production. Renew Sustain Energy Rev 16(6): 3744–3753

Manzano-Agugliaro F, Alcalyde A, Montoya FG, Zapata-Sierra A, Gil C (2013) Scientific production of renewable energies worldwide: an overview. Renew Sustain Energy Rev 18:134–143

Michalena E, Angeon V (2009) Local challenges in the promotion of renewable energy sources: the case of Crete. Energy Policy 37:2016–2018

Molnarova K, Sklenicka P, Stiborek J, Svobodova K, Salek M, Brabec E (2012) Visual preferences for wind turbines: location, numbers and respondent characteristics. Appl Energy 92:269–278

Panwar N, Kaushik SC, Kothari S (2011) Role of renewable energy sources in environmental protection: a review. Renew Sustain Energy Rev 15(3):1513–1524

Pasqualetti MJ, Gipe P, Righter RW (2002) A landscape of power. In: Pasqualetti MJ, Gipe P, Righter RW (eds.) Wind power in view: energy landscapes in a CrowdedWorld. Academic Press, San Diego, pp 3–16

Pedersen E, Larsman P (2008) The impact of visual factors on noise annoyance among people living in the vicinity of wind turbines. J Environ Psychol 28:379–389

Skanavis C, Koumouris G, Petreniti V (2005a) Public participation mechanisms in environmental disasters. Environ Manag 35(6):821–837

Skanavis C, Sakellari M, Petreniti V (2005b) The potential of free-choice learning for environmental participation in Greece. Environ Educ Res 11(3):321–333

Skanavis C, Sakellari M (2007) Assessment of environmental intention of journalists. Appl Environ Educ Commun 6(3):233–240

Skanavis C, Giannoulis C (2009) A training model for environmental educators and interpreters employed in Greek protected areas and ecotourism settings. Int J Sustain Dev World Ecol 16(3):164–176

Skanavis C, Sakellari M (2011) International tourism, domestic tourism and environmental change: environmental education can find the balance. TOURISMOS Int Multi J Tourism 6(1):233–249

Skanavis C, Sakellari M (2012) Free-choice learning suited to women's participation needs in environmental decision-making processes. Environ Educ Res 8(1):1–17

Slovic P (1993) Perceived risk, trust and democracy. Risk Anal 13:675–682

Soderholm P, Ek K, Pettersen M (2007) Wind power deployment in Sweden: global policies and local obstacles. Renew Sustain Energy Rev 11:365–400

St. Denis G, Parker P (2009) Community energy planning in Canada: the role of renewable energy. Renew Sustain Energy Rev 13:2088–2095

Thayer RL, Freeman CM (1978) Altamont: public perceptions of a wind energy landscape. Landscape Urban Plann 14:379–398

Thomas BG, Urquhart J (1996) Energy for the 1990s and beyond. J Energy Convers Manag 37:1741–1752

Tolon-Becerra A, Lastra-Bravo X, Bienvenido-Ba′rcena F (2011) Proposal for territorial distribution of the EU 2020 political renewable energy goal. Renew Energy 36(8):2067–2077

Tsoutsos T, Tsouchlaraki A, Tsiropoulos M, Serpetsidakis M (2009) Visual impact evaluation of a wind park in a Greek island. Appl Energy 86:546–553

Wolsink M (2007) Planning of renewable schemes: deliberative and fair decision-making on landscape issues instead of reproachful accusations of non-cooperation. Energy Policy 35:2692–2704

Wustenhagena R, Wolsink M, Burer MJ (2007) Social acceptance of renewable energy innovation: an introduction to the concept. Energy Policy 35:2683–2691

Zakhidov RA (2008) Central Asian countries energy system and role of renewable energy sources. Appl Solar Energy 44(3):218–223

The Political-Economics of the Green Industrial Revolution: Renewable Energy as the Key to National Sustainable Communities

Woodrow W. Clark II and Xing Li

Abstract A Green Industrial Revolution (GIR) consists of renewable energy, smart green sustainable communities, water, and waste along with advanced technologies that are cost-effective in implementing it. The GIR started in China and the EU, but has taken the USA by surprise. The EU, South Korea, and Japan had started a GIR over two decades ago. The GIR can be reflected in the significant paradigm change from the fossil fuels and nuclear power plants of the Second Industrial Revolution (2IR), which had dominated global economics since the late 1890s, to renewable energy in the late 1990s, which has grown, and is growing at an extraordinarily rapid rate into the twenty-first century. While the US had invested and even began to commercialize some of the technologies developed into mass markets by the EU and Japan, it failed in the last two decades to move ahead of corporate vested interests in maintaining the 2IR in large part due to ignoring and even politicizing the science about climate change being caused by humankind. The problem today in the GIR is that the economic stimuli that helped the 2IR emerge into being dominant around the world was based on government basic economic aid such as land grants (oil drilling), mass transportation (rail roads to transport coal and highways today) infrastructures, tax breaks that are still in place today from over a 100 years ago, and finally the entire auto industry that was once hybrid, electric, and even agriculture oil juice-based (Henry Ford, as the farmer) to dependency on fossil fuels especially oil and now natural gas. The GIR needs these same kinds of government financial support mechanisms. The United States of America (USA) needs to make these significant economic paradigm changes now to become a part of the GIR.

W. W. Clark II (✉)
Cross Disciplinary Scholars in Science and Technology (CSST), UCLA, Los Angeles, USA
e-mail: wwclark13@gmail.com

X. Li
Department of Culture and Global Studies, Aalborg University, Aalborg, Denmark
e-mail: xing@cgs.aau.dk

E. Michalena and J. M. Hills (eds.), *Renewable Energy Governance*,
Lecture Notes in Energy 23, DOI: 10.1007/978-1-4471-5595-9_22,

1 Introduction: Cases of Corporate and Business Influences

Corporate interests and impacts on public policy are extremely significant around the world. In the USA, the Supreme Court ruled in 2012, about 6 months before the USA national presidential and Congressional election that anyone (defined now as even companies and corporations) could contribute any amount of money to people running for election for any office. The USA Supreme court ruled 2 years before, that an individual or group could contribute any sum of money from any source to any other person or group since it was protected by the USA Constitution under "free speech." This put the USA national election in the hands of the wealthy and corporate interests. In total, over USD $1 billion was put into the presidential and Congressional elections. About 70–80 % of the money went to Republican candidates. The results were different from what was hoped by these wealthy people and even predicted because the American public voted to reelect the Democrats, with President Obama, and therefore increase the Democratic majority in the USA Senate and increase its minority position in the USA House of Representatives.

The future impact of money focused on certain locations and candidates is uncertain. In particular, the USA national and local elections, targeted specific funds toward candidates who would support traditional 2IR energy resources. Funds were targeted against those who supported controls and environmental restrictions on fossil fuels, lower greenhouse gases, eliminate pollution, and who supported environmental regulatory programs or increase in renewable energy systems. One case in point with the USA National election concerned the Obama administration's financial support of a solar energy company for over USD $500 million in USA government support of debt loans. In the year before the national election and less than 2 years after receiving the USA debt funds, the company declared bankruptcy. There was no investigation into the solar company, owners or its supporters, some of whom were associated with the Obama administration through the USA Department of Energy.

The USA national government is not alone. Such cases arise and are common in developing countries, but also in other western developed countries. For example, California at the local level, where the public votes were for funds to rebuild and modernize its schools, a superintendent and the facilities manager in a wealthy Southern California School District were both convicted and jailed for fraud for taking public money for themselves. Furthermore, a large national American corporation settled out of court for over USD $6.5 million in payments back to the school district.

This kind of problem is not new in the 2IR or to the USA. The basic problems are those people and companies who had made their wealth (and hence political power) from the 2IR for over a 100 years through fossil fuels production, distribution, and use. With that economic power, the leaders and their companies today have "deep pockets" to finance the political leaders who will continue to promote their future. The 2IR leaders, most economics, and corporations argue that economics is critical. And there is a need today to let "market forces," that is, businesses and companies compete to determine the outcomes or final results of energy supplies and their demands for use in society.

Another example of how this is not correct comes from Ukraine that wanted to increase its support of oil being piped from Russia through the Ukraine to western European nations. A former Prime Minister (2011) accepted personally 10 % of natural gas funds from Russia, which were piped through the Ukraine. She was voted out of office; went on trial and convicted; then put in jail by the new Prime Minister.

Similar situations occur all over the world.

Consider China, which had "leapfrogged" into the GIR from being dependent upon fossil fuel, which dominated its rapid growth into to the 2IR with very little regard to the climate, pollution, and urban infrastructure needs. The last Chinese Central Government Administration (ending in November 2012) had a high level official whose wife was convicted of supporting a western businessman and then killing him (this high-level official was recently sentenced to life prison). That resulted in a dramatic change in the Chinese Central Government leadership, which took office in March 2013 after the Chinese New Year. Meanwhile, China for most of the last decade at the turn of the twenty-first century grew large with developed cities and communities in the 2IR as its economy was rapidly beginning to pass the USA in becoming #1 in the world (Leung 2010).

China became then the number one carbon emissions nation in the world, overtaking the USA that held that distinction for over three decades. China over the last decade had used the 2IR to leapfrog into the GIR, but at terrible leadership and environmental costs. What can western nations learn by this? And then, how can all nations work together collaboratively to get off dependency on fossil fuels from the 2IR and into the GIR that use renewable energy to stop, mitigate, and reverse climate change?

2 RE: The Gap Between the West and the East

China, however, has developed its own form of non-western government and economics. There are some similarities of political influences between the USA and other western nations to China. Clark and Li (2004) made this point a decade ago by noting that the Chinese government used a "social capitalism" political model whereby governments are responsible for social issues including the environment and renewable energy, as well as health and retirement issues for its people. On the contrary, Western capitalism looks at what profits are best for the corporation, shareholders, and senior managers first and foremost. This economic ideology lacks both ethics and legitimate use of economics, calling into question that economics is not a science but simply a way to use numbers to justify political positions and decisions (Clark 2012).

The Chinese national governments have led Chinese modernization into the GIR by creating and providing financing through the 11th Five Year Plan. In the last 12th Five Year Plan and now in the plans for the 13th Five Year Plan, renewable energy systems, water protection, and waste recycling are key components (Lo 2011). Each Five Year plan provides clear and formulated policies along with funding to address environmental issues and their solutions through sustainable communities, all of which are key elements in the GIR (Ma et al. 2011). With the new Chinese central government in power for 5 years, the 13th Five Year Plan will be very different from plans in the past two decades.

The projections are that the new government will be far less western in its "economic and business" policies. Instead, rather than just build and expand cities and communities with marginal concern for the environmental impacts, just the opposite is predicted from the current government (CMST 2011). Even more critical is the concern about energy demand and where the supply of energy will come from and thus how China can remain energy dependent so as not to get tempted with natural gas supplies from Russia, such as what happened with the Ukraine.

Denmark (and other Nordic nations, including Germany) also has national plans that include renewable energy, water, and waste infrastructures. The Danish current plan (2012) calls for the nation to be 100 % energy independent through renewable energy by 2025 (Lund et al. 2011). Denmark, as a nation, is already 45 % today, according to some reports (Gipe 2011–2013).

Meanwhile, the USA and other nations have no such national plans, especially in energy and renewable energy in particular. Instead, the USA leaves the decisions on energy to "market forces" which are focused primarily on their past business models that are usually rooted in fossil fuels from the Second Industrial Revolution (2IR). It is no wonder that the USA remains behind in the Green Industrial Revolution (GIR). The problem with the USA in comparison to the rest of the developed and developing world has now been well documented but denied by the USA companies and politicians who continue to support the 2IR.

2.1 China has "Leapfrogged" into the Green Industrial Revolution

China is perhaps one of the few countries that regard energy security as a vital component of their *national interests* (Leung in press). Energy security has become the essential premise for China to achieve its national goal of quadrupling its gross domestic product in 2020 (Cai et al. 2008). There is a genuine consensus among Chinese leaders and scholars that energy has become a key strategic issue for China's economic development, social stability, and national security. There is also the realization that China's key national interests[1] are highly dependent on the access to sufficient energy resources. Hence China's energy demands for "fossil-fired generation of electricity grew 0.3 % last year, wind-generated electricity grew 35.5 %. Solar-generated electricity grew 414 % while nuclear generation grew 12.6 %". Trabish notes in Green Media (2013: 1) about China the surprising scale of the shift in renewable energy, taking into consideration the huge pre-existing base of traditional generation:

- Fossil-fired generation, which still accounted for 79 % of China's electricity generation in 2012, added 12 terawatt-hours for a cumulative total of 3,910.2 terawatt-hours of generation.
- Wind's 35.5 % growth meant an addition of 26 terawatt-hours of generation. But it achieved a much smaller cumulative 100.4 terawatt-hours of electricity production. Its growth did, however, put it ahead of nuclear energy's cumulative 98 terawatt-hours of generation.
- Solar reached a cumulative 3.5 terawatt-hours of electricity production.
- Hydroelectric power had the biggest total growth of generation in China, adding 196 terawatt-hours to reach 864 terawatt-hours of electricity production.

From 2000 to 2005, China's energy consumption rose by 60 %, accounting for almost half of the growth in world energy consumption. Between 2000 and 2003, China's share of the increase in global demand for aluminum, steel, nickel, and copper was, respectively, 76, 95, 99, and 100 % (Konnan and Jian 2008). Chinese energy demand has more than doubled during the past decade. China will consume about 41 % of global coal consumption and 17 % of global energy supply by 2050 (China Statistics Press 2013).

However, in the past two decades the liberal market economic policy is making the Chinese economic development highly dependent on stable supplies of energy and resources, which are primarily fossil fuels from the 2IR. When viewing the global scale, any crisis to China's access to overseas resource and maritime shipping routes will have a negative impact on China's growth and trade-dependent economy. The lesson China has learned is prompting Beijing to seriously undertake the indispensable green industrial revolution. China's new Five year plan has indicated a great leap forward with the GIR. Today China is integrating renewable energy systems with mass transportation including the use of energy storage devices such as hydrogen fuel cells (Liu 2012).

In order to avoid the mistakes of the western developed nations, China has moved ahead in a variety of infrastructure areas through its national government,

[1] There are other perspectives and viewpoints about energy and the role of renewable energy. Each of these perspectives are rooted in the economics of groups (companies) and nations who have build strong vested financial interest in fossil fuels and nuclear power. Today, there are growing vested interests in renewable energy. Even within the last two decades, fossil fuel companies started making investments in the renewable energy sector. However, they are very limited and rarely a main part of the corporate portfolio. See Brown 2008, Shields 2010 and McNall et al. 2011.

large cities, and regions (Clark and Isherwood 2007 and 2010). Yet there are problems in China as it seeks to lead globally through renewable energy and GIR. The New York Times presented several articles in March 2013, which presented how the negative impact on China's environment recently killed over 16,000 pigs, who were then floating in rivers that supply drinking water to Shanghai. On the other side, SunTech one of the world's leading solar energy companies early in 2013 had a major division (Wuxi) go bankrupt which had at its peak, a market capitalization of over USD $16 billion, but owed eight Chinese banks over USD 1.1 billion that it did not pay on time, causing the "insolvency." As Keith Bradsher reported in 2013:

> *More than any other country, China had leaned heavily on renewable energy to solve its problems of severe air pollution and dependence on energy imports from politically unstable countries in the Middle East and Africa…*
>
> *Suntech, a centerpiece of the country's efforts, had grown to 10,000 employees in its hometown, Wuxi, on China's east coast, and even set up a small factory in Arizona to assemble panels. But a tenfold expansion of Chinese solar panel manufacturing capacity from 2008 to 2012 pushed down the price of solar panels about 75 %, undermining the economics of the business* (2013; 1).

3 The Western Economic Paradigm Must Change

According to Jeremy Rifkin (2004 and 2012), the EU is in a Third Industrial Revolution (3IR). While Clark and Cook (2011 and 2013) argue that the (3IR) is really a GIR that concerns renewable energy, smart green communities, advanced technologies that produce, store, and transmit energy for infrastructures. However, with further investigation, the various industrial revolutions were not only built upon energy sources. They included massive manufacturing, building, and home structures that all needed infrastructures to service them. The challenge was to make these communities "sustainable" so as not to pollute the environment, water, and create greenhouse gases and carbon which are the root causes of climate change (Bruntland 1983).

The development of the USA into a powerful world leader had a lot to do with its military strength but also its economic development of fossil fuels for over a century in the Second Industrial Revolution (2IR) including the technologies that support them, such as combustion engines and related technologies including atom bomb and nuclear power (Chomsky 2012). And the fossil fuels continue to be there. Hence, climate change on the environment is continuing to be damaged in order to produce more oil and natural gas now through increased supplies of shale oil and "fracking" for gas, which has massive amounts of environmental problems and concerns. But this 2IR retards and places the USA back by decades when compared to emerging economics and even other western developed nations.

As historians have documented the development of the 2IR in the USA was primarily based on "state capitalism" since oil companies got land grants, funding, and even trains or pipelines for transporting their fossil fuels. That governmental support continues today. Consider the issue of the USA getting shale oil from Alberta, Canada, and the massive pipelines installed through the USA to get the oil to the USA. Furthermore, these same companies get tax breaks and credits such that their economic responsibility to the USA is minimal.

The Western paradigm much change: The USA needs to look comprehensively into the corporate and political reactions to the 2011 Japanese tsunami and ensuing nuclear power plant explosions, as well as the 2010 BP oil spill in the Gulf of Mexico off Louisiana. The USA and other countries cannot ignore the environmental consequences and economic costs of the 2IR that have handicapped it from moving into the GIR. The end result is not good for the American people, let alone the rest of the world since more pollution, emissions, and environmental consequences create abnormal and unusual intense storms (tsunamis. Hurricanes and tornadoes) destroy communities, infrastructures, and kill people in their path. Today, these climate change problems are costing America thousands of lives and billions of US Dollars.

4 Background to the Green Industrial Revolution

The Green Industrial Revolution (GIR) emerged at the end of the twentieth century due in large part to the end of the Cold War that dominated the globe since the end of World War II. The Second Industrial Revolution (2IR) had dominated the twentieth century because it was primarily based on fossil fuels and technologies that used primarily mechanical and combustion technologies. On the other hand, the GIR is one of the renewable energy power and fuel systems, and smart "green" sustainable communities that use more wireless, virtual communications, and advanced storage devices like fuel cells (Clark and Cooke 2011). The GIR is a major philosophical paradigm change in both thinking and implementation of environmentally sound technologies that requires a new and different approach to economics (Clark 2011).

The USA during the 1970s and then again in the early 1990s dismissed that humankind was responsible for climate change (UNIPCC 2000 and 2007) which became later apparent for both Democrat and Republican Presidential Administrations in their lack of pro-active polices globally through the Kyoto Accords and subsequent UN Conferences in København (December 2009) and Cancun (2010) and others. On the other hand, in the early 1990s, economic changes in Europe and Asia were made due to the end of the Cold War to meet the new global economy. The Asian and EU conversions from military and defense programs to peacetime business activities, was much smoother than that of the USA. When environmental economist, Jeremy Rifkin, recognized this change and developed the concept of a "Third Industrial Revolution" in his book *"The European Dream (2004),"* he did not recognize that Japan and South Korea had been in a GIR decades before 2011 (Clark and Li 2004; Clark and Cooke 2011).

However, Clark et al. 2006 published a paper on the "Green Hydrogen Economy" that made the distinction between "clean" and "green" technologies when related to hydrogen and other energy sources. Clean technology is used to describe fossil fuels in an environmentally friendly manner, such as "natural gas" and even "clean coal." Green, on the other hand, means specifically renewable sources such as the sun, wind, water, wave, and ocean power. In short, there is a dividing line between what technologies were part of the 2IR (that is clean technologies such as clean coal and natural gas) and the GIR of which focus is on renewable energy such as solar, wind, ocean, and wave power and geothermal as well as the integration of these technologies, and the use of storage devices like fuel cells. The GIR focused on climate change and changing the technologies and fuels that caused it; or could at least mitigate and change the negative pollution and emission problems that impacted the earth.

Clark and Fast (2008) in founding "qualitative economics" made the point about economics needing to become a science. Qualitative economics provides the scientific basis for the GIR because it focuses on the need to provide hypotheses, test them, and then repeat the tests needed to make scientific predictable conclusions. While quantitative data is important, there must also be the definition and meaning given from observations of ideas, numbers, words, symbols, and even sentences. Hence the misuse of "clean" tech as natural gas being good for the environment is incorrect. Natural gas in real scientific terms is a fossil fuel and that is not good for the environment.

4.1 Key Elements of the GIR

So far The Green Industrial Revolution impacts America on the local level rather than at the regional, state, or national levels. Infrastructures of energy, water, waste, transportation, and IT among others and how they are integrated as the core to the GIR (op.cit. Clark and Cooke 2011). These infrastructure systems need to be compatible yet integrated with one another. For example, renewable energy power generation must be used for homes, businesses, hospitals, and non-profit organizations (government, education, and others) that are metered and monitored as "smart on-site grids" and also used for the energy in vehicles, mass train, and buses among other transportation infrastructures (Knakmuhs 2011). Such "agile energy" or "flexible systems" (Clark and Bradshaw 2004) allow people to generate their own power while also being connected to a central power grid. However, both the local power and central power in the GIR need to be generated from renewable energy sources, with stand by and backup storage capacity.

There are five key basic elements for The Green Industrial Revolution: (1) Energy efficiency and conservation, (2) Renewable power generation systems, (3) Smart grid connections within sustainable communities, (4) Advanced technologies like fuel cells, flywheels, high-speed rail, and (5) Education, training, and certification of professionals and programs.

First, communities and individuals all need to conserve and be efficient in the use of energy as well as other natural resources like land, water, oceans, and the atmosphere. Second, renewable energy generated from wind, sun, ocean waves, geothermal, water, and bio-waste must be the top priority for power on-site and for central plants.

The third element is the need for smart girds on the local and regional levels in which both the monitoring and control of energy can be done in real time. Meters in buildings and homes need to establish base load on-site power use so that conservation can be done (systems put on hold or turned off, if not used) and then renewable energy power is generated when demand is needed. The fourth element needs to be advanced storage technologies such as fuel cells, batteries, regenerative brakes, and ultra-capacitors. These devices can store energy from renewable sources, like wind and solar that produced electricity intermittently, unlike the constant supply of carbon-based fuel sources. Finally, the fifth element is education and training for a workforce, entrepreneurial, and business sector that is growing and providing employment opportunities in the GIR.

5 The Role of Governance in Smart Renewable Energy Systems

In general, the GIR concerns homes, businesses, government and large office, and shopping areas, which can all, monitor their use of natural resources like energy and water. For example, communities need devices that capture unused water, and that can transform waste into energy so that they can send any excess power that is generated to other homes or neighbors. Best cases from around the world of sustainable communities that follow these elements of the GIR exist today around the world (Ma et al. 2011).

Essentially the GIR has started from governments who were concerned about the current and near-future societal impact of businesses and industries in their countries. The EU and Asian nations in particular have had long cultural and historical concerns over environmental issues. The Nordic nations, for example, have created Eco-cities as well as the reuse of waste for more than three decades. Sweden, Denmark, and Norway have all either eliminated dependency on fossil fuels now for power generation or will be in the near future. Most of the Nordic countries also have started to shut down nuclear plants, especially after Fukushima. Finland is the only country in decades that has started a nuclear power plants due to their international need for energy. Above all, Finland did not want to become dependent up Russia for energy fuels. France is the only country in the EU, which since the 1980s, has built and operated nuclear plants that now account for over 75 % of its energy supply. However, today there is increasing concern over nuclear power and more renewable energy systems being installed and operated.

6 Economics is the Problem and the Solution: Because the World is Round, not Flat

The "market force" neo-classical economic paradigm[2] represented by American and UK are economic policies put into action for over the last four decades. Prime Minister Thatcher and then President Reagan were the embodiment and champions of this economic paradigm derived from Adam Smith (Clark and Fast 2008). Much of the supply and demand methods are based on the fossil fuels used for creating energy and power for buildings and transportation. Then either the supply or demand are not balanced, then the market forces (private sector companies) must step in to provide resources for either area so that the economic balance is achieved and maintained. Market force economics had some influence on the EU and Asia, but then that 2IR economic paradigm demonstrated its failure in October 2008 with the global economic collapse that started in the US on Wall Street. That failure meant some of the government programs in the EU and Asia, which had succeeded, now needed to be given more economic attention because they differed greatly from the USA and UK economic models.

As said above, these Asian and other nations have been in the GIR now themselves for several decades, which succeeded and continued to do so with a different economic model. Northern EU, Japan, South Korea, and China are clear documented examples of a different economic model. For example, a key economic government programs representing the GIR in the EU is the Feed-in-Tariff (FiT), which started in Germany during the early 1990s and was successfully taking route in Italy, Spain, and Canada as well as nations in the EU and Asia. Still, there are problems in Spain and Germany (Gipe 2011–2013) with the over use of FiT for solar and wind farms in a bad global economic situation. Thus, these and other nations have decided to cut this measure back. On he other hand, the USA have not incorporated yet a FiT in any significant, long-term planned policy programs on a national, let alone state and city levels. Some American communities and states only have started very restrictive and modest FiT programs.

7 European Union Economic Policies: The Nordic Countries + Germany

Germany jumped out in the lead of the GIR in the EU with its FiT legislation in 1990. Basically the FiT is an incentive economic and financial structure to encourage the adoption of renewable energy through government legislation. The FiT policy obligates regional or national electricity utilities to buy renewable

[2] This term is rooted in Adam Smith and those economists (the vast majority) that followed his thinking, which argues that there are supply and demands which need to be balanced in order for any economic system to function (Clark and Fast 2008).

electricity at above-market rates. Successful models like that exist such as the EU tax on fuels and the California cigarette tax, both of which cut smoking dramatically in California and for people to use mass transit and trains rather than drive their cars as much in the EU. But also provide incentives and metering mechanisms to sell excess power generated back to the power grid. Other EU nations, especially Spain, followed and the policy is slowly being developed in Canada and some US states and cities.

7.1 Germany Was World's Leading Producer of Solar Systems

The reason for German leadership was that it has more solar systems installed than any other nation based upon also on the creation of world leading solar manufacturing companies, solar units sold which are installed and measured by sales, amount of kilowatts per site with records kept by the local and national governments (Gipe 2012). The extensive use of solar by Germany is despite the fact that the nation has many cloudy and rainy days along with significant snow in the winter which is common to northern Europe. Over 250,000 "green" jobs were created in Germany alone. Germany led the world in the solar and wind industries until 2011 when China took over that position as global leader.

Japan implemented in 2010 a similar aggressive FiT system in order to stimulate its renewable energy sector and regain renewable energy technological (solar and system companies and installations) leadership that it held in the early part of the twenty-first century. Measurements were kept by the solar companies as well as local and national governments. MITI, the Japanese national research organization measures the use of renewable energy systems on a quarterly basis. However, the aftermath of the Japanese earthquake and destruction of the nuclear power plants in April 2011 could actually expedite renewable energy growth and installation through a number of government programs and incentives that are being proposed.

7.2 Denmark is Another Case: Wind

Other European countries have implemented the GIR through programs as well. Denmark, for example, is committed to be generating 100 % of its energy from renewable power sources by 2050. While trying to meet that goal, the country has created new industries, educational program, and therefore careers. One good example of where this policy has accomplished dramatic results is the City of Frederikshavn in the Northern Jutland region of Denmark. The City has 45 % renewable energy power now and by 2015, it will have 100 % power from renewable energy sources (Lund 2009). In terms of corporate development in the renewable energy sector, for example, one Danish company, Vestas is now the world's leading wind power turbine manufacturer with partner companies all over the world because of its partnership and joint ventures in China. Vestas continues to introduce improved third-generation turbines that are lighter, stronger, and more efficient and reliable. They also continue to design new systems, like those that can be installed offshore away from impacted urban areas.

7.3 Other EU Nations

Germany, Spain, Finland, France, UK, Luxembourg, Norway, Denmark, and Sweden are on track to achieve their renewable energy generation goals. Italy is fast approaching the same goals when in 2010, it took the distinction as having the most MW of solar installed from Germany. Other EU countries are lagging behind, especially in Central and Eastern Europe. The EU has required all its member nations to implement programs like those in Western EU in order to be energy independent from getting oil and gas, especially now since most of these supplies come from North Africa, the Middle East, and Russia.

Individual EU nations have widely different starting positions in terms of resource availability and energy policy stipulations. France and Finland were already noted as supporting nuclear power. Meanwhile Russia along with Ukraine and other eastern EU nations seek natural gas to sell to other nations. The UK and the Netherlands have offshore gas deposits, although with reduced output predictions. And in 2013, the Danish government is investigating the use of fracking to get natural gas from last areas of land in that country. In Germany, lignite offers a competitive foundation for base-load power generation although hard coal from German deposits is not internationally competitive. These 2IR sources of energy are debatable for these and other countries. Even Austria and Norway are looking to expand hydropower, which is the dominating energy source in those countries for generating power; though expansion is limited and had environmental impacts as well.

Other EU directives toward energy efficiency improvement and greenhouse gas emission reductions also impact electricity generation demand. Many EU members have taken additional measures to limit GHG emissions at the national level. Since the EU-15 is likely to miss its pledged reduction target without the inclusion of additional tools, the European Parliament and the Council enacted a system for trading GHG emission allowances in the Community under the terms of Directive 2003/87/EC dated 13 October 2003. CO2 emissions trading started in January 2005 but have not produced the desired results due limitations of "cap and trade" economic measures and the use of auctions over credits given for climate reduction.

After cap and trade auction systems were established for 3 years, by 2011, the results are not good. However, despite the economics and "markets" not performing as predicted even by 2012, the commitment to this 2IR economic model continues. Basically the carbon exchanges have performed poorly and not as promised to either buyer or seller of carbon credits (or other exchange mechanisms). The initial issues are emission caps not being tight enough and lack of significant EU or local government oversight (EU 2009). By 2010, many of the exchanges had closed or combined with others. The problem was often cited at the lack of supporting governmental (EU or by nation) policies but the real issue is that the economics do not work as well as the control over carbon emissions. The trading and auction mechanisms furthermore, do not provide direct and measurable solutions to the problem of emissions and its impact on climate change. A far more direct finance and economic mechanism as proposed by several EU nations and China would be to have a "carbon tax."

In short, far more aggressive use of the FiT type of financing and/or direct carbon taxes needs to be made. On its own the solar industry would not move fast enough into the GIR. In many ways, this is the lesson for other nations. In fact, the reality of the 2IR historically has been to have strong and continuous government incentives from the late nineteenth century to the present day. The definition and model of economics of a market remains critical in understanding how the USA can move into the GIR. Consider now how Japan and South Korea did just that: moved into the GIR with strong government leadership and financial support.

8 Asia Takes the Lead: Japan and South Korea are Leaders in the Green Industrial Revolution

While it took an extraordinary political transition to prompt Europe to open the door to the Green Industrial Revolution, Japan and South Korea in particular has taken a completely different path. And now China is moving aggressively ahead in the GIR. Most of the information and data below will be focused on China. For example, China led the US and the other G-20 nations in 2009 for annual clean energy investments and finance, according to a new study by "The Pew Charitable Trusts" (Lillian 2010: 4)

In April 2011, China became the world leader of financial investment in "clean tech" with $54 billion invested which was over $10 billion from second place Germany and almost double third place, USA (San Jose 2011: 8). Wind was the favorite sector of renewable energy with $79 billion invested globally.

This article noted in particular a comment by a senior partner in a venture capital firm, "a lot of the clean technologies are dependent on policy and government support to scale up. In some other parts of the world (not USA), you have more consistency in the way these types of funds are appropriated" (San Jose 2011: 8).

8.1 The Case of Japan with the Culture and Historical Need for Energy

The Japanese have had a long cultural and business history in commercializing environmental technologies. The 2011 earthquake made Japan focus back on that historical tradition. The future has yet to become clear and will not be defined for some months and years ahead. However, in Japan, the environment took a backseat to industrial development during the drive toward modernization and economic development that began in the latter half of the nineteenth century. After nearly 300 years of self-imposed isolation from the world, Japan was determined to catch up to the industrialized West in a fraction of the time it took Europe and the US to make their transitions, eventually emerging as a great power in the beginning of the twentieth century.

Economic development continued unabated until World War II, when its capacity was destroyed by American bombings. Economic growth restarted again in the postwar period at a rapid pace but with a distinctive orientation and concern for the limited nature resources of the island nation.

By the 1970s, on the strength of its industry and manufacturing capabilities, Japan had attained its present status as an economic powerhouse. Companies like TOTO (concerned with water and waste conservation and technologies) along with the automakers concerned with atmospheric pollution emerged as global leaders. A large part of that success was the need for the government to invest in research and development organizations (e.g., METI) to support companies and business growth, what would now be called the GIR. For example, high-speed rail was started in Japan in the mid-1980s and expanded. The transportation systems were economically efficient as well as being environmentally sound and provided for the public at reasonable rates.

While this incredibly successful period of development left many parts of the country wealthy, it also resulted in serious environmental problems. In addition, the oil crisis had hit Japan particularly hard because of its lack of natural resources, making it difficult for the industrial and manufacturing sectors to keep working at full capacity. To respond to the effects of pollution, municipalities began working in earnest on ways to reduce emissions and clean up the environment, while Japanese industry responded to the oil crisis by pushing for an increase in energy efficiency.

At the same time, Japan's economy was evolving more toward information processing and high technology, which held the promise of further increases in energy efficiency. After WWII, Japan created new innovative management "team" systems that were copied in 1980s by the USA and some EU countries. Many manufacturing firms saw value in establishing plants in other developed countries in part to create a market for their products, employ local workers, and establish firm and solid roots. For example, Toyota and Honda established their Western Hemisphere Headquarters in Torrance, California. Other high tech companies established large operations throughout the US. In this way, the Japanese in government, industry, and academia have worked collaboratively with local and regional communities to reincorporate traditional Japanese ideas about conservation and respect for the environment in order to create sustainable lifestyles compatible with modern living.

Community-level government efforts in Japan, supported by national government initiatives, have led to unique advancements in energy efficiency and sustainable lifestyles, including novel ways of preventing and eliminating pollution. As it stands, Japan is responsible for some 4 % of global CO_2 emissions from fuel combustion, and though this is the lowest percentage among major industrialized nations, it is still something the country intends to reduce, with a long-term goal of reducing emissions by 60–80 % by 2050. With the majority of energy still coming from coal, Japan also is attempting a large shift toward renewable energy.

As of November 2008, residential-use solar power generation systems have been put in place in around 380,000 homes in Japan. A close examination of the data on shipments domestically in Japan shows that 80–90 % are intended for residential use, and such shipments are likely to increase, as the government aims to have solar panel equipment installed in more than 70 % of newly built houses by 2020 to meet its long-term goals for reductions in emissions. Current goals for solar power generation in Japan are to increase its use 10-fold by 2020 and 40-fold by 2030, and large proposed subsidies for the installation of solar—9 billion yen or $99.6 million total in the first quarter 2009—along with tax breaks for consumers will continue the acceleration of solar adoption by Japanese households.

8.2 Efficiency and Conservation Need to Merge with Renewable Energy

In recent years, Europe, China, Southeast Asia, and Taiwan, saw tremendous growth in energy generation almost entirely from solar power installations. However, these have mostly involved large-scale solar-concentrated power facilities not fit for individual households. In Japan, however, as solar power generation systems for residential use become increasingly commonplace, the renewable energy Governance system has become concentrated in creating sustainable communities through use on roofs of local homes and businesses.

The same is true with the LED light bulbs, and now solar panels. Today, LED bulbs may cost a few pennies more but they last far longer than a regular light bulb and can be recycled without issues of mercury and other waste contamination. The result is better lighting for homes and offices but more significantly less costs in terms of the systems and the environment. Some LED bulbs are guaranteed to last from 6 to 8 years (Nularis 2012). While energy demands in homes and offices continues to rise due to the Internet, computers, and video systems, the installation of energy efficient and now cost saving systems if very much in demand. Some states are even requiring by law the change over from the older light bulbs to the newer LED ones.

9 Smart Green Integrated Grids

The grid of the future has to be 'smart' and flexible and based on the principles of sustainable development. As the Brundtland Report said in 1987, "as a minimum, sustainable development must not endanger the natural systems that support life on Earth: the atmosphere, the waters, the soils and the living beings" (Bruntland 1983: Introduction). With that definition in mind, a number of communities sought to become sustainable over the last three decades.

Integrated "agile" (flexible) strategies applied to infrastructures are needed for creating and implementing 'on-site' power systems in all urban areas that often contain systems in common with small rural systems. The difference in scale and size of central power plants (the utility size for thousands of customers) with on-site or distributed power can be seen in the economic costs to produce and sell energy. Historically the larger systems could produce power and sell it for far less than the local power generated locally for buildings. Those economic factors have changed in the last decade (Li and Clark 2009). Now, on-site power particularly from renewable energy power (e.g., solar, wind, geothermal and biomass) has become far more competitive and is often better for the environment. Large-scale wind farms and solar concentrated systems are costly and lose efficiency due to transmission of power over long distances (Martinot and Droege 2010).

9.1 China is the Case

Asia's shift to renewable energy will require extensive retraining. Consider the case of wind power generation in China. In the early 1990s, Vestas saw Asia and China as the new emerging big market. Vestas agreed to China's "social capitalist" business model (Clark and Li 2004; Clark and Jensen 2002), where the central government sets a national plan, provides financing, and gives companies direction for business projects over 5-year time frames, which are then repeated and updated. Business plans are critical to any company, especially when set and followed by national governments.

A major part of the Chinese economic model required that foreign businesses be co-located in China with at least a 50 % Chinese ownership. This meant that in the late twentieth and early twenty-first century, the Chinese government owned companies or were the majority owners of the new spin-off government-owned ventures, established international companies or businesses started in China. Additionally, China required that the "profits" or money made by the new ventures to be kept in China for reinvestments.

Next to the renewable energy companies like wind and solar industries, are the entire ancillary supporting businesses also needed to support the companies from mechanics, software, plumbing and electricity to installation, repair and maintenance, and other areas. Supporting industries were also needed such as law, economics, accounting, and planning, especially since the Chinese government began to create sustainable communities that required all these skill sets (Clark and Isherwood 2009 and 2010). Hence these businesses grew and became located in China.

9.2 China and Social Capitalism

The Chinese social capitalism model is not rigid with the government owning controlling percentage (over 50 %) of a company. Many businesses were started by the Chinese government with its holding of 25–33 % of shares, while the other firms were owned by the former government employees, until the companies went public (Li and Clark 2009). Yet in almost all cases, the companies are competitive globally and are performing remarkably well as demonstrated again in the renewable energy sector where in early 2011, SunTech, a Chinese-based publically traded company, became the world's largest manufacturer and seller of solar panels (Chan 2011). According to a press release by the company in February 2011, it has delivered more than 13 million PV panels to customers in more than 80 countries.

Today, China is a (if not the) world leader in wind energy production and manufacturing with over 3,000 MW installed in China alone (Vestas 2011). The Chinese are now following a similar business model in the solar industry (Martinot et al. 2007: 10). As such, China and Inner Mongolia (IMAR) have contracted Vestas to install 50 MW in IMAR (op. cit. Vestas 2011), according to a Report from the Asian Development Bank (Clark and Isherwood 2007 and 2010) which argues for targeted needs to:

- Create international collaborations between universities and industry
- Conduct research and development of renewable energy technologies
- Build and operate science parks to commercialize new technologies into businesses.
- Provide and promote international exchanges and partnerships in public education, government, and private sector businesses

10 Comparing Renewable Energy Globally

The end results for the EU are smart homes and communities through "distributed" or "on-site" integrated smart renewable energy systems. The Green Industrial Revolution starts in the home so that energy efficiency and conservation are a significant part of everyone's daily life. The home is the place to start. But it is also the place to start with the other elements of the GIR: renewable energy generation, storage devices, smart green communities, new fuel sources for the home, and transportation.

This lack of planning has both long-term and short-term impacts. The finance of new energy technologies and systems (like any new technology) is often dependent on government leadership through programs in public policy and finance. Fossil fuel energy systems in the 2IR have been funded and supported by the governments of western nations through tax reductions and rebates that continue today. For the GIR, it is only logical and equitable that such economic and financial support continues. That means the American national government should provide competitive long-term tax incentives, grants, and purchase orders for renewable energy sources rather than just fossil fuels.

Meanwhile, the EU, South Korea, and Japan took the leadership in the planning, finance, and creation of renewable energy companies, while other nations including the USA did not (Li and Clark 2009). For example, because of the national policy on energy demand and use, Japan has one of the lowest energy consumption measurements in the developed world. This has been made possible by its continued investment in the long-term energy conservation while developing renewable sources of energy and companies that make these products. Japan's per capita energy consumption is 172.2 million Btu versus 341.8 million Btu in USA.

11 How to Pay For Renewable Energy Systems: The Next Economics

One critical of a long-term economic plan is the need for life cycle analysis (LCA) versus cost benefit analysis (CBA). While not discussed much in this paper about these two very different accounting processes that cover the topic in-depth as the systems apply to government spending. Each approach is critical in how businesses learn what their cash flow is and their return on investment (ROI). The CBA model only provides for 2–3 year ROI since that is what most companies (public or government) require for quarterly and annual reports. However, for new technologies (like renewable energy, but also even wireless and WIFI technologies) more than a few years are needed on the ROI. The same was true in the 2IR when oil and gas were first discovered and sold. Now in the GIR, longer economic and financial ROIs are needed.

LCA covers longer time periods, such as 3–6 years and within renewable energy systems some as long as 10–20 years depending on the product and/or service. Furthermore, LCA includes externalities such as environment, health, and climate change factors, all of which have financial and economic information associated with them. The point is that cost benefit analyses are limited. The basic concept is that the LCA consists of one long-term finance model in the USA today for solar systems is called a Power Purchase Agreement (PPA) that contracts with the solar installer or manufacturer for 20–30 years. PPA is a financial arrangement between the user "host customer" of solar energy and a third party developer, owner, and operator of the photovoltaic system (Clark and Isherwood 2010).

11.1 Economic and Finance Issues

The customer purchases the solar energy generated by the contractor's system at or below the retail electric rate from the owner, who in turn along with the investor receives federal and state tax benefits for which the system is eligible on an annual basis. These LCA financial agreements can range from 6 months to 25 years and hence allow for a longer ROI. However, there are other ways to finance new technologies especially if they are installed on homes, office, and apartment buildings. Today, financial institutions and investors can see a ROI that is attractive when the solar system on a home, for example, is financed as a lease, part of tax on the home, or included in the mortgage itself like plumbing, lighting, and air-conditioning are today.

What is interesting are some newer economic ideas on how to finance technologies that reduce "global climate change." One way to describe the GIR financial mechanisms is by looking at the analytical economic models that financed the 2IR. For example, the 2IR was based upon the theory of abundance. The earth had abundant water and ability to treat waste. Hence buildings, businesses, homes, and shopping complexes all had plumbing for fresh water and drainage for waste. The same scenario occurred in electrical systems that took power from a central grid for use in the local community buildings. Locally and globally, people have found that systems work but now with climate change need to conserve resources and be more efficient.

When these economic considerations are factored into even the CBA rather than a LCA financial methodology, the numbers do not work (Sullivan and Schellenberg 2011). The financial consideration for energy transmission and then monitored by smart systems are needed, but costly. Long distances make them even more costly because then the impact of the climate (storms, tornadoes, floods, etc.) with required operation and maintenance are added today with security factors. The actual "smart" grid is at the local level where these and other uncontrolled costs can be eliminated and monitored.

11.2 New Financing Mechanisms

The financing of water, waste, electrical, and other systems for buildings was over time incorporated into the basic mortgage for that building. In short, modern 2IR infrastructure systems were no longer outside (e.g., the out house or water faucet) but inside the building. What this 2IR financial model does is set the stage for the GIR financial model. Much of the 2IR financing for fossil fuels and their technologies came about as leases or building mortgages.

A variation of the 2IR model which is a bridge to the GIR is the PACE (Property Assessed Clean Energy) program started in 2008 in Berkeley, California whereby home owners can install solar systems on their buildings, for example, but pay for them from a long-term supplemental tax that is transferred with the sale of the property assessment on their property taxes. The financing is secured with a lien on the property taxes, which acquires a priority lien over existing mortgages. The program was put on hold in July 2010 when the Federal Housing Finance Agency (FHFA) expressed concerns about the regulatory challenge and risk posed by the priority lien established by PACE loans. Nevertheless, the US Department of Energy continues to support PACE.

The dramatic change to the GIR, however, moves past that financial barrier. Mortgages are part of the long-term cost for owning a property. Therefore, in the GIR, the conservation and efficiency for the 2IR technologies in buildings can be enhanced with the renewable energy power, smart green grids, storage devices, and other technologies through mortgages that can be financed from one owner to another over decades (20–30 years or more).

This sustainable finance mortgage model is long term as a life cycle analysis (LCA) framework since it provides for new technologies and installation costs to the consumer that makes the GIR attainable within short time. Changes in technologies (modernization, lower costs, higher efficiency, etc.), need to be updated since the new technologies can easily be substituted and replaced for the earlier ones. For example, changes in the water and waste systems in buildings have evolved and become more significant in the last decade alone due to concerns about water, waste, and energy efficiency. What needs to happen is that the banking and lending industries try this GIR finance model with LCA on selected areas, like building mortgages. After some case studies are done, the LCA finance model can be replicated or changed as needed.

12 Conclusions and Future Research Recommendations

The basic point of this chapter is to highlight the need for economics to be scientific in its hypothesis statements, measurements, and data collection so that like a science, economics can predict the future. For example, the creation and use of life cycle analysis can provide a more exact and precise understanding of sustainable communities in terms of their energy use, conservation, and power generation. Also on the local level, which is within buildings themselves, the use of mortgage financing to include renewable energy power systems can provide cost savings in the near term into decades for the future. Above all these changes will reduce carbon emissions and pollutions in order to mitigate and control climate change.

Furthermore, the economics of the 2IR and the GIR are not very similar. Nor should they be. However, there are certain parallel conditions that need to be used in both revolutions. That is, for example, the role of government since it must often take the first steps in directing, creating, and financing technologies. As the 2IR needed government to help drill for oil and gas as well as mine for coal, the government needed to build rail and road transportation systems to transport the fuels from one place to another.

The problem with the 2IR was that the neo-classical economic model came to dominate it in terms of supply and demand economics, which even the Economist has seriously questioned (Economist 2009 as well as January 23 2012b, and April 21 2012a). Since the global economic crisis in October 2008, there has been a serious rethinking and search for how economics will be done in the future. A key is making the "field" of economics into a science (Clark and Fast 2008 and Clark 2012). To do that requires people who consider themselves to be economists take a look at other sciences as well as what they teach, publish, and do in their careers today.

When economics is seen from a global perspective and not just the western (that is UK and USA perspective) it is very different and needs to learn from these other perspectives. One specific issue is that economics in many parts of the world is not based on "market forces", but instead on close relationship with government and public needs such as health, environment, and infrastructures like sustainable communities (Clark and Li 2012).

The GIR is very much in the same economic need for direct government involvement. The evidence can be seen in Asia and the EU. And especially now in China, the central government plans for environment and related technologies help a nation move into the GIR. Moreover, there is a strong need for financial support that are not tax breaks or incentives alone, but investments, grants and purchasing of GIR green technologies, such as renewable energy, fuel cells, and integrated infrastructure systems.

This can be seen in the US today with the debate over smart grids. What are those? And who pays for them? When the smart grid is defined as a utility, then the government must pay for them since they are part of the transmission of energy, for example, over long distances that must be secure and dependable. But as the GIR moves much more into local on-site power, the costs of the smart grid are at the home, office buildings, schools and colleges, shopping mall, and entertainment centers. Local governments are also involved as they are often one of the largest consumers of energy in any region and hence emitters of carbon and pollution. Within any building, a smart grid must know when to regulate and control meters and measurement of power usage and conservation. The consumer on the other hand needs the new advanced technologies, but the government must support these additional costs and their use of energy as they impact the local community and larger regions residential and business needs.

Economics has changed in the GIR. And yet, economics has a basis of success in the 2IR. Historically, 2IR economics was successful because the government was needed to support its technologies along with goods and services. The evolution into the neo-classical form of economics was far more a political strategy backed by companies who wanted control of infrastructure sectors. But the reality was that "greed" took over and has now forced a rethinking of economics as nations now move into the GIR.

Acknowledgments The authors thank Jon McCarthy for both economic details and edits as well as Lucas Adams, Kentaro Funaki, Claus Habermeier, and Russell Vare for international perspectives and data. Namrita, Singh, and Jerry Jin checked on data contributions and review of this paper, which is derived from one published in Contemporary Economic Policy Journal.

References

Brown, LR (2008) Plan B 3.0: mobilizing to save Civilization. W.W. Norton & Company, NY

Bruntland Commission (1983) Report. Our common future. UN commission general assembly resolution #38/161 for process of preparation of the environmental perspective to the year 2000 and beyond. Oxford University Press, Oxford, 1987

Cai W, Wang C, Chen J, Wang K, Zhang Y, Lu X (2008) Comparison of CO2 emission scenarios and mitigation opportunities in China's five sectors in 2020. Energy Policy 36:1181–1194

Chan S (2011) President of SunTech, Global Solar Industry Prospects in 2011. SolarTech Conference, Santa Clara, California

China Ministry of Science and Technology (CMST) (2011) The 12th Five-Year plan of electric vehicle in China has been implemented, p 2

China Statistics Press (2013) Yearbook of China transportation and communication, First Edition ed. Yearbook house of China transportation and communication, Beijing

Chomsky N (2012) Plutonomy and the precariat: on the History of the US economy in decline, tompatch.com, pp 1–5. http://truth-out.org/news/item/8986-plutonomy-and-the-precariat-on-the-history-of-the-us-economy-in-decline76. Accessed 8 May 2012

Clark II WW, Grant C, Jerry J, Ching-Fuh L (2014) The Green industrial revolution (in English and Mandarin). Publisher to be determined in China

Clark II WW et al. (2014) Global sustainable communities design handbook. Elsevier Press, NY

Clark II WW (2012) The next economics: a science in energy, environment, and climate change. Springer Press, NY

Clark II WW, Bradshaw T (2004) Agile energy systems: global solutions to the California energy crisis. Elsevier Press, NY

Clark II WW, Cooke G (2011) Global energy innovations. Praeger Press, NY

Clark II WW Author and Editor (2011) Sustainable communities design handbook. Elsevier Press, NY

Clark II WW, Isherwood W (2007) Special report on energy infrastructures in the west: lessons learned for inner Mongolia, PRC. Asian Development Bank

Clark II WW, Isherwood W (2009) Report on energy strategies for inner Mongolia autonomous region, Utilities Pol J, doi: 10.1016/j.jup.2007.07.003

Clark II WW, Isherwood W co-authors (2010) Leapfrogging energy infrastructure mistakes for inner Mongolia, Utilities Pol J, Special issue

Clark II WW, Jensen JD (2002) Capitalization of environmental technologies in companies: economic schemes in a business perspective, Int J Energy Technol Pol, vol 1, nos 1/2, Inderscience, London, UK

Clark II WW, Michael F (2008) Qualitative economics: toward a science of economics. Coxmoor Press, Oxford, UK

Clark II WW, Rifkin J et al. (2006) A green hydrogen economy. Special issue on hydrogen, Energy Policy, Elsevier, vol.34, Fall 06, (34):2630–2639

Clark II WW, Li X (2004) Social capitalism: transfer of technology for developing nations. Int J Technol Transf Commercialization 3(1), Inderscience, London

Clark II WW, Li X (2012) Social capitalism: China's economic rise, (chapter 7). In: Clark WW (ed.) The next economics. becoming a science in energy, environment, and climate change. Springer, NY

Economist (2012) The third industrial revolution, special issue, April 2012

Economist (2012) The rise of state capitalism: the emerging world's new model. Special issue with Lenin smoking a cigar on the cover. Two articles. 23 Jan 2012

Economist (2009) Modern economic theory: where it went wrong—and how the crisis is changing it, special issue with Bible melting on cover. Editorial and three articles. 18 July 2009

Gipe P (2011–2013) Feed-in-Tariff Monthly Reports, March 2011

Konan DE, Jian Z (2008) China's Quest for Energy Resources on Global Markets, Pac Focus 23(3)

Knakmuhs H (2011) Smart Transmission: making the pieces fit, RenGrid 2(3). www.renewgridmag. Accessed April 2011, p 1+
Leung GCK (2010) China's oil use, 1990–2008. Energy Policy 38:932–944
Li X, Clark II WW (2009) Crises, opportunities and alternatives globalization and the next economy: a theoretical and critical review, (chapter 4). In: Xing L, Winther G (eds) Globalization and transnational capitalism. Aalborg University Press, Denmark
Lillian J (2010) New and Noteworthy. Sun Dial, Solar Industry, May 2010, pp 3–4
Liu W (2012) The integration of sustainable transportation into the future renewable energy systems in China, Ph.D Thesis, Aalborg University, Denmark
Lo V (2011) China's Role in Global Economic Development, Speech by Chairman of Shui On Land, given at Asian Society, Los Angeles, CA, April 25, 2011
Lund H (2009) Sustainable Towns: the case of Frederikshavn. 100 % Renewable Energy, (chapter 10). Sustainable Communities. Springer Press, New York
Lund H, Hvelplund F, Mathiesenan BV, Østergaard PA (2011) Main report of Coherent Energy and Environmental System Analysis, Multiscience Publishing, UK
Ma L, Liu P, Fu F, Li Z, Ni W (2011) Integrated energy strategy for the sustainable development of China. Energy 36(2):1143–1154
Martinot E, Droege P (2010) Renewable energy for cities: opportunities, policies, and visions. Series of Reports from 2007–2010 http://www.martinot.info/Martinot_Otago_Apr01_cities_excerpt.pdf
McNall SG, James H, George B (eds) (2011) Global challenges and opportunities. vol 1. Praeger Press, New York
Nularis (2012) Data and Information. www.nularis.com
Rifkin J (2012) The third industrial revolution. Penguin Putnam, NY
Rifkin J (2004) The european dream. Penguin Putnam, NY
San Jose Business Journal (2011) Clean energy financing jumps to record $243B, April 1, 2011: p 8
Shields C (2010) Renewable energy: facts and fantasies. 2GreenEnergy Publications, USA
Sullivan M, Josh S (2011) Smart grid economics: the cost-benefit analysis, RenGrid, 2(3). www.renewgridmag. Accessed April 2011, pp 12–13+
Trabish HK (2013) Growth in China's wind sector eclipses growth in coal. GreenTech Media. Stat of the day, March 22, 2013, pp.1–2. http://www.greentechmedia.com/articles/read/Growth-in-Chinas-Wind-Sector-Eclipses-Growth-in-Coal?ut...3/27/2013
United Nations Intergovernmental Panel on Climate Change (UNIPCC) (2007) United Nations, NY: Report Three (2000) and Report Four (2007)
Vestas (2011) From www.vestas.com

Conclusions

Answering the "Mega-What" Question: Who is Finally Renewable Energy For

Evanthie Michalena and Jeremy Maxwell Hills

1 Introduction: The RE Global Phenomena

The chapters have demonstrated intensive and diverse efforts in the development and management of RE across the world, and in both developing and developed countries. This "herd behavior" is unusual in a global sense as countries are in different stages of development and facing differing priorities which maybe vary from stopping parts of the population from starving, to reducing economic in-migration. However, RE has a broad and rather effective instigator and promoter through the UNFCCC and the Kyoto Protocol which provides a common driver for RE development in many of the countries covered in the chapters of this book. For developing countries, which may be struggling with disease and malnutrition, and in which energy security may be less of an issue when a majority of rural dwellers do not have electricity to meet their basic needs, the Kyoto Protocol may seem somewhat of a lower priority. Some might even argue that energy-guzzling western countries are outsourcing RE implementation to developing countries so as to subsidise the development of a market for western RE machinery.

It is apparent in the RE sector that there are international or multilateral powers at work. International institutions boast in having a great influence in facilitating the functioning of energy markets, lowering the transaction costs and setting certain rules and standards. There is indeed the need for multi-country analysis and assessment of a number of RE aspects which these organisations are well positioned to deliver. For example, the changing and growing portfolio of RE

E. Michalena (✉)
ENeC Laboratory, UMR 8185, Paris-Sorbonne (Paris IV) and CNRS INSHS, Paris, France
e-mail: evamich@gmail.com

J. M. Hills
CTL Consult, Newcastle upon Tyne, UK
e-mail: j.hills@ctl-consult.com

E. Michalena and J. M. Hills (eds.), *Renewable Energy Governance*,
Lecture Notes in Energy 23, DOI: 10.1007/978-1-4471-5595-9_23,

technologies and inherent efficiency and production costs (Kosmadakis et al.) which would benefit from consistent cross-country multi-annual assessments. Or, the assessment of impacts of RE technologies which are greater than equivalent fossil fuel generation, as identified for wind farms in the UK and the terrestrial ecotoxicity and abiotic depletion associated with the RE machinery materials (Kouloumpis et al.). This collaborative and supportive approach to international facilitation of the RE industry should mean that improvements in governance are top priorities of all institutions of all levels. However, this requires institutions to work together towards common goals. An example in relation to regional players in the Black Sea area is provided by Hills and Michalena.

However, in the Chapter of Kottari and Roumeliotis on international agencies such as IEA, IRENA, RENI and REEE, it is commented that functions of international agencies are fragmented and lack cohesion and thus compromise the public and private interests in RE. The authors also elaborate "jealously" between two of those agencies and apparent jostling for the RE agenda. This does suggest that on the international RE governance stage there are a number of interests which can constrain or even block the path of RE deployment. Within the global agreements in which the benefits of reducing CO_2 are projected to benefit all the world, there would seem to be a number of claims from countries which will particularly benefit from RE; these claims seem to be aligned to geopolitical, trade and diplomatic interests.

2 National Motives for RE

Much has been made of the value of RE for energy security and at a national level this might seem a valuable asset of a RE government-directed approach. If fossil fuel import is reduced from potentially volatile and distanced areas and more national effort is made to collect natural and renewable energy from within the national borders, then more national control is supposed to be established over energy supply. In the Chapter of Matthewman and Byrd, New Zealand—a relatively small and isolated country—seems to have good energy security with about 75 % of electricity from RE; maybe this would appear as an enviable position. However, cheap energy prices mask the illusion of "infinite energy". There are limited initiatives on reduction of energy use, limited investment in improving the supply grid and consumers are subject to national energy regulation. Declining national fossil fuel sources, coupled to increasing variation in RE supply (linked to more erratic rainfall and thus hydropower generation), are creating potential electricity supply issues in New Zealand. Yet New Zealanders are disinterested and intolerant of anything else but the maintenance of cheap and plentiful energy. This example displays a national portrait or character in terms of RE use as well as specific national challenges to maintain/grow the RE component of the energy mix in the light of climate change.

In general, there are many aspects which have significant effects on the way in which RE is deployed and managed. Koster and Anderies considered energy transitions towards RE in their Chapter and they conclude that regardless of the kind of RE available, RE implementation rests on a handful of institutional drivers; the most essential one being government commitment to increase the RE share in its energy portfolio. However, in some cases, what may superficially seem as the passive desire to increase RE's contribution and meet targets (e.g. EU or Kyoto), may rather downplay the wider strategic and nationalistic motives. In Germany, for example, political decisions led to a move away from importing fossil fuels to developing RE technology, building in-country plant for in-country generation and building in-country plant for export. This means that RE was formulated around a political trade mechanism to promote German interests of which energy is a just by-product (Buschmann and Lauber). However, this trade promotion effort is being internally challenged by public uneasiness about expensive financial support promoting RE generation in Germany as well as externally by price-competitive Chinese imports of RE plant.

Like in Germany, Norway has also nationally been directing RE developments. Norway is well blessed not only with vast oil reserves but also significant and exploited hydropower (which dominates electricity generation already). Thus, besides moving to wider use of RE for power needs other than electricity, Norway could rest contentedly if national energy security and global energy targets are met. However, the country has invested in innovative RE R&D centres across Norway through the "climate settlement", it has created energy certificates in a joint market with Sweden to promote further RE incentives, and there is ongoing discussion on making Norway a hydropower storage facility to provide intermittent generation at peak times to the wider European electricity market (Skjølsvold et al.). Thus, Norway's "goldmine" of energy has a RE footprint which is potentially growing in a number of ways beyond national boundaries: helping to create financial inflows through RE energy certificates or even the potential of rapid RE generation when the wider European population demand might peak.

Other countries have taken a rather less strategic and entrepreneurial course on energy. In Cyprus, for example, an accident in traditional generation made apparent the vulnerability of the energy generation, however this did not change the endless, costly and time-consuming bureaucratic procedures necessary to obtain a permit for RE generation. As a consequence, the opportunity provided by the 8th largest explosion in the world and losing 60 % of the national generation capacity, the country failed to lead to any rapid national response to RE expansion (Fokaides et al.).

Moving from a small country to the largest CO_2 emitter in the world, China has taken an approach which is much aligned with national energy production and security and which appears to be the most unified, top-down approach to implementing RE policies and programmes (Koster and Andereis). Obviously, as a consequence of the massive RE drive, it seems that creation of this large national manufacturing base in RE, coupled to cheap labour and existing market links, has inevitably led to export opportunities. The relatively late adopter but massive entry of an authoritarian administration which can progress on high level decisions, obviating time-consuming multi-level pluralistic stakeholder consultations and discourses, may yet undermine strategic motives for RE advancement in other countries. However, such countries may not lose face entirely as the energy security and climate mitigation benefits may suffice; still, as demonstrated in the next section, this has come at quite a cost to us, the consumers.

2.1 Instruments of RE Governance

Whilst setting RE generation targets at the global or national level maybe admirable, underneath these headlines there needs to be some degree of organisation to allow these targets to be met. However, it is noticeable in the case of RE that even with government championing of their targets, they tend not to pay directly for the infrastructure or operation of the RE systems which help meet the targets. This reflects the approach often used in fossil fuel energy generation sector, but less so in transport (roads and rail) and water supply in which government invests directly in infrastructure. Leaving RE to the private sector, in most cases, means that some form of functional interface between governmental targets and private sector investment must be created.

A common approach for meeting RE targets is for the government to set out one or more financial instruments which attract in entrepreneurs and investors. Government financial promotion of RE may come through tax relief on certain RE plants, but more commonly financial instruments provide payment per unit of RE generated. These instruments may have a number of additional features; for example they may define a minimum amount or RE to be supplied by each operator, RE quotas may be traded through "green energy certificates" or similar and/or financial bandings for different types of RE technology generation may be used to favour certain technologies (e.g. Lauber and Buschmann, McCormack and Norton, Morkens et al.). This basic model is quite simple with the government setting targets and then providing adequate financial incentives for investors to be attracted to RE investment. For governments, especially those severely affected by the economic downturn (Boemi and Papadopoulos), this kicks the need to pay for the high up-front infrastructure costs of RE plant into the future, and helps balances the present government finances. There is however a need to provide a degree of guarantee and longevity of financial pay-backs to the investors, or through a more political lens, to ensure progress to RE targets in the present administration whilst offsetting the costs to future governments and maybe alterative political parties.

Whilst the government target/investor incentive mode of operation may appear simple, there are three interrelated facets of it that have been repeatedly touched on in these chapters.

2.1.1 Setting the Rate of the Financial Instrument

There is a clear trade-off between setting-up financial instruments, say a FIT, which is so low that it does not attract investment and setting it so high that there is goldrush of investment. For example, Boemi and Papadopoulos note the progressive system of RE in Spain where RE reaches 10 % electricity demand, however, they also note that it only came at "significant cost". McCormack and Norton note the "boom" in the UK RE market based on a generous FIT and subsequent court action delaying the UK Government rapidly reducing the FIT as it tried to do.

2.1.2 Restricting Access to Financial Instruments

Other chapters note that access to these financial instruments is restricted which leads to weaker RE penetration. The main blockage in the path to the financial instruments for the investor/developer seems to be the government. For Greece, Boemi and Papadopoulos identify the long and tedious licensing procedures which lead to more than 2 years' delay before the first investments began to be implemented. Also for Greece, Metaxas and Tsinisizelis note the frequent delays in obtaining a grid connection quotation from the grid operator (in law 4 months, up to 2 years in practice) and destabilisation of the RE sector through imposition of a retroactive tax measure on RE which arguably infringes EU Law. Blockages can also be planning laws, for example Caralis et al. describe the extensive areas excluded from wind farm development of the island of Lesvos (Greece) to landscape (urban areas, roads, beaches, protected areas e.g. Natura2000, forests), wind generation feasibility (e.g. slope and wind potential) and aesthetic considerations. Blocking the access to pro-RE financial instruments is common in many of the case studies presented; this is mainly due to onerous governmental procedures. In the case of an investor, the delay in obtaining a permit to develop a site may be critical in the break-even point of the investment and the accrued profit over time. Time is money for an investor, thus the financial pay-off (e.g. FIT) needs to be set higher to accommodate inefficiencies or delays in governmental processing.

2.1.3 Weak Feedback and Monitoring Mechanisms

When looking at policies and administrative procedures, it can be easy to consider the electricity system as a static system. However, there are constant changes in electricity demand which must be met with changes in supply through a combination of base-load and rapid onset electric generation. RE must take part in this system, and its own dynamics accounted for, for example negligible wind generation during low wind periods (for example, the winter high pressure weather system over Northern Europe in 2007) or no PV generation in late evening and during the night.

On the contrary, the financial instruments tend to be highly static in comparison to the constant generation-supply fluctuations. FIT's for example can be set for a number of years, and their legacy can be for 15 or 20 years into the future (sometimes with inflation taken into account). Thus, FIT's in many cases are so static that they do not follow the energy market but just provide an incentive for investors to access the market to generate electricity using a technology which fails to attain grid parity.

FIT's, or similar financial instruments, can thus lead to boom or bust, if they are not orientated to the influencing dynamics, such as energy demand curves, costs of borrowing, duration of permitting procedures etc. A more dynamic form of financial instrument, more linked to the energy market than investment incentive, has been tried in Flanders (Flemish Green Electricity Certificate System FGECS; Moorkens et al.) which has efficacy (i.e. delivered the Flemish RE targets) but the effectiveness and efficiency of the system is still in doubt. There does seem to be a need for financial instruments which incentivise investors as well as operate in rhythm with the energy markets, but which have monitoring and feedback loops to allow ongoing management of boom and bust scenarios of any possible change in government RE policy.

2.2 The Hidden Costs

The consequences of inappropriate RE rates, blockages to access financial instruments and limited feedback and monitoring are now quite obvious to the consumer. Over-generous financial instruments mean that the consumer pays more per unit of electricity than necessary with a lean FIT; although both are above grid parity rates paid previously. Blockages in government licensing/planning system lead to increased costs and uncertainty for the investor, which has to be taken into account through increasing the FIT incentive; again at a cost to the consumer. The efficiency of the governmental regulatory authorities and relevant planning departments in affecting RE penetration are sparsely evaluated to ensure they are "fit for purpose" and rarely taken into account on the consumers bills.

Additionally, the onerous permitting process increases the value of the permit once it is finally obtained, and this means that a "market of permits" can be developed which increases possibilities for corruption of consenting officials; again ultimately the costs of those practices will be paid by the consumer. The financial legacy of RE for the consumer will be decadal and in many countries RE is still growing so costs will increasingly be layered onto consumers bills. Many chapters note the growing uneasiness of RE costs (e.g. Boemi and Papadopoulos, Lauber and Buschmann, McCormack and Norton, Moorkens et al.). As worked out on the case of RE in Spain and quoted in the Economist *"..it is not hard to think of better ways of spending such large sums of taxpayers' money"*.[1]

What the financial instruments have provided for, is many new entrants into the energy markets through the production of small and localised RE generation. However, this might be not benefit of the monolithic monsters of fossil fuel electricity generation, who operate on plant orders of magnitude bigger than many RE generation, even if this looks as being the case: Yes, wind generators are growing in size and wind farms are growing bigger, which helps to accommodate the interests of the traditional players (e.g. UK Forewind—Dogger Bank offshore wind farm project of a proposed 1.2 GW). Yes, carbon neutral production may also be possible through traditional fossil fuel generation coupled to large-scale Carbon Capture and Storage (Kosmadakis et al.). Yes, combined use of wind energy with pumped storage could increase RE penetration and help moderate supply for use at peak times (Caralis et al.). However, governmental management of consents and financial instruments seem to have many inefficient aspects, even in the simple RE generation/FIT payout model, so that moving to more innovative models may be challenging. Especially, when effective and efficient articulation of RE policy is required by government departments, but those very departments fail to ensure effective and joined-up policy and process. As the consumer, and not the government operatives, ultimately pay, then such inefficiencies can remain unaccounted and obscured.

3 The Multiple Levels of RE Governance

A focus on the national targets and financial tools does not uncover all of the dimensions of governance. Hamilton and Kellett, use Australia as an example to discuss which institutional level or levels are necessary to have a functional energy policy. Australia seems to provide an example which is organised at the Federal level but the State/local level administrations are not clear on what to do; thus, there seems a lack of vertical integration. The Federal-level assumption seems to be that the market, with a FIT, is adequate to meet the national targets. However, the unsubstantive or inconsistent local level government and governance means that RE penetration is variable.

In contrast to Australia, France's FIT instruments provided a strong enough incentive for mass private sector involvement which tended to override any local concerns (Burger and Mancebo). The commune's Mayor seems to be the key-player in the process, through supporting or allowing developers access to the local planning board for planning permission; although the Mayor is often a landowner/farmer as well. Developers pay for commune-level endowments or sponsorship or pro-wind farm initiatives by way of compensation. It seems as if the wind farm

[1] The Economist, July 20th, 2013, p. 54.

penetration is very much of a local and parochial issue with only the FIT being the national-level driver.

All of this discussion seems to be based around national financial instruments. However, there are signs that RE can be produced outwith this frame if there is directed and cohesive local governance. Frantzeskaki et al. assess two case studies (Texel, Netherlands and Udny and Urgha in Scotland, UK) in which communities have developed RE without centralised support. The authors identify a number of characteristics of these front-runners: ability to share risks, creation of new energy governance arrangements in the absence of a policy agenda or prior to policy agendas (Udny and Urgha) and skills, expertise and innovative potential for pursuing community needs such as security and energy independence. Parpairis and Lagos also identify situations where local level RE has developed outside any national incentive, for example in tourist resorts in Fiji. This case is about a cost effective alternative electricity supply and thus, although it may count towards the national RE generation capacity, is done for hotel operators profit and not sustainability.

Koster and Anderies have identified eight institutional drivers for national-level energy transitions: government commitment, polycentricity, stakeholders and community involvement, transparency of information, pilot programmes and technology innovation, compliance to Kyoto protocol, grid connectivity and compliance monitoring. The national case studies presented in their chapter tended to have most if not all of these institutional drivers. However, as RE becomes increasingly vertically integrated and starts to develop along more local/community needs, then there is shift in institutional drivers. For example, for local communities undertaking RE initiatives, as outlined in the previous paragraph, aspects such as compliance to Kyoto protocols and polycentricity are less important. This lack of connectivity between the motives and drivers of national and local, means that vertical integration of policy is necessary.

When RE is considered at lower and lower spatial scales, down to the scale of the actual generating RE plant, there is a conceptual transformation. At the most local scale the issues of energy supply or security, or even the degree, to which the government will achieve Kyoto targets, become less relevant or nearly irrelevant. These issues are overridden by local concerns such as declining aesthetics, increased traffic disturbance for construction and operation, land or conservation impacts etc. Skanavis et al. note that there should be a focus on responsible environmental behaviour of citizens through concentration on strengthening the participants' decision-making skills and critical analysis abilities. Only then can local dialogue be achieved with developers and local administrations to identify mutually beneficial ways forward. However, it is not clear which sector should undertake the mobilisation and upgrading of citizens skills. The tendency seems to be for payments from the generation profits or some type of community endowment to be used to offset the negative effects, but that also defers any further capacity upgrading to local communities.

Consider a RE policy which is built up by parish, commune, province or country, region and then national policies. Would that local-aggregated policy match with the nationally set policy for RE, or would there be differences of focus, accruement of benefits, motivation and delivery tools? In many of the chapters, governance forms an awkward interface between these domains.

4 Renewable Energy for People and Societies

The RE target setting of Kyoto, international agencies (such as EU) and national governments has established the focus of RE onto CO_2 reduction. However, we have also heard of strategic underpinning of these targets to provide national benefits, for example for trade and export purposes by Germany and Norway. The penetration of RE is one of a gamete of approaches which moves towards climate compatible development in which reduction of the carbon intensity of economic growth works alongside improving human development and societal wellbeing. *Climate compatible development (CCD) aims to both reduce climate change impacts through mitigation and increase the resilience of society through climate change adaptation.*

Within the auspices of CCD the aim is to optimise these benefits using the array of possible mitigation and adaptation tools and technologies. With the inefficiencies in expenditure on RE and the ineffectiveness of government management and monitoring of RE, then there is some doubt that RE is one of the best tools for CCD. In addition, local negativities to RE deployment are not in harmony with the improvement in societal wellbeing achieved through a cohesive and holistic approach to coping with climate change.

If RE focusses on just energy, and energy targets, then many of the potential gains are lost. If we go back to Jane Wanjiku, the fruit seller in Kenya, energy for her means increased personal security to extend her selling into the night and smoke-free light for her children to do their homework (Mwangi et al.). The benefits accrued go beyond just CO_2 negligible energy to economic stability and wellbeing as well as future medical legacies for her children. Energy, and especially RE as presented in these chapters, seems to exist in a narrow and sectoral frame; a focus on generation and supply and not benefits.

5 Final Conclusion: The "Mega-What" Question: is Finally Renewable Energy for Targets, for RE Developers or for People?

The focus on the penetration of renewable energy (RE) to meet targets, misses the value-added dimensions of RE, and thus fails to optimise societal value. For example, solar- powered irrigation pumps in Niger or Yemen might be of negligible

CO_2 relevance to the national or global carbon budget and associated targets, but they can potentially bring income, productivity, environment, cultural, community and health benefits. If the focus is just on targets, then small-scale deployment of RE in a way tailored to local wellbeing and economic development would tend be displaced for large-scale RE generation deployment. Certainly effective and efficient RE has a part in climate compatible development, but the present preoccupation of RE governance on the scale of generation might miss the needs of the future generations.

Writing on the Green Industrial Revolution, Clark and Li note the inadequacy of formulating action just on "market forces" and note the need for a close linkage between the government and public needs such as health, environment and infrastructures like sustainable communities. Parpairis and Lagos concluded from case studies across the world that RE tends not to lead to endogenous development. These Chapters question the fundamental market-based instigators for RE as well as the local/community benefits that seem to flow from RE. At the level of the users of electricity, it is not clear if their electricity coming out of their socket is RE, fossil fuel or nuclear generated; the users cannot distinguish. For people, RE is thus totally indistinguishable from other forms of electricity in the grid. Thus, at a local scale, RE is not connected or interlinked with most of the livelihoods or economic activities which take place.

It is clear that RE has the potential to promote climate compatible development but it has to be interlinked to other activities which can be broadly classified as the green economy. As stated by UNEP[2] *"Increasing investments in renewable energy, as part of a green economy strategy spanning all major sectors, can contribute to reducing health and environmental impacts from energy production and use, while ensuring the basis for long-term economic growth"*. However, this views RE as one *interlinked* tool to promote a green economy. Furthermore, UNEP note that RE needs to play a role in an "integrated strategy" especially in rural areas with vulnerable populations. The example of solar powered pumps used above demonstrates such an approach in which irrigation is powered by RE thus creating a mitigation benefit, but it also builds climate resilience of the local people and thus has an adaptation benefit.

This book has covered many examples of governmental incentives that promote RE interests in meeting RE targets; and a number of reasons and examples where this has come at some considerable cost to consumers. What is missing however, is a more holistic and integrated approach to RE as part of green economy development. Examples analysed in this book, have not demonstrated, for example, selection criteria for RE permits that include improvements in social-equity and human wellbeing of citizens to provide the socially value-added dimension. Why not—are these not the responsibility of the consenting institutions of our

---

[2] United Nations Environment Programme (2011) Towards a Green Economy: pathways to sustainable development and poverty eradication. UNEP.

governments? Clearly, these societal aspects are less important to RE developers than the guaranteed tariffs and profit accrual from RE deployment.

So this journey into RE governance concludes with the "Mega-What" question of "who is Renewable Energy for"? At present it seems that RE penetration is much wanted for accomplishment of governmental and international targets to serve RE developers interests; targets are on the way to being met in many cases and numerous RE developers have had a boom decade. But consumers pay, not only for the electricity but also for the many identified inefficiencies in the planning and deployment systems; while benefits are not yet clear and increasing fuel poverty seems to be the most likely result. This concept of benefits becomes even more "abstract" to final end-users as RE operates in relative isolation and not as a part of integrated "green" development. The consequences of this sector-specific planning are likely to become increasingly felt with further future climate change and economic turbulence.

The vacuum in RE governance at the citizen, community group or societal level (over and above NIMBY type concerns) may yet emerge, however it has to compete with the powers of international negotiations and agreements, national RE targets accompanied in some cases by punitive non-compliance punishments, national government geopolitical and trade trajectories and lobbying from RE developers for a guaranteed and sizeable return on investments.

Maybe it is time that RE development is for people rather than for Feed-In-Tariffs, trade and targets?

Erratum to: Renewable Energy Governance

Complexities and Challenges

Evanthie Michalena and Jeremy Maxwell Hills

Erratum to:
E. Michalena and J. M. Hills (eds.), *Renewable Energy Governance*, DOI 10.1007/978-1-4471-5595-9

The volume number for this book has been changed from 57 to 23.

The online version of the original book can be found under DOI 10.1007/978-1-4471-5595-9

E. Michalena (✉)
ENeC Laboratory, UMR 8185, Paris-Sorbonne (Paris IV) and CNRS INSHS, Paris, France
e-mail: evamich@gmail.com

J. M. Hills
CTL Consult, Newcastle upon Tyne, UK
e-mail: j.hills@ctl-consult.com

E. Michalena and J. M. Hills (eds.), *Renewable Energy Governance*,
Lecture Notes in Energy 23, DOI: 10.1007/978-1-4471-5595-9_24,

MIX
Papier aus verantwortungsvollen Quellen
Paper from responsible sources
FSC® C105338

If you have any concerns about our products,
you can contact us on
ProductSafety@springernature.com

In case Publisher is established outside the EU,
the EU authorized representative is:
Springer Nature Customer Service Center GmbH
Europaplatz 3, 69115 Heidelberg, Germany

Printed by Libri Plureos GmbH
in Hamburg, Germany